OPTIMIZATION METHODS FOR LARGE-SCALE SYSTEMS

...with applications

McGRAW-HILL BOOK COMPANY

New York St. Louis San Francisco Düsseldorf Johannesburg
Kuala Lumpur London Mexico Montreal New Delhi
Panama Rio de Janeiro Singapore Sydney Toronto

OPTIMIZATION METHODS FOR LARGE-SCALE SYSTEMS

...with applications

DAVID A. WISMER, Editor

Department of Engineering Systems
School of Engineering and Applied Science
University of California, Los Angeles

OPTIMIZATION METHODS
FOR LARGE-SCALE SYSTEMS

 Library of Congress Catalog Card Number 73–122272

07–071154–2

1234567890 MAMM 754321

This book was set in Imprint Old Face 101 by The St Catherine Press, Ltd., and printed on permanent paper and bound by The Maple Press Company. The designer was Ernest W. Blau. The editors were Dale L. Dutton and Karen Kesti. Stephen J. Boldish supervised production.

PREFACE

Recently an increasing amount of research has been directed toward the solution of optimization problems where the set of equations describing the system is of high dimension, i.e., a large-scale system. This statement is true both for static system optimization where the system is described by linear or nonlinear algebraic or transcendental equations and for dynamic system optimization where the system is described by ordinary or partial differential equations which may be linear or nonlinear. Research in the former area is usually categorized as mathematical programming and in the latter area as optimal control.

The purpose of this book is to bring together a representative selection of the contributions to the theory and application of large-scale system optimization methods. We have tried to balance the selection between theoretical and applied contributions to both static and dynamic systems. This admittedly broad goal has precluded the inclusion of much introductory material normally found in optimal control or mathematical programming texts. We feel justified however in including work from both fields since the dynamic system optimization problem, although more general, can always be viewed (approximately) as a large mathematical programming problem by discretizing the independent variable. Thus the solution to the optimal control problem is a set of functions in some appropriate function space while the solution to the mathematical programming problem is a set of numbers in an appropriate finite-dimensional space. Because of this difference in problem definition, the actual dimensionality implied by the term *large-scale system* may be vastly different for these two problems. Given the state of the art as it now exists for static systems, we can expect to solve problems of much higher dimensions than for the corresponding dynamic system case.

Since the separate topics of mathematical programming and optimal control are each highly developed, it is uncommon to find them both treated in a single book as they are here. In order to do this and still keep a unifying thread throughout the book, we have confined our attention to those developments which are applicable to large-scale systems. As a rule such developments are characterized by (1) *decomposition* of the large system into many smaller subsystems which are later *composed* or *coordinated* to reconstruct the original system, or by (2) *aggregation* of variables in the large system thereby reducing its dimensionality. The ultimate gain from these methods is not only to obtain a computer solution in a reasonable time (or at all) but also to aid in the conceptualization and understanding of large-scale system interactions.

Much of the work using the decomposition and coordination approach, especially for nonlinear systems, has originated at the Systems Research Center at Case Western Reserve University under the heading of *multilevel control*. This approach involves decomposing the Lagrangian or Hamiltonian functional into independent parts by the introduction of auxiliary variables. The smaller subproblems thereby obtained are solved separately and iteration of the subproblem solutions is required to obtain an overall solution which satisfies the original problem. Generally speaking, two different approaches are available for solving these iterative problems; namely (1) model coordination or feasible methods which decompose the system model equations, and (2) goal coordination or dual-feasible methods which decompose the system objective function. Multilevel optimization methods have been developed for both static and dynamic systems. About half the chapters of the book pertain to the theory and application of these methods and include many numerical examples. The remaining chapters deal with the large-scale optimization problem from different points of view but with the same objective in mind.

We have tried to balance the book between theory and applications. Most of the theory-oriented chapters contain numerous examples and the applications-oriented chapters also contain contributions to the theory.

In Chapter 1, J. D. Schoeffler gives a detailed introduction to multilevel systems theory as developed for static system optimization. This chapter reviews the classical saddle value and Kuhn-Tucker conditions from which the multilevel theory stems and includes several varied examples of applications.

In Chapter 2, A. M. Geoffrion introduces some fundamental concepts in large-scale mathematical programming and their application. These concepts serve to unify the existing mathematical programming theory as well as to characterize most algorithms and are called by the author *problem manipulation* and *solution strategies*. A variety of examples from diverse areas of application serve well to illustrate the ideas presented.

Chapter 3, by G. B. Dantzig and R. M. Van Slyke, treats the optimization of linear static and dynamic systems. Following a brief review of linear programming concepts and terminology, the authors explain the decomposition technique for linear programming and then apply it to the solution of linear optimal control problems.

Chapter 4, by J. D. Pearson, presents a detailed treatment of multilevel systems theory as developed for dynamic systems including the application to smoothing and filtering problems. The duality arising from decomposition of the problem is exploited and careful attention is given to computational considerations including the use of a conjugate gradient method for the coordination of subproblems. The chapter also includes several numerical examples.

In Chapter 5, M. Aoki discusses the aggregation of variables to reduce the dimensionality of linear control systems. This concept, long in use in economics, has only recently been applied to optimal control problems. This chapter also presents some results for system identification and state estimation problems.

Chapter 6, by D. A. Wismer, discusses the usefulness of multilevel concepts for the optimization of systems described by partial differential equations. When approximated by ordinary differential equations or algebraic equations these systems are particularly well-suited to multilevel methods. An efficient and simple coordination algorithm is derived and several numerical examples are given.

In Chapter 7, E. J. Bauman shows how the independent variable (presumably time) in a dynamic system can be partitioned into subintervals and the resulting subsystems solved independently. A new coordination algorithm is derived and applied to the solution of an optimal rocket trajectory problem.

Finally in Chapter 8, J. D. Schoeffler treats the special problems encountered when multilevel techniques are applied to industrial processes in an on-line environment. This important class of problems arises whenever a control computer is connected directly to the physical plant being controlled. In a practical, nonmathematical way, this chapter discusses three different approaches to the decomposition of actual industrial processes.

This book contains many important results which have only recently appeared in the open literature. It is hoped that it will serve both as a graduate text and reference book in all cases where the question of high system dimensionality arises. This may occur in the fields of optimal control theory, systems theory, or mathematical programming. Because of the book's broad scope, different portions of it may be used as a particularly cohesive group. The four groupings most readily apparent are:

1. Static system optimization—Chapters 1, 2, 3, and portions of 6 and 8
2. Dynamic system optimization—Chapters 3, 4, 5, 6, 7, and 8
3. Multilevel optimization—Chapters 1, 4, 6, 7, and 8
4. Optimization theory for large-scale systems—Chapters 1, 2, 3, 4, and 5.

In order to master all the theorems in the book, one would probably need a background in analysis, linear algebra, random processes, and the calculus of variations. However, for the specialized topical groupings listed above, only a subset of these tools is needed. In addition, by skipping the proofs, much can be learned about computational algorithms for large-scale systems with very little prior mathematical sophistication.

In conclusion, it is a pleasure to acknowledge the friendly cooperation afforded by all the contributors to this book.

David A. Wismer

CONTENTS

OPTIMIZATION METHODS FOR LARGE-SCALE SYSTEMS

...with applications

1

STATIC MULTILEVEL SYSTEMS

JAMES D. SCHOEFFLER

1-1 MULTILEVEL SYSTEMS AND DECOMPOSITION

Successful design of large, complex systems invariably involves decomposition of the system into a number of smaller subsystems each with its own goals and constraints.[7,9] The resulting interconnection of subsystems may take on many forms, but one of the most common is the hierarchical form in which a given-level unit controls or coordinates the units on the level below it and in turn is controlled by the units on the level above it. The information available to a unit on a given level and the way such a unit can make use of the information to influence or control another unit has been the object of much study. Of particular importance is the consideration of whether or not such an arrangement is capable of acting in an overall system-optimal manner and if so, how to ensure that all units, acting according to their own goals, will somehow achieve this overall goal.[4,10]

In the case of static optimization of a large process (an industrial process for the production of steel, paper, or oil is a typical example), these questions are far from academic, for considerable effort has been extended toward the use of large real-time digital computers to attempt to perform this optimization. It is clear in these situations that the complexity of the processes involved is such that it is impossible to design a single integrated optimizing system for the entire process. This is because the number of skilled people required and the time required to install such a system are excessive—for example, a process changes from time to time as technology advances and in fact changes faster than such a control system can be designed and implemented. Consequently, it is necessary to decompose the problem into a number of smaller problems, the optimization of subsystems of the process for example, and to solve these problems one at a time. Yet there exist overall process goals which must be attained, and it is important to solve the individual problems in such a way that the overall goal will be achieved.

Unfortunately, it is clear that an integrated optimization problem involving many subsystems and variables cannot be decomposed or partitioned into independent subproblems which can be independently optimized. In fact it is often true that because of an abnormal situation in one part of a plant, the rest of the process must be operated in a far from optimal manner in order to achieve good overall performance.[13] Optimizing a single subsystem in a large system without regard to the effects of interactions can lead to such degraded performance in other subsystems that overall process performance is worse than without any optimization.

The objective of this section is to introduce multilevel systems which do permit decomposition of static optimization problems into independent subproblems each of which when solved independently yields the overall system optimum. This is achieved by ensuring that certain variables or parameters in the subproblems called *coordinating variables* are free to be manipulated by a second-level controller whose task it is to choose the coordinating variables in such a way that the independent first-level systems are forced to choose solutions which in fact correspond to an overall system optimum.

The decentralization of the static optimization problem is carried out in a two-step sequence. First, the integrated problem (objective function and constraints) is converted to a two-level or multilevel form with separate and distinct tasks assigned to the levels.

Second, those parts of the first-level task or problem which do not interact with other parts are split apart, forming a decomposition or partition of the first-level problem. Success of the method of course depends upon the method by which the original problem is converted into a two-level problem for this must be done in such a way that the resulting first-level problem has a number of noninteracting parts which can be treated separately.

Although there may be an infinity of different ways of transforming a given constrained optimization problem into a multilevel problem, they are all essentially combinations of two different approaches which may be termed the *model-coordination method* (or the *feasible method*) and the *goal-coordination method* (or the *dual-feasible method*).[8] The next two sections discuss each of these approaches by means of a simple example involving two coupled subsystems and without regard to proof of existence or convergence.

1-1-1 The Model-coordination Method

Consider the integrated optimization problem associated with the system diagrammed in Fig. 1-1. The variables in the system are defined below:

$\mathbf{m}$ = vector of manipulated variables

$\mathbf{m}^i$ = vector of manipulated variables for subsystem i

$\mathbf{y}$ = vector of output variables for the system

$\mathbf{y}^i$ = vector of output variables for subsystem i

$\mathbf{x}^1$ = vector of interaction variables from subsystem 1 to subsystem 2

$\mathbf{x}^2$ = vector of interaction variables from subsystem 2 to subsystem 1

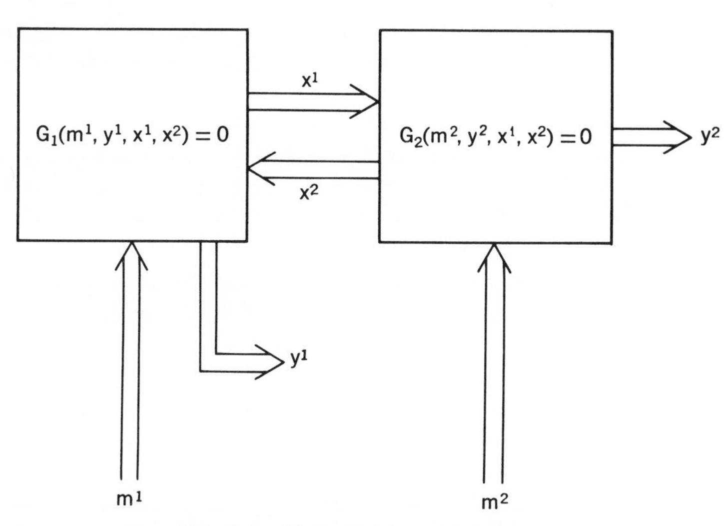

FIG. 1-1. Example of coupled system.

Let the static system equations be

$$\mathbf{G}(\mathbf{m}, \mathbf{y}, \mathbf{x}) = 0$$

where the vector $\mathbf{G}$ contains the system equations for each subsystem:

$$\mathbf{G}^i(\mathbf{m}^i, \mathbf{y}^i, \mathbf{x}^1, \mathbf{x}^2) = 0 \qquad i = 1, 2$$

Let the objective function which is to be minimized be

$$P(\mathbf{m}, \mathbf{y}, \mathbf{x}) = P_1(\mathbf{m}^1, \mathbf{y}^1, \mathbf{x}^1) + P_2(\mathbf{m}^2, \mathbf{y}^2, \mathbf{x}^2)$$

Notice that the performance function has been selected so that actually each individual subsystem has its own performance function and the overall system performance is the sum of the two subsystem performances. The minimization is to take place over all allowable **m**, **y**, and **x** which satisfy the system equations:

$$\min_{\mathbf{m},\mathbf{y},\mathbf{x}} P(\mathbf{m}, \mathbf{y}, \mathbf{z})$$

$$\text{subject to } \mathbf{G}(\mathbf{m}, \mathbf{y}, \mathbf{x}) = 0$$

Depending on the problem, the set of allowable variables may be finite or infinite, real or integer, constrained or unconstrained.

Although the performance function may be separated into two non-interacting functions, one for each subsystem, there is interaction because of the interaction variables **x** which affect both subsystems. The model-coordination method converts this integrated optimization problem into a two-level problem by *fixing the interaction variables*.[2] That is, fix the interaction variables at some value, say **z**:

$$\text{constrain } \mathbf{x} = \mathbf{z}$$

Then under these conditions, the integrated problem may be split into a first-level problem and a second-level problem as follows:

FIRST-LEVEL PROBLEM:

$$\text{determine } H(\mathbf{z}) = \min_{\mathbf{m},\mathbf{y}} P(\mathbf{m}, \mathbf{y}, \mathbf{z})$$

$$\text{subject to } \mathbf{G}(\mathbf{m}, \mathbf{y}, \mathbf{z}) = 0$$

SECOND-LEVEL PROBLEM:

$$\min_{\mathbf{z}} H(\mathbf{z})$$

That is, if **x** is considered to be a constant **z**, the first level optimizes the objective function by choosing the variables **m** and **y** so that P is minimized and the system equations are satisfied for the given **z**. If

$$S_0 = \{(\mathbf{m}, \mathbf{y}) \mid \mathbf{G}(\mathbf{m}, \mathbf{y}, \mathbf{z}) = 0\}$$

then the minimization of P is over the set S_0. Notice that this set may be empty in some problems for some choices of **z**. Define

$$S_1 = \{\mathbf{z} \mid H(\mathbf{z}) \text{ exists}\}$$

That is, S_1 is the set of all **z** such that the system equations may be satisfied for this value of **z** and the minimum of the objective function exists (is finite). If the original integrated optimization problem has a solution, then S_1 is not empty for it contains (at least) the point $\mathbf{z} = \mathbf{x}_{\text{opt}}$, that is, the optimal values for the interaction variables.

The actual implementation of this two-level solution to the problem is not usually done simultaneously as indicated here but rather sequentially. That is, the second level produces an estimate $\mathbf{z}$ of $\mathbf{x}_{\text{opt}}$, the optimal values for the interaction variables, and transmits this estimate to the first-level unit which determines the optimal values for $\mathbf{m}$ and $\mathbf{y}$ assuming that $\mathbf{x} = \mathbf{z}$. The first-level unit in turn transmits the values for $\mathbf{m}$ and $\mathbf{y}$ to the second-level unit which then produces a better estimate for the interaction variables, etc. Thus the solution of the optimization problem proceeds in an iterative fashion with the effort divided between the first- and second-level units.

This method of producing a multilevel form of the problem leads directly to a method of decomposition of the first-level problem because the first-level problem partitions immediately into two independent problems:

FIRST-LEVEL PROBLEM FOR SUBSYSTEM i ($i = 1, 2$):

$$H_i(\mathbf{z}) = \min_{\mathbf{m}^i, \mathbf{y}^i} P_i(\mathbf{m}^i, \mathbf{y}^i, \mathbf{z}^i)$$

$$\text{subject to } \mathbf{G}_i(\mathbf{m}^i, \mathbf{y}^i, \mathbf{z}^1, \mathbf{z}^2) = 0$$

The second-level problem is to choose $\mathbf{z}$ so as to minimize $H(\mathbf{z}) = H_1(\mathbf{z}) + H_2(\mathbf{z})$. The various minimizations are to be done over the appropriate sets, so that the minima exist. The resulting multilevel decentralized solution to this optimization problem is diagrammed in Fig. 1-2.

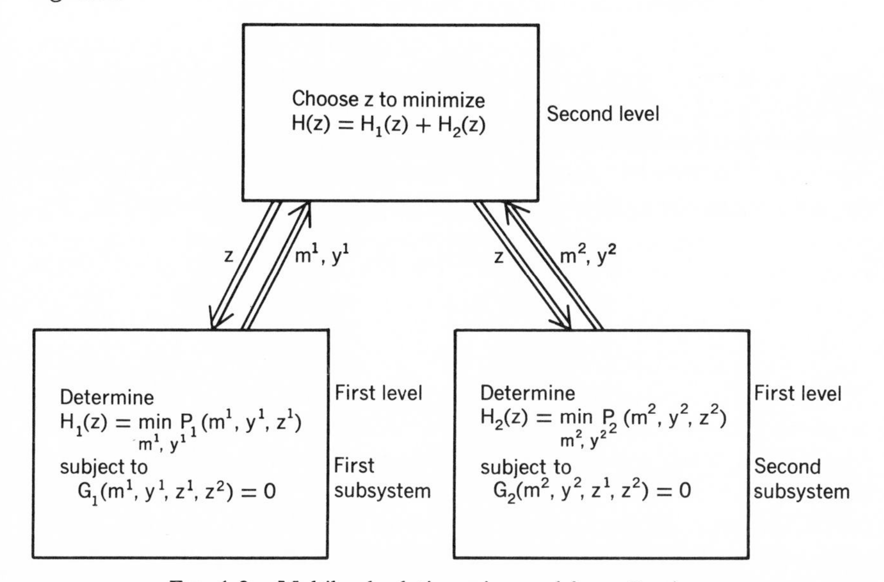

FIG. 1-2. Multilevel solution using model coordination.

The set of variables which are fixed $\mathbf{x}$ are termed the *coordinating variables*. The term model coordination arises from the fact that decomposition is made possible by adding a constraint to the mathematical models of the system, namely, that certain internal interacting variables be fixed. The alternate name, the feasible method, arises from the fact that throughout the iterative calculation, all intermediate values for the variables $\mathbf{m}$, $\mathbf{y}$, and $\mathbf{x}$ are feasible or allowed. Thus the system could actually operate with any of these intermediate values, although with less than optimal performance. For this method to be useful in the interacting subsystem problem defined here, it is of course necessary that the set S_0 be nonempty for reasonable values of $\mathbf{z}$. This means that the dimension of the manipulated variables $\mathbf{m}^i$ must be high enough to allow a solution of the subsystem equations

$$\mathbf{G}_i(\mathbf{m}^i, \mathbf{y}^i, \mathbf{z}^1, \mathbf{z}^2) = 0$$

for reasonable values of $\mathbf{z}$. In other words, the subsystems must be output-controllable considering $\mathbf{z}$ to be their output.

The important generalization here is that the first-level problem was created by fixing certain variables in the original optimization problem and assigning to the second level the task of determining these coordinating variables. It is important to consider means of doing this coordination or second-level problem but this question is sidestepped until the mathematical basis for decomposition is more carefully laid.

1-1-2 The Goal-coordination Method

Consider again the two coupled subsystems and the optimization problem defined in Sec. 1-1-1. The goal-coordination method argues that a coupled problem cannot merely be separated into two parts and solved separately if overall optimality is to be achieved. Hence it is first necessary to literally remove the interactions by "cutting" all links between subsystems.[3,12,13] This may be done as indicated in Fig. 1-3, where the outputs of the subsystem which are inputs to another subsystem are labeled $\mathbf{x}^i$ as before, but the corresponding inputs to the subsystems are now labeled $\mathbf{z}^i$ and the meaning of the word "cut" is literal; that is, $\mathbf{z}^i$ and $\mathbf{x}^i$ need not be equal. Moreover, the $\mathbf{z}^i$ now act like arbitrary manipulated variables and must be selected, like $\mathbf{m}$, $\mathbf{y}$, and $\mathbf{x}$, by the optimizing subsystems. This of course decouples the two subsystems completely and since the objective function was already decoupled, it is clear that there is no interaction in the system at all. However, in order to ensure that the independent subsystem problems yield the overall system optimality, it is necessary to ensure that the *interaction-balance principle* be satisfied, namely, that the independently selected $\mathbf{x}^i$ and $\mathbf{z}^i$ actually be equal.[8]

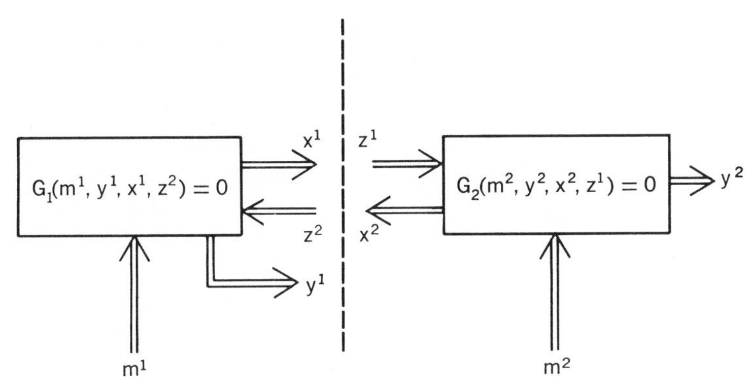

FIG. 1-3. Decoupled system.

The multilevel formulation of this problem then involves cutting the interacting variables in order to create a first-level problem which can easily be decentralized or decomposed into independent subproblems. The second level attempts to force the first-level independent subproblems to arrive at a solution for which the interaction-balance principle holds. This may be done by modifying the goals of the first-level problems.

Consider the addition of a penalty term which penalizes the performance of the system if interactions do not balance:

$$P(\mathbf{m}, \mathbf{y}, \mathbf{x}, \mathbf{z}, \boldsymbol{\lambda}) = P_1(\mathbf{m}^1, \mathbf{y}^1, \mathbf{x}^1) + P_2(\mathbf{m}^2, \mathbf{y}^2, \mathbf{x}^2) + \boldsymbol{\lambda}'(\mathbf{x} - \mathbf{z})$$

where $\boldsymbol{\lambda}$ is a vector of weighting parameters (positive or negative as needed) which causes any interaction unbalance $(\mathbf{x} - \mathbf{z})$ to affect performance. With the introduction of the $\mathbf{z}$ variables, the system equations are now

$$\mathbf{G}_1(\mathbf{m}^1, \mathbf{y}^1, \mathbf{x}^1, \mathbf{z}^2) = 0$$

$$\mathbf{G}_2(\mathbf{m}^2, \mathbf{y}^2, \mathbf{x}^2, \mathbf{z}^1) = 0$$

Define the set S_0 as values of those variables $\mathbf{m}$, $\mathbf{x}$, $\mathbf{y}$, and $\mathbf{z}$ which satisfy the constraints

$$S_0 = \{(\mathbf{m}, \mathbf{y}, \mathbf{x}, \mathbf{z}) \mid \mathbf{G}_1 = \mathbf{G}_2 = 0\}$$

Minimizing the objective function (with penalty term) over the set of allowable system variables results in a function of $\boldsymbol{\lambda}$:

$$H(\boldsymbol{\lambda}) = \min_{(\mathbf{m},\mathbf{y},\mathbf{x},\mathbf{z}) \in S_0} P(\mathbf{m}, \mathbf{y}, \mathbf{x}, \mathbf{z}, \boldsymbol{\lambda})$$

Define the set S_1 as the domain of $H(\boldsymbol{\lambda})$:

$$S_1 = \{\boldsymbol{\lambda} \mid H(\boldsymbol{\lambda}) \text{ exists}\}$$

That is, S_1 is the set of all $\boldsymbol{\lambda}$ such that the minimum of P exists. Assuming there exists some vector $\boldsymbol{\lambda}$ such that solving the above optimization problem with the penalty term results in the interactions being in balance, the set S_1 is not empty.

First expand the penalty term into

$$\boldsymbol{\lambda}'(\mathbf{x} - \mathbf{z}) = \boldsymbol{\lambda}_1'(\mathbf{x}^1 - \mathbf{z}^1) + \boldsymbol{\lambda}_2'(\mathbf{x}^2 - \mathbf{z}^2)$$

Then the first-level problems separate into

SUBSYSTEM 1:

$$\min_{\mathbf{m}^1,\mathbf{y}^1,\mathbf{x}^1,\mathbf{z}^2} P_1(\mathbf{m}^1, \mathbf{y}^1, \mathbf{x}^1) + \boldsymbol{\lambda}_1'\mathbf{x}^1 - \boldsymbol{\lambda}_2'\mathbf{z}^2$$

$$\text{subject to } \mathbf{G}_1(\mathbf{m}^1, \mathbf{y}^1, \mathbf{x}^1, \mathbf{z}^2) = 0$$

SUBSYSTEM 2:

$$\min_{\mathbf{m}^2,\mathbf{y}^2,\mathbf{x}^2,\mathbf{z}^1} P_2(\mathbf{m}^2, \mathbf{y}^2, \mathbf{x}^2) - \boldsymbol{\lambda}_1'\mathbf{z}^1 + \boldsymbol{\lambda}_2'\mathbf{x}^2$$

$$\text{subject to } \mathbf{G}_2(\mathbf{m}^2, \mathbf{y}^2, \mathbf{x}^2, \mathbf{z}^1) = 0$$

Notice that the goals of the individual subsystems have been modified in that the coordinating variables $\boldsymbol{\lambda}$ enter into each subsystem goal. Since the objective of the second-level unit is to choose $\boldsymbol{\lambda}$, it is clear that the coordination of the independent first-level problems is taking place by manipulating the first-level problem objectives, thus the term goal coordination. The alternate term, dual-feasible method, arises from further consideration of the function $H(\boldsymbol{\lambda})$ in Sec. 1-2-6.

Whether the coordinating parameter $\boldsymbol{\lambda}$ is considered to be a penalty weight or a Lagrange multiplier is immaterial. The important generalization to be made here is that just as in the case of the model-coordination scheme, the multilevel formulation of the problem was created by fixing certain variables (interaction variables in Sec. 1-1-1 and Lagrange multiplier variables in this section). Just as before, the goal or task of the second level is to determine the optimal values for these fixed or coordinating variables. Again it must be emphasized that the numerical solution is iterative with the two levels alternating solution of their associated problems.

Once having reformulated the problem in the multilevel format, the decomposition of the first-level problems into independent subproblems is straightforward. The subproblems are defined below and diagrammed in Fig. 1-4.

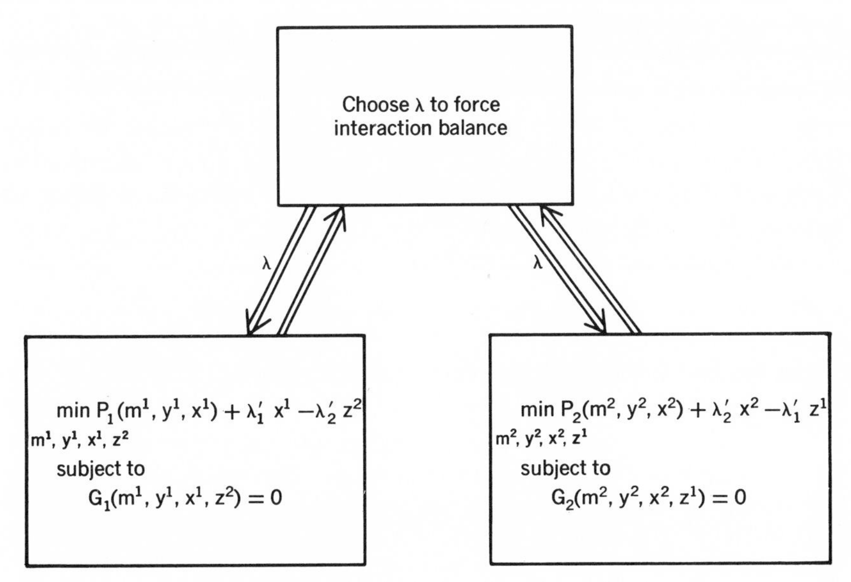

FIG. 1-4. Multilevel solution via goal coordination.

FIRST-LEVEL PROBLEM:

Determine $H(\boldsymbol{\lambda})$ by minimizing $P(\mathbf{m}, \mathbf{y}, \mathbf{x}, \boldsymbol{\lambda})$ over the set S_0.

SECOND-LEVEL PROBLEM:

Choose $\boldsymbol{\lambda}$ such that solution to the first-level problem results in satisfaction of the interaction-balance principle; that is, $\mathbf{x} = \mathbf{z}$.

By approaching the formulation of the optimization problem from a Lagrange multiplier point of view, the penalty term may be viewed as a constraint (that the interactions be in balance) and the conditions under which a $\boldsymbol{\lambda}$ exists which solves the integrated optimization problem may be determined. Furthermore, the second-level problem may then be shown to be a well-behaved optimization problem in its own right. Again these results must await the formulation of the mathematical basis of the optimization problem in Sec. 1-2.

1-1-3 Multilevel Formulation of Optimization Problems

From the two approaches to multilevel, decentralized solutions of optimization problems presented in the previous two sections, general conclusions may be drawn. Consider an objective function and a set of constraints. The constraints may be on the allowed values of the variables in the problem (for example, all variables nonnegative) or may be in the form of equality or inequality relations among the variables (constraints which often arise in practical situations because of limits on system operation). If the constraints arise because of the requirement of interaction balance, they are of course simple equality constraints. In any case, if the integrated optimization problem is formulated by including the constraints through the use of either Lagrange multipliers or penalty terms, conversion to a multilevel form is then done by

1. Choosing a set of coordinating variables (actual variables, Lagrange multipliers, penalty weights, or any combination of these) and assuming these to be *fixed*, thereby producing a first-level optimization problem with certain fixed variables
2. Assigning to the second level the task of determining the optimal values of the coordinating variables
3. Deriving an algorithm by which the second-level unit may successively improve its estimate of the optimal interaction variables

The choice of coordinating variables is of course made on the basis of decomposition of the first-level problem. That is, a set of coordinating variables must be selected so that the resulting first-level problem, which is the original problem with the coordinating variables assumed fixed, may be decomposed into independent subproblems. From an efficiency point of view, it is desirable that the solutions to the independent first-level problems be very simple since in the iterative solution they are solved many times.

In general, optimization problems may involve functions which are differentiable or not differentiable, convex or not convex, and defined on infinite or finite sets. Similarly, constraints may be of any form. In order to formulate the multilevel decentralized form for the solution of the constrained optimization problem, it is necessary to be able to first formulate it on the basis of knowledge about its solutions (or lack thereof), and then to formulate algorithms for solution of the first- and second-level problems. Thus before proceeding to examples of multilevel solutions of useful real problems, it is first necessary to consider carefully the mathematical basis for the solution of nonlinear optimization problems. This area is called *mathematical programming*.

1-2 MATHEMATICAL PROGRAMMING BASIS FOR DECOMPOSITION

Many physical problems can be formulated as optimization problems, that is, as the problem of determining the maximum or minimum of a function subject to some constraints. If both the function to be optimized and the constraints are static, the problem is called a mathematical programming problem. Many dynamic problems can be reformulated as static problems through parameterization as will be shown in an example later. The characteristics of solutions to such problems vary considerably depending upon the type of function to be optimized (e.g., linearity, differentiability, convexity), the characteristics of the constraints imposed on the solution (e.g., linearity, convexity, equality), and the characteristics of the variables involved (e.g., real, integer). Peculiar situations are the rule rather than the exception in complex problems and can cause numerical solution techniques to fail in a nonapparent manner. Consequently, different numerical techniques are required for solution depending upon the specific problem at hand. To choose a solution technique or derive a new method for solving such problems, it is important to know of the existence, uniqueness, and characteristics of solutions as well as the behavior of the functions in any area of the space involved in which a search for solutions is to take place. This is especially true for large optimization problems in which success depends critically upon the method chosen to solve the problem. Since the multilevel decomposition technique is essentially directed toward the solution of large nonlinear optimization problems with many constraints, the techniques and results of mathematical programming are directly applicable. A number of important results are summarized here which provide the basis for the determination of the characteristics of problems which can be solved by multilevel decomposition, and more important, also discussed are the conditions under which the method is assured of efficiently providing solutions.

Whatever the source of the optimization problem (design, operation, control) a given set of variables or parameters, called $\mathbf{x}_1, \mathbf{x}_2, \ldots, \mathbf{x}_n$, must be chosen to optimize some performance measure. Allowed values for the $\mathbf{x}$'s are usually restricted for practical reasons (variables must be positive perhaps, or production greater than some minimum, or total cost equal to available funds, etc.). The problem is to select from the set of permissible $\mathbf{x}$'s that set or sets which optimize the performance function. This may be formalized to the following:

Consider a static optimization problem, assumed for convenience to be minimization, involving a vector $\mathbf{x}$ of n real variables and real-valued functions $f(\mathbf{x})$, $w_i(\mathbf{x})$, and $g_i(\mathbf{x})$ defined on Euclidean n space E^n:

$$\min_{\mathbf{x}} f(\mathbf{x})$$

$$\text{subject to } \mathbf{x} \in S_1 \cap S_2 \cap S_3$$

where

$$S_1 \subseteq E^n$$

$$S_2 = \{\mathbf{x} \mid w_i(\mathbf{x}) = 0, i = 1, 2, \ldots, n_w\}$$

$$S_3 = \{\mathbf{x} \mid g_i(\mathbf{x}) \leqslant 0, i = 1, 2, \ldots, n_0\}$$

Here the $w_i(\mathbf{x})$ are a set of equality constraints, and the $g_i(\mathbf{x})$ a set of inequality constraints. The set S_1 is often taken to be $(E^n)^+$ and it is usually convenient to handle it separately rather than embedding it in the set S_3. Any of the constraints may be absent. The characteristics of this problem, called the ***primal problem***, are primarily determined by the convexity and differentiability of the functions involved. These terms as well as others are defined below.[14]

A point set S in E^n is called *convex* if for any two points $\mathbf{x}_1$ and $\mathbf{x}_2 \in S$, all points

$$\lambda \mathbf{x}_1 + (1 - \lambda)\mathbf{x}_2 \qquad 0 \leqslant \lambda \leqslant 1$$

belong to S. Thus all points on a line connecting any two points in a convex set are also in the set. If S is closed and convex, it is called a *convex domain.* If $f(\mathbf{x})$ is a real-valued function of n variables defined on a convex set $S \subseteq E^n$, then $f(\mathbf{x})$ is said to be a *convex function* if for every pair of points $\mathbf{x}_1$ and $\mathbf{x}_2 \in S$

$$f[\lambda \mathbf{x}_1 + (1 - \lambda)\mathbf{x}_2] \leqslant \lambda f(\mathbf{x}_1) + (1 - \lambda) f(\mathbf{x}_2) \qquad 0 \leqslant \lambda \leqslant 1$$

If $\mathbf{x}_1 \neq \mathbf{x}_2$ and the $\leqslant$ symbol can be replaced by a strict inequality, the function is *strictly convex.* Geometrically, a convex function cannot assume a larger value along the line joining $\mathbf{x}_1$ and $\mathbf{x}_2$ than a linear function which takes on the values $f(\mathbf{x}_1)$ and $f(\mathbf{x}_2)$ at $\mathbf{x} = \mathbf{x}_1$ and $\mathbf{x}_2$ respectively.

The importance of convex functions in mathematical programming is essentially that they are "the next generalization from linearity" with the following characteristics:[14,11]

1. A linear function is convex but not strictly convex.
2. A convex function is continuous in the interior of its domain.
3. A positive linear combination of convex functions is convex.
4. Each local minimum of a convex function in a convex domain S is a global minimum over this domain.
5. A local minimum of a strictly convex function is actually unique and the global minimum.
6. If $f(\mathbf{x})$ is convex, then $f(\mathbf{x} - a)$ is convex.
7. If $f(\mathbf{x})$ is convex, then the set $S = \{\mathbf{x} \mid f(\mathbf{x}) \leqslant 0\}$ is also convex. This result is important because many constraints are of this form.
8. The intersection of convex sets is convex.

The last four properties ensure that a minimization problem involving a strictly convex objective function and a convex constraint set has a unique minimum, an important result.

Clearly, convexity of the functions involved will permit specific statements about the existence and uniqueness of minima of the primal problem.

Stationarity Conditions are necessary but not sufficient conditions for a local minimum of a function. If $f(\mathbf{x})$ is a real-valued function of n variables $\mathbf{x} = [x_1, x_2, \ldots, x_n]$ defined on E^n and if $f(\mathbf{x})$ is differentiable (all partial derivatives exist everywhere and are continuous) then in order for $\mathbf{x} = \mathbf{x}^0$ to be a minimum, all partial derivatives there must be zero. Since this is also true at maxima and inflection points of the function, the conditions are necessary but not sufficient. If the partial derivatives of $f(\mathbf{x})$ are arranged in a vector, the vector is called the *gradient* of $f(\mathbf{x})$ and is written

$$\nabla f(\mathbf{x}) = \left[\frac{\partial f}{\partial x_1} \cdots \frac{\partial f}{\partial x_n}\right]'$$

and the stationarity condition is simply that $\nabla f(\mathbf{x}^0) = \mathbf{0}$. If $\mathbf{x} \in S \subset E^n$ and the minimum occurs on the boundary then stationarity is no longer necessary. If the matrix of second partials exists, then at a minimum of the function $f(\mathbf{x})$ not on a boundary, the matrix of second partials must be positive semidefinite. At any point $\mathbf{x}_1$, $\nabla f(\mathbf{x}_1)$ is normal to the surface $f(\mathbf{x}) = f(\mathbf{x}_1)$ and points in the direction of steepest ascent, a fact that is important and useful in numerical procedures.

The concept of a *saddle point* is of central importance in constrained optimization problems. If $L(\mathbf{x}, \mathbf{u})$ is a real-valued function of the n-vector $\mathbf{x}$, $\mathbf{x} \in S_1 \subseteq E^n$ and the m-vector $\mathbf{u} \in S_2 \subseteq E^m$, then the point $(\mathbf{x}^0, \mathbf{u}^0)$ is called a saddle point for L if

$$L(\mathbf{x}^0, \mathbf{u}) \leqslant L(\mathbf{x}^0, \mathbf{u}^0) \leqslant L(\mathbf{x}, \mathbf{u}^0)$$

for all $\mathbf{x} \in S_1$ and $\mathbf{u} \in S_2$. Thus $(\mathbf{x}^0, \mathbf{u}^0)$ is a minimum in the $\mathbf{x}$ variables and a maximum in the $\mathbf{u}$ variables, resulting in the usual saddlelike shape conceptually. If $L(\mathbf{x}, \mathbf{u})$ is differentiable in all variables, then at a saddle point interior to S_1 and S_2, all partial derivatives must be zero (a stationary point). If second partials exist, then the matrix of second partials with respect to the $\mathbf{x}$ variables must be positive semidefinite and the matrix of second partials with respect to the $\mathbf{u}$ variables must be negative semidefinite.

Two approaches to solving the primal problem are through stationarity conditions and through saddle-point conditions. Both are discussed in the next sections.

1-2-1 Stationarity Conditions for Minima

If all of the functions involved in the primal problem are differentiable, that is, if partial derivatives with respect to all of the x_i exist and are continuous throughout the space of definition of the functions, then stationarity conditions which the minimum point must satisfy may be derived. The primal problem for this case is summarized in Table 1-1. The allowable values for the variables lie in the intersection of three sets. Sets S_2 and S_3 are determined by equality and inequality constraints respectively. The set S_1 is the set of nonnegative real numbers, a constraint which is more easily handled separately from the general inequality constraints in S_3.

Provided a minimal regularity condition (to be discussed) is fulfilled, Lagrange multipliers may be used to enforce the constraints in the following way. Form the Lagrange-like function as shown in Table 1-1. Then necessary conditions in order for the point $\mathbf{x}^0$, $\boldsymbol{\lambda}^0$, $\boldsymbol{\mu}^0$ to be the solution to the primal problem are the Kuhn-Tucker stationarity conditions[5] as shown in Table 1-1 (assuming that the regularity condition is fulfilled). The conditions are grouped into three parts. The first set

Table 1-1. Stationarity Conditions for the Minimum

Primal Problem

$$\min_{\mathbf{x}} f(\mathbf{x})$$

$$\text{subject to } \mathbf{x} \in S_1 \cap S_2 \cap S_3$$

where

$$\begin{aligned} S_1 &= (E^n)^+ \\ S_2 &= \{\mathbf{x} \mid w_i(\mathbf{x}) = 0, i = 1, 2, \ldots, n_w\} \\ S_3 &= \{\mathbf{x} \mid g_i(\mathbf{x}) \leqslant 0, i = 1, 2, \ldots, n_g\} \\ \mathbf{x} &= n\text{-vector of real variables} \\ f(\mathbf{x}), w_i(\mathbf{x}), g_i(\mathbf{x}) &= \text{real-valued functions} \end{aligned}$$

defined on S_1 and *differentiable* everywhere

Lagrange Function

$$L(\mathbf{x}, \boldsymbol{\lambda}, \boldsymbol{\mu}) = f(\mathbf{x}) + \sum_{i=1}^{n_w} \lambda_i w_i(\mathbf{x}) + \sum_{i=1}^{n_g} \mu_i g_i(\mathbf{x})$$

Stationarity Conditions for the point $(\mathbf{x}^0, \boldsymbol{\lambda}^0, \boldsymbol{\mu}^0)$

1. $\nabla_{\mathbf{x}} L(\mathbf{x}^0, \boldsymbol{\lambda}^0, \boldsymbol{\mu}^0) \geqslant 0$

   $(\mathbf{x}^0)' \nabla_{\mathbf{x}} L(\mathbf{x}^0, \boldsymbol{\lambda}^0, \boldsymbol{\mu}^0) = 0$

   $\mathbf{x}^0 \geqslant 0$

2. $w_i(\mathbf{x}^0) = 0 \qquad i = 1, 2, \ldots, n_w$

3. $g_i(\mathbf{x}^0) \leqslant 0$

   $\mu_i^0 g_i(\mathbf{x}^0) = 0$

   $\mu_i \geqslant 0 \qquad i = 1, 2, \ldots, n_g$

requires that the partial derivatives with respect to each of the $\mathbf{x}_i$ be positive or zero. If the set of allowable $\mathbf{x}$ were the entire E^n rather than just the first quadrant, this would cause the first condition to be simply all partial derivatives equal to zero and the second and third parts of the first condition would be superfluous. These parts are present in order to enforce the conditions $\mathbf{x}_i \geqslant 0$. Notice that if any element of the gradient vector is nonzero, the corresponding $\mathbf{x}_i$ must be zero, that is, on the boundary of S_1.

The second condition is present to handle any equality constraints. Notice that condition 2 is equivalent to $\nabla_\lambda L(\mathbf{x}^0, \boldsymbol{\lambda}^0, \boldsymbol{\mu}^0) = 0$, which makes it identical to the first condition when $S_1 = E^n$. This condition requires that the equality constraints be used to determine the $\mathbf{x}^0$, $\boldsymbol{\lambda}^0$, and $\boldsymbol{\mu}^0$ in the first condition. Finally the third condition ensures that the inequality conditions hold. If it is noted that the g_i are the partial derivatives of L with respect to the μ_i, the third condition becomes

$$\begin{aligned} \nabla_\mu L(\mathbf{x}^0, \boldsymbol{\lambda}^0, \boldsymbol{\mu}^0) &\leqslant 0 \\ (\boldsymbol{\mu}^0)' \nabla_\mu L(\mathbf{x}^0, \boldsymbol{\lambda}^0, \boldsymbol{\mu}^0) &= 0 \\ \boldsymbol{\mu}^0 &\geqslant 0 \end{aligned}$$

which is identical in form to the first condition with the $\mathbf{x}$ and $\boldsymbol{\mu}$ interchanged. This condition behaves similarly also. That is, if the solution lies on the constraint boundary, the multiplier $\boldsymbol{\mu}_i^0$ is nonzero whereas if the constraint is not binding, the multiplier is zero. Notice that the only essential difference between the equality and inequality constraints is that the multipliers are constrained to be nonnegative for inequality constraints.

These conditions were first derived by Kuhn-Tucker in 1951 and represent one of the most important results in the area of mathematical programming. To see the problem of applying these results, consider the example problem:

$$\begin{aligned} \min f(\mathbf{x}) &= (x_1 - 1)^2 + (x_2 - 2)^2 \\ \text{subject to } x_2 - x_1 &= 1 \\ x_1 + x_2 &\leqslant 2 \\ \mathbf{x} &\in (E^2)^+ \end{aligned}$$

Forming the Lagrangian gives $L(\mathbf{x}, \boldsymbol{\lambda}, \boldsymbol{\mu}) = (x_1 - 1)^2 + (x_2 - 2)^2 + \lambda(x_2 - x_1 - 1) + \mu(x_1 + x_2 - 2)$.

The stationarity conditions are (numbering corresponds to that of Table 1-1):

1.
$$2x_1 - \lambda + \mu - 2 \geqslant 0$$
$$2x_2 + \lambda + \mu - 4 \geqslant 0$$
$$x_1(2x_1 - \lambda + \mu - 2) = 0$$
$$x_2(2x_2 + \lambda + \mu - 4) = 0$$
$$x_1 \geqslant 0$$
$$x_2 \geqslant 0$$

2.
$$x_2 - x_1 = 1$$

3.
$$x_1 + x_2 \leqslant 2$$
$$\mu(x_1 + x_2 - 2) = 0$$
$$\mu \geqslant 0$$

It is easily verified that $x_1 = 0.5$, $x_2 = 1.5$, $\lambda = 0$, and $\mu = 1.0$ satisfy the stationarity conditions and correspond to the solution to the primal problem. It is of course not at all apparent how one would solve the set of stationarity conditions to determine $\mathbf{x}^0$, $\boldsymbol{\lambda}^0$, and $\boldsymbol{\mu}^0$ in a complex problem. Nonetheless they are valuable.

Although the stationarity conditions in Table 1-1 are only necessary for differentiable functions in general, it is possible to prove that they are in fact necessary and sufficient if the function $f(\mathbf{x})$ is convex and the sets S_1, S_2, and S_3 are convex.[5] This is indicated in Table 1-4, which lists three sets of conditions for optimality depending on various combinations of convexity and differentiability, and the conditions under which each apply. From the listed properties of convex functions, it follows that these three sets are convex if the inequality functions $g_i(\mathbf{x})$ are convex and if the equality constraints are linear or absent.

1-2-2 The Regularity Assumption

The stationarity conditions of Kuhn and Tucker are restricted by a regularity assumption or constraint condition as it is termed.[5,1] This means, essentially, that singularities on the boundaries of the constraint space (such as cusps) cannot be present if the Lagrange multiplier technique is to be used. A convenient form for the constraint condition is the following one given by Beltrami.[1]

Constraint Condition or Regularity Assumption: Let $\mathbf{x}^0$ solve the primal problem. There must exist a vector $\mathbf{h} \in E^n$ such that at each equality constraint, the inner product $(\nabla w_i(\mathbf{x}^0), \mathbf{h}) < 0$ and at each *binding* inequality constraint, $(\nabla g_i(\mathbf{x}^0), \mathbf{h}) < 0$. That is, a *single* $\mathbf{h}$ must exist for all binding constraints.

This regularity assumption ensures that the multipliers associated with the binding constraints are finite or bounded. To see how the stationarity conditions fail to apply in the absence of regularity, consider the following primal problem:

$$\min f(\mathbf{x}) = -x_1$$

$$\text{subject to } w(\mathbf{x}) = (x_2 - 2) + (x_1 - 1)^3 = 0$$

$$g(\mathbf{x}) = (x_1 - 1)^3 - (x_2 - 2) \leqslant 0$$

$$x_1 \geqslant 0 \qquad x_2 \geqslant 0$$

These constraints are diagrammed in Fig. 1-5, where it is self-evident that the optimum lies at $x_1 = 1$ and $x_2 = 2$ and that a cusp exists there.

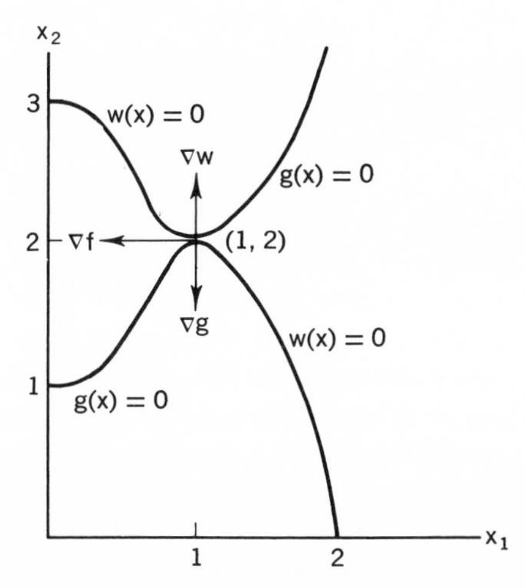

FIG. 1-5. Example of Sec. 1-2.

But at the optimum, the gradients of the constraints (both of which are binding) are

$$\nabla w(\mathbf{x}^0) = (0, 1)$$

$$\nabla g(\mathbf{x}^0) = (0, -1)$$

Hence in applying the regularity assumption conditions there results

$$\nabla w'\mathbf{h} = h_2 < 0$$

$$\nabla g'\mathbf{h} = -h_2 < 0$$

Clearly no $\mathbf{h}$ exists and the regularity condition is violated. This result may also be seen from consideration of the stationarity conditions in Table 1-1.

Since the optimum does not occur on the boundary of S_1, the first condition becomes

$$\nabla L(\mathbf{x}^0, \boldsymbol{\lambda}^0, \boldsymbol{\mu}^0) = \mathbf{0}$$

Hence at the optimum, this gives the gradient of the function in terms of the gradients of the constraints:

$$\nabla f(\mathbf{x}^0) = -\sum_{i=1}^{n_w} \boldsymbol{\lambda}_i^0 \nabla w_i(\mathbf{x}^0) - \sum_{i=1}^{n_g} \boldsymbol{\mu}_i^0 \nabla g_i(\mathbf{x}^0)$$

But observe from Fig. 1-5 that the gradients of the constraints are in the $\mathbf{x}_2$ direction whereas the gradient of the function itself is in the $\mathbf{x}_1$ direction. Clearly, no finite values of the multipliers exist which permit the above relation to hold and consequently the stationarity conditions cannot hold at this point.

As pointed out by Beltrami,[1] it is clear from the above example that the regularity condition is necessary and in addition, if the rank of the matrix of gradients of the binding constraints is maximal, the multipliers are uniquely determined but not otherwise.

Several important cases exist in which the regularity condition is automatically satisfied:

1. If there are no equality constraints and all inequality constraints are convex in S_1 and if there exists at least one $\mathbf{x}$ which is inside the constraint set (not on the boundary), then the regularity conditions hold.
2. If the rank of the matrix of gradients of all constraints at $\mathbf{x}^0$ is maximal, the regularity conditions are fulfilled.
3. If the constraints are linear, the regularity conditions are automatically satisfied.

The Lagrange formulation of the constrained optimization problem and the regularity conditions are closely related to another approach to these problems, namely, the penalty function method. The penalty function approach changes a constrained optimization problem into an unconstrained problem, or sequence of problems, by adding "penalties" to the performance function which penalizes the solution when the constraints are violated. In terms of the primal problem, a penalty function formulation might yield

$$\phi(\mathbf{x}, \boldsymbol{\lambda}, \boldsymbol{\mu}) = f(\mathbf{x}) + \sum_{i=1}^{n_w} \lambda_i w_i^2(\mathbf{x}) + \sum_{i=1}^{n_g} \mu_i g_i^2(\mathbf{x}) u_i^2(\mathbf{x})$$

Here the inequality constraints have been converted to equality constraints by multiplying them by $u_i(\mathbf{x})$ defined by

$$u_i(\mathbf{x}) = \begin{cases} 1 & \text{if } g_i(\mathbf{x}) > 0 \\ 0 & \text{if } g_i(\mathbf{x}) \leqslant 0 \end{cases}$$

and the λ_i and μ_i are positive constants or weights which cause violations of the constraints to affect the performance function. In this method, the constraints are satisfied when the penalty terms are zero, and this is forced by adjusting the coefficients λ_i and μ_i, increasing them until the constraints are satisfied. The difference between this and the Lagrange approach is actually quite small and Beltrami has shown that in fact as the coefficient of the penalty term of a binding constraint approaches infinity, the product of the coefficient and the constraint function (which of course is approaching zero) approaches the value of the multiplier in the Lagrange formulation.[1] He uses this result to prove the Kuhn-Tucker stationarity conditions from this different point of view.

1-2-3 Saddle-point Conditions

In the case of differentiable functions with the optimum occurring at an interior point, it is evident that the stationarity conditions also imply that the point $\mathbf{x}^0$, $\boldsymbol{\lambda}^0$, $\boldsymbol{\mu}^0$ is a saddle point of the Lagrange function. If the optimum occurs on the boundary of the set S_1 or the constraint set S_3, then the signs of the slopes of the Lagrangian imply that the optimum is a constrained saddle point. This result is very important because a saddle point is defined in terms of minimizations and maximizations rather than stationarity conditions and there is a computational advantage to searching for a minimum rather than a stationary point. Moreover, in the case of multilevel problems, the decomposition leads to subproblems which are minimizations and which have physical meaning.

The saddle-point condition is equivalent to the stationarity conditions for differentiable functions and consequently is sufficient for nonconvex problems and both necessary and sufficient for convex differentiable problems (that is, for primal problems in which all functions are differentiable at all points in the appropriate sets). These results are summarized in Table 1-4. But the saddle-point condition is actually more general than this and is applicable to nonconvex and nondifferentiable functions under a wide variety of conditions, providing a powerful basis for the solution of nonlinear optimization problems. The material here is based primarily on the paper of L. S. Lasdon.[6]

To derive these, first consider the essential characteristics of a constrained saddle point (that is, a saddle point in a constrained space $S_1 \subseteq E^n$):

Theorem 1-1: A point $\mathbf{x}^0$, $\boldsymbol{\lambda}^0$, $\boldsymbol{\mu}^0$ with $\boldsymbol{\mu}^0 \geqslant 0$ is a constrained saddle point of the Lagrangian $L(\mathbf{x}, \boldsymbol{\lambda}, \boldsymbol{\mu})$ associated with the primal problem (Table 1-2) *if and only if* conditions 1 to 4 below hold:

1. $\mathbf{x}^0$ minimizes $L(\mathbf{x}, \boldsymbol{\lambda}^0, \boldsymbol{\mu}^0)$ over S_1
2. $w_i(\mathbf{x}^0) = 0 \qquad i = 1, 2,..., n_w$
3. $g_i(\mathbf{x}^0) \leqslant 0 \qquad i = 1, 2,..., n_g$
4. $\mu_i^0 g_i(\mathbf{x}^0) = 0 \qquad i = 1, 2,..., n_g$

Notice the lack of assumptions about the set S_1 and the functions $f(\mathbf{x})$, $w_i(\mathbf{x})$, or $g_i(\mathbf{x})$. Thus this theorem includes finite sets S_1, and nonconvex and nondifferentiable functions. Notice that condition 1 calls for a *global*, not local, minimization. These conditions are summarized in Table 1-2.

The proof of this theorem is straightforward and involves only the

Table 1-2. Saddle-point Condition for a Minimum

Primal Problem

$$\min_{\mathbf{x}} f(\mathbf{x})$$

$$\text{subject to } \mathbf{x} \in S_1 \cap S_2 \cap S_3$$

where

$$\begin{aligned} S_1 &\subseteq E^n \\ S_2 &= \{\mathbf{x} \mid w_i(\mathbf{x}) = 0, i = 1, 2,..., n_w\} \\ S_3 &= \{\mathbf{x} \mid g_i(\mathbf{x}) \leqslant 0, i = 1, 2,..., n_g\} \\ \mathbf{x} &= n\text{-vector of real variables.} \\ f(\mathbf{x}), w_i(\mathbf{x}), g_i(\mathbf{x}) &= \text{real-valued} \\ &\quad\ \text{functions defined on } S_1 \end{aligned}$$

Lagrange Function

$$L(\mathbf{x}, \boldsymbol{\lambda}, \boldsymbol{\mu}) = f(\mathbf{x}) + \sum_{i=1}^{n_w} \lambda_i w_i(\mathbf{x}) + \sum_{i=1}^{n_g} \mu_i g_i(\mathbf{x})$$

Saddle-Point Condition for point $(\mathbf{x}^0, \boldsymbol{\lambda}^0, \boldsymbol{\mu}^0)$

1. $\mathbf{x}^0$ minimizes $L(\mathbf{x}, \boldsymbol{\lambda}^0, \boldsymbol{\mu}^0)$ over S_1
2. $w_i(\mathbf{x}^0) = 0 \qquad i = 1, 2,..., n_w$
3. $g_i(\mathbf{x}^0) \leqslant 0$

   $\mu_i^0 g_i(\mathbf{x}^0) = 0$

   $\mu_i^0 \geqslant 0 \qquad i = 1, 2,..., n_g$

definitions of a saddle point and Lagrangian. To show necessity, assume that the point $\mathbf{x}^0$, $\boldsymbol{\lambda}^0$, $\boldsymbol{\mu}^0$ is a saddle point of the Lagrangian of the primal problem. Then by definition of a saddle point,

$$L(\mathbf{x}^0, \boldsymbol{\lambda}^0, \boldsymbol{\mu}^0) \leqslant L(\mathbf{x}, \boldsymbol{\lambda}^0, \boldsymbol{\mu}^0) \qquad \text{for all} \quad \mathbf{x} \in S_1$$

which is condition 1 of the theorem. For a saddle point,

$$L(\mathbf{x}^0, \boldsymbol{\lambda}, \boldsymbol{\mu}) \leqslant L(\mathbf{x}^0, \boldsymbol{\lambda}^0, \boldsymbol{\mu}^0) \qquad \text{for all} \quad \boldsymbol{\mu} \geqslant 0 \quad \text{and all} \quad \boldsymbol{\lambda}$$

Expand this in terms of its definition, giving

$$f(\mathbf{x}^0) + \sum_{i=1}^{n_w} \lambda_i w_i(\mathbf{x}^0) + \sum_{i=1}^{n_g} \mu_i g_i(\mathbf{x}^0) \leqslant f(\mathbf{x}^0) + \sum_{i=1}^{n_w} \lambda_i^0 w_i(\mathbf{x}^0) + \sum_{i=1}^{n_g} \mu_i^0 g_i(\mathbf{x}^0)$$

or

$$\sum_{i=1}^{n_w} (\lambda_i - \lambda_i^0) w_i(\mathbf{x}^0) + \sum_{i=1}^{n_g} (\mu_i - \mu_i^0) g_i(\mathbf{x}^0) \leqslant 0$$

Now if any $w_i(\mathbf{x}^0)$ is nonzero, a value for λ_i (which is unconstrained as to sign) may be selected so that the inequality would be violated. Hence condition 2 above must hold. Similarly, if any $g_i(\mathbf{x}^0)$ is positive, a positive μ_i can be selected so that the inequality is violated. Hence condition 3 above must hold. Since the $w_i(\mathbf{x}^0) = 0$, the inequality for $\boldsymbol{\mu} = 0$ becomes

$$\sum_{i=1}^{n_g} \mu_i^0 g_i(\mathbf{x}^0) \geqslant 0$$

However, the μ_i^0 are positive and the $g_i(\mathbf{x}^0) \leqslant 0$. Hence it is also true that

$$\sum_{i=1}^{n_g} \mu_i^0 g_i(\mathbf{x}^0) \leqslant 0$$

Clearly then

$$\mu_i^0 g_i(\mathbf{x}^0) = 0 \qquad i = 1, 2, \ldots, n_g$$

and condition 4 must hold and the necessity is proved.

Sufficiency is even more straightforward to prove. Assuming the conditions listed in the theorem, condition 1 immediately implies that

$$L(\mathbf{x}^0, \boldsymbol{\lambda}^0, \boldsymbol{\mu}^0) \leqslant L(\mathbf{x}, \boldsymbol{\lambda}^0, \boldsymbol{\mu}^0) \qquad \text{for all} \quad \mathbf{x} \in S_1$$

which is half of the saddle-point definition. The other half is shown by writing

$$L(\mathbf{x}^0, \boldsymbol{\lambda}, \boldsymbol{\mu}) = f(\mathbf{x}^0) + \sum_{i=1}^{n_w} \lambda_i w_i(\mathbf{x}^0) + \sum_{i=1}^{n_g} \mu_i g_i(\mathbf{x}^0)$$

Since $w_i(\mathbf{x}^0) = 0$ and $g_i(\mathbf{x}^0) \leqslant 0$ and $\mu_i \geqslant 0$, the second term is zero and the third term nonnegative. Consequently,

$$L(\mathbf{x}^0, \boldsymbol{\lambda}, \boldsymbol{\mu}) \leqslant f(\mathbf{x}^0)$$

But

$$L(\mathbf{x}^0, \boldsymbol{\lambda}^0, \boldsymbol{\mu}^0) = f(\mathbf{x}^0)$$

and hence

$$L(\mathbf{x}^0, \boldsymbol{\lambda}, \boldsymbol{\mu}) \leqslant L(\mathbf{x}^0, \boldsymbol{\lambda}^0, \boldsymbol{\mu}^0)$$

and the sufficiency is proved also.

The result that makes the preceding theorem powerful and important is the following:

Theorem 1-2: If the point $\mathbf{x}^0, \boldsymbol{\lambda}^0, \boldsymbol{\mu}^0$ is a saddle point for the Lagrangian of the primal problem, then $\mathbf{x}^0$ is the solution to the primal problem.

Thus under very general conditions, a sufficient condition for the solution to the constrained optimization problem exists, namely, that a constrained saddle point exists. This theorem is proved by noting that if the point $\mathbf{x}^0, \boldsymbol{\lambda}^0, \boldsymbol{\mu}^0$ is a saddle point, the conditions listed in the previous theorem must hold. Hence the constraints of the primal problem are satisfied by $\mathbf{x}^0$ and it is only necessary to show that $f(\mathbf{x}^0) \leqslant f(\mathbf{x})$ for all $\mathbf{x}$ which satisfy the constraints. Since the saddle point exists, it follows that

$$f(\mathbf{x}^0) + \sum_{i=1}^{n_w} \lambda_i^0 w_i(\mathbf{x}^0) + \sum_{i=1}^{n_g} \mu_i^0 g_i(\mathbf{x}^0) \leqslant f(\mathbf{x}) + \sum_{i=1}^{n_w} \lambda_i^0 w_i(\mathbf{x}) + \sum_{i=1}^{n_g} \mu_i^0 g_i(\mathbf{x}) \qquad \text{for all} \quad \mathbf{x} \in S_1$$

By conditions 3 and 4 of the previous theorem, the left-hand side of this inequality reduces to $f(\mathbf{x}^0)$. Now for any $\mathbf{x} \in S_2$, $w_i(\mathbf{x}) = 0$ and for any $x \in S_3$, $g_i(\mathbf{x}) \leqslant 0$. Hence for any $x \in S_1 \cap S_2 \cap S_3$,

$$\sum_{i=1}^{n_w} \lambda_i^0 w_i(\mathbf{x}) = 0$$

and since $\mu_i^0 \geqslant 0$,

$$\sum_{i=1}^{n_g} \mu_i^0 g_i(\mathbf{x}) \leqslant 0$$

and consequently the above inequality reduces to

$$f(\mathbf{x}^0) \leqslant f(\mathbf{x}) \qquad \text{for all} \quad \mathbf{x} \in S_1 \cap S_2 \cap S_3$$

and the theorem is proved.

If the sets S_1, S_2, and S_3 are all convex [and recall that S_2 is the set of $\mathbf{x}$ satisfying the equality constraints and this set is convex only if the $w_i(\mathbf{x})$ are linear or absent] then the saddle point necessarily exists and $\mathbf{x}^0$ solves the primal problem provided that there exists at least one $\mathbf{x}$ satisfying the inequality constraints strictly. That is, if there exists at least one $\mathbf{x}$ such that $g_i(\mathbf{x}) < 0$ for $i = 1, 2,..., n_g$ and simultaneously satisfying the $w_i(\mathbf{x}) = 0$, then the saddle-point condition is both necessary and sufficient. These results are summarized in Table 1-4 although the above regularity condition on $\mathbf{x}$ is not listed there. This result is proved by Karlin.[15]

The importance of the saddle point is not simply that it is necessary and sufficient for convex functions, but that it is sufficient for any optimization problem with constraints and provides a method for solving such problems, namely, *search for a saddle point of the Lagrange function using minimization techniques.*

The only difficulty with these results is that the minimization of the Lagrangian is assumed done with $\boldsymbol{\lambda} = \boldsymbol{\lambda}^0$ and $\boldsymbol{\mu} = \boldsymbol{\mu}^0$ which generally are not known in advance. In order to provide a basis for decomposition and a practical solution of these problems, it is necessary to consider the relation between the Lagrangian with nonoptimal $\boldsymbol{\lambda}$ and $\boldsymbol{\mu}$ and the primal problem.

1-2-4 Dual Formulation of the Optimization Problem

The strategy for finding the solution to the optimization problem is simply to search for the saddle point of the Lagrange function for if successful, the saddle point is sufficient to solve the problem, whether or not the functions involved are convex, differentiable, etc.[6] In order to find the saddle point, however, the Lagrangian must be examined for values of the Lagrange multipliers other than the optimal ones. Because these multipliers enter linearly into the Lagrangian, much can in fact be said about solutions by introducing the concept of the dual function.

Define the function $h(\boldsymbol{\lambda}, \boldsymbol{\mu})$ as shown and call it the *dual* function:

$$h(\boldsymbol{\lambda}, \boldsymbol{\mu}) = \min_{\mathbf{x} \in S_1} L(\mathbf{x}, \boldsymbol{\lambda}, \boldsymbol{\mu})$$

Notice that the minimization of the Lagrangian is taken over all allowable $\mathbf{x}$ with $\boldsymbol{\lambda}$ and $\boldsymbol{\mu}$ fixed as parameters but that these $\mathbf{x}$ may not satisfy the constraint sets S_2 and S_3. Furthermore, the minimum of the Lagrangian may not exist for all $\boldsymbol{\lambda}$ and $\boldsymbol{\mu}$ so that $h(\boldsymbol{\lambda}, \boldsymbol{\mu})$ is not defined for all allowed values of $\boldsymbol{\lambda}$ and $\boldsymbol{\mu}$. Define

$$D = \{(\boldsymbol{\lambda}, \boldsymbol{\mu}) \mid h(\boldsymbol{\lambda}, \boldsymbol{\mu}) \text{ exists and } \boldsymbol{\mu} \geqslant 0\}$$

Thus D is the set of all multipliers for which the Lagrangian has a finite minimum.

The set D may be empty in some instances but is very well behaved if the set S_1 is bounded and the Lagrangian is continuous in $\mathbf{x}$ for all $\mathbf{x} \in S_1$ and all allowed $\boldsymbol{\lambda}$ and $\boldsymbol{\mu}$ (that is, for all $\boldsymbol{\mu} \geqslant 0$ and all $\boldsymbol{\lambda}_i$). Introducing the dual function is no restriction on the original problem because the point $(\boldsymbol{\lambda}^0, \boldsymbol{\mu}^0)$ must belong to D since if the saddle point exists, the Lagrangian is a minimum there according to condition 1 of Table 1-2.

But how does $\mathbf{h}(\boldsymbol{\lambda}, \boldsymbol{\mu})$ permit the determination of $(\boldsymbol{\lambda}^0, \boldsymbol{\mu}^0)$? Consider the following theorem using the same definitions as in Table 1-2:

Theorem 1-3: The dual function $h(\boldsymbol{\lambda}, \boldsymbol{\mu}) \leqslant f(\mathbf{x})$ for all $\mathbf{x} \in S = S_1 \cap S_2 \cap S_3$ and $(\boldsymbol{\lambda}, \boldsymbol{\mu}) \in D$.

The proof that the dual function provides a lower bound on the minimum value of $f(\mathbf{x})$ is simply to assume $\mathbf{x} \in S_1 \cap S_2 \cap S_3$ and write that

$$h(\boldsymbol{\lambda}, \boldsymbol{\mu}) = \min_{\mathbf{x} \in S_1} L(\mathbf{x}, \boldsymbol{\lambda}, \boldsymbol{\mu}) \leqslant L(\mathbf{x}, \boldsymbol{\lambda}, \boldsymbol{\mu}) \qquad \mathbf{x} \in S_1$$

and expand this using the definition of the Lagrangian:

$$h(\boldsymbol{\lambda}, \boldsymbol{\mu}) \leqslant f(\mathbf{x}) + \sum_{i=1}^{n_w} \lambda_i w_i(\mathbf{x}) + \sum_{i=1}^{n_g} \mu_i g_i(\mathbf{x}) \qquad \mathbf{x} \in S_1$$

Notice that if $\mathbf{x} \in S_2$, then $w_i(\mathbf{x}) = 0$, $i = 1, 2, \ldots, n_w$ and hence

$$h(\boldsymbol{\lambda}, \boldsymbol{\mu}) \leqslant f(\mathbf{x}) + \sum_{i=1}^{n_g} \mu_i g_i(\mathbf{x}) \qquad \mathbf{x} \in S_1 \cap S_2$$

Furthermore, if $\mathbf{x} \in S_3$ (as it does by hypothesis), then $g_i(\mathbf{x}) \leqslant 0$, $i = 1, 2, \ldots, n_g$, and if $(\boldsymbol{\lambda}, \boldsymbol{\mu}) \in D$, then $\boldsymbol{\mu} \geqslant 0$. Thus

$$\sum_{i=1}^{n_g} \mu_i g_i(\mathbf{x}) \leqslant 0 \qquad \mathbf{x} \in S$$

and

$$h(\boldsymbol{\lambda}, \boldsymbol{\mu}) \leqslant f(\mathbf{x}) \qquad \mathbf{x} \in S \qquad (\boldsymbol{\lambda}, \boldsymbol{\mu}) \in D$$

and the theorem is proved.

Since the dual function provides a lower bound on $f(\mathbf{x})$, the greatest lower bound must occur at the maximum value of $h(\boldsymbol{\lambda}, \boldsymbol{\mu})$ in D and leads us to define the *dual problem*:

$$\max_{(\boldsymbol{\lambda}, \boldsymbol{\mu}) \in D} h(\boldsymbol{\lambda}, \boldsymbol{\mu})$$

The primal and dual problems are summarized in Table 1-3 and are important because of the theorem:

Theorem 1-4: The point $\mathbf{x}^0$, $\boldsymbol{\lambda}^0$, $\boldsymbol{\mu}^0$ is a constrained saddle point of the optimization problem defined in Table 1-3 *if and only if*

1. $\mathbf{x}^0$ solves the primal problem.
2. $(\boldsymbol{\lambda}^0, \boldsymbol{\mu}^0)$ solves the dual problem.
3. $f(\mathbf{x}^0) = h(\boldsymbol{\lambda}^0, \boldsymbol{\mu}^0)$.

Table 1-3. Dual Formulation of the Minimality Condition

Primal Problem

$$\min_{\mathbf{x}} f(\mathbf{x})$$

$$\text{subject to } \mathbf{x} \in S_1 \cap S_2 \cap S_3$$

where

$$S_1 \subseteq E^n$$
$$S_2 = \{\mathbf{x} \mid w_i(\mathbf{x}) = 0, i = 1, 2,\ldots, n_w\}$$
$$S_3 = \{\mathbf{x} \mid g_i(\mathbf{x}) \leqslant 0, i = 1, 2,\ldots, n_g\}$$
$$\mathbf{x} = n\text{-vector of real variables}$$
$$f(\mathbf{x}), w_i(\mathbf{x}), g_i(\mathbf{x}) = \text{real-valued functions defined on } S_1$$

Lagrangian

$$L(\mathbf{x}, \boldsymbol{\lambda}, \boldsymbol{\mu}) = f(\mathbf{x}) + \sum_{i=1}^{n_w} \lambda_i w_i(\mathbf{x}) + \sum_{i=1}^{n_g} \mu_i g_i(\mathbf{x})$$

Dual Function

$$h(\boldsymbol{\lambda}, \boldsymbol{\mu}) = \min_{\mathbf{x} \in S_1} L(\mathbf{x}, \boldsymbol{\lambda}, \boldsymbol{\mu})$$

$$D = \{\boldsymbol{\lambda}, \boldsymbol{\mu} \mid \mathbf{h}(\boldsymbol{\lambda}, \boldsymbol{\mu}) \text{ exists}\}$$

Dual Problem

$$\max_{(\boldsymbol{\lambda}, \boldsymbol{\mu}) \in D} h(\boldsymbol{\lambda}, \boldsymbol{\mu})$$

Conditions for Optimality of the point$(\mathbf{x}^0, \boldsymbol{\lambda}^0, \boldsymbol{\mu}^0)$

1. $\mathbf{x}^0$ solves the primal problem
2. $(\boldsymbol{\lambda}^0, \mu^0)$ maximizes $h(\boldsymbol{\lambda}, \boldsymbol{\mu})$ over D
3. $f(\mathbf{x}^0) = h(\boldsymbol{\lambda}^0, \boldsymbol{\mu}^0)$

This theorem states the remarkable result that a constrained saddle point has been found (and hence solves the desired optimization problem according to Table 1-2) provided that after maximizing the dual function $h(\boldsymbol{\lambda}, \boldsymbol{\mu})$ and finding $\mathbf{x}^0$ which corresponds to the maximizing $\boldsymbol{\lambda}$ and $\boldsymbol{\mu}$, that is, $\boldsymbol{\lambda}^0$, $\boldsymbol{\mu}^0$, the corresponding values of the performance function and the dual function are equal. By the previous theorem, $h(\boldsymbol{\lambda}, \boldsymbol{\mu}) \leqslant f(\mathbf{x}^0)$, and this theorem then says that if this lower bound is in fact equal to the function $f(\mathbf{x}^0)$, a saddle point and consequently the global optimum for the primal problem have been found. This result is important enough to justify its careful proof.

First consider the necessity. Assume the point $\mathbf{x}^0$, $\boldsymbol{\lambda}^0$, $\boldsymbol{\mu}^0$ is a constrained saddle point of the primal problem in Table 1-3. Then by Table 1-2, $\mathbf{x}^0$ solves the primal problem (condition 1 above) and it is necessary only to show that conditions 2 and 3 hold. Because $\mathbf{x}^0$, $\boldsymbol{\lambda}^0$, $\boldsymbol{\mu}^0$ is a constrained saddle point, conditions 1 to 3 in Table 1-2 hold. Hence $\mathbf{x}^0$

Table 1-4. Summary of Conditions on Primal Problem

All Functions Differentiable	All Functions Convex	Stationarity Conditions (Table 1-1)	Saddle-point Conditions (Table 1-2)	Duality Conditions (Table 1-3)
Yes	Yes	Necessary and sufficient	Necessary and sufficient	Necessary and sufficient
Yes	No	Necessary only	Sufficient only	Sufficient only
No	Yes	Not applicable	Necessary and sufficient	Necessary and sufficient
No	No	Not applicable	Sufficient only	Sufficient only

minimizes the Lagrangian $L(\mathbf{x}, \boldsymbol{\lambda}^0, \boldsymbol{\mu}^0)$ over S_1. Thus by the definition of $h(\boldsymbol{\lambda}, \boldsymbol{\mu})$,

$$h(\boldsymbol{\lambda}^0, \boldsymbol{\mu}^0) = f(\mathbf{x}^0) + \sum_{i=1}^{n_w} \lambda_i{}^0 w_i(\mathbf{x}^0) + \sum_{i=1}^{n_g} \mu_i{}^0 g_i(\mathbf{x}^0)$$

The terms involving the constraints are identically zero because when $\mathbf{x}^0$ solves the primal problem each $w_i(\mathbf{x}^0) = 0$ and each $\mu_i{}^0 g_i(\mathbf{x}^0) = 0$ (Table 1-2). Thus

$$h(\boldsymbol{\lambda}^0, \boldsymbol{\mu}^0) = f(\mathbf{x}^0)$$

or condition 3 holds.

By Theorem 1-3,

$$h(\boldsymbol{\lambda}, \boldsymbol{\mu}) \leqslant f(\mathbf{x}^0) = h(\lambda^0, \boldsymbol{\mu}^0)$$

for all $(\boldsymbol{\lambda}, \boldsymbol{\mu}) \in D$. Thus the point $(\boldsymbol{\lambda}^0, \boldsymbol{\mu}^0)$ maximizes $h(\boldsymbol{\lambda}, \boldsymbol{\mu})$ over D and condition 2 is shown to follow. The necessity is proved.

Sufficiency of the theorem is shown by assuming conditions 1 to 3 hold and showing that the point $\mathbf{x}^0$, $\boldsymbol{\lambda}^0$, $\boldsymbol{\mu}^0$ is a constrained saddle point by showing that the three conditions of Table 1-2 must hold.

First, since $\mathbf{x}^0$ solves the primal problem, all constraints are satisfied:

$$w_i(\mathbf{x}^0) = 0 \qquad i = 1, 2, \ldots, n_w$$

$$g_i(\mathbf{x}^0) \leqslant 0 \qquad i = 1, 2, \ldots, n_g$$

and hence condition 2 and part of condition 3 of Table 1-2 are satisfied.

By definition,

$$h(\boldsymbol{\lambda}^0, \boldsymbol{\mu}^0) = \min_{\mathbf{x} \in S_1} L(\mathbf{x}, \boldsymbol{\lambda}^0, \boldsymbol{\mu}^0)$$

but it is necessary to prove that $\mathbf{x}^0$ is actually the $\mathbf{x}$ which minimizes the Lagrangian for $\boldsymbol{\lambda} = \boldsymbol{\lambda}^0$ and $\boldsymbol{\mu} = \boldsymbol{\mu}^0$.

Let $\mathbf{x}^*$ be the $\mathbf{x}$ which minimizes $L(\mathbf{x}, \boldsymbol{\lambda}^0, \boldsymbol{\mu}^0)$ such that if $\mathbf{x}^* \neq \mathbf{x}^0$,

$$L(\mathbf{x}^*, \boldsymbol{\lambda}^0, \boldsymbol{\mu}^0) < L(\mathbf{x}^0, \boldsymbol{\lambda}^0, \boldsymbol{\mu}^0)$$

A strict inequality is written here since $\mathbf{x}^*$ minimizes over S_1 and if the equality case arose, $\mathbf{x}^0$ would be taken as the minimizing $\mathbf{x}$. We now look for a contradiction. By definition,

$$h(\boldsymbol{\lambda}^0, \boldsymbol{\mu}^0) = L(\mathbf{x}^*, \boldsymbol{\lambda}^0, \boldsymbol{\mu}^0) = f(\mathbf{x}^*) + \sum_{i=1}^{n_w} \lambda_i^0 w_i(\mathbf{x}^*) + \sum_{i=1}^{n_g} \mu_i^0 g_i(\mathbf{x}^*)$$

$$< f(\mathbf{x}^0) + \sum_{i=1}^{n_w} \lambda_i^0 w_i(\mathbf{x}^0) + \sum_{i=1}^{n_g} \mu_i^0 g_i(\mathbf{x}^0)$$

But $h(\boldsymbol{\lambda}^0, \boldsymbol{\mu}^0) = f(\mathbf{x}^0)$ by hypothesis, so that

$$\sum_{i=1}^{n_w} \lambda_i^0 w_i(\mathbf{x}^0) + \sum_{i=1}^{n_g} \mu_i^0 g_i(\mathbf{x}^0) > 0$$

where again the inequality is strictly greater than zero. But $\mathbf{x}^0$ solves the primal and hence $w_i(\mathbf{x}^0) = 0$ and $g_i(\mathbf{x}^0) \leqslant 0$ for all i. Since $(\boldsymbol{\lambda}^0, \boldsymbol{\mu}^0)$ solves the dual, $\boldsymbol{\mu}^0 \geqslant 0$, and since $\mathbf{x}^0$ solves the primal,

$$\mu_i^0 g_i(\mathbf{x}^0) = 0 \qquad i = 1, 2, \ldots, n_g$$

and this contradicts the above strict inequality (and means that condition 3 of Table 1-2 is fulfilled). Hence $\mathbf{x}^* = \mathbf{x}^0$ or $\mathbf{x}^0$ is actually the $\mathbf{x}$ which minimizes the Lagrangian when $\boldsymbol{\lambda} = \boldsymbol{\lambda}^0$ and $\boldsymbol{\mu} = \boldsymbol{\mu}^0$. Thus condition 1 of Table 1-2 is satisfied and the sufficiency is proved.

The dual problem is actually well behaved in the sense that it is a straightforward maximization problem. In fact, the problem of multiple maxima will not usually exist as the following theorem indicates:

Theorem 1-5: The dual function $h(\boldsymbol{\lambda}, \mu)$ defined in Table 1-3 is concave over any convex subset of its domain D.

To show this, let D^* be a convex subset of D and consider any two points in D^*, $(\boldsymbol{\lambda}^1, \boldsymbol{\mu}^1)$ and $(\boldsymbol{\lambda}^2, \boldsymbol{\mu}^2)$. Any point on the line connecting them may be written as

$$\boldsymbol{\lambda} = \alpha\boldsymbol{\lambda}^1 + (1 - \alpha)\boldsymbol{\lambda}^2$$

$$\boldsymbol{\mu} = \alpha\mu^1 + (1 - \alpha)\mu^2 \qquad 0 \leqslant \alpha \leqslant 1$$

and lies in D^* since D^* is convex. By definition of the dual function (Table 1-3),

$$h(\boldsymbol{\lambda}, \boldsymbol{\mu}) = \min_{\mathbf{x} \in S_1} L(\mathbf{x}, \boldsymbol{\lambda}, \boldsymbol{\mu})$$

But the Lagrangian is linear in $\boldsymbol{\lambda}$ and $\boldsymbol{\mu}$ and therefore

$$h(\boldsymbol{\lambda}, \boldsymbol{\mu}) = \min_{\mathbf{x} \in S_1} [\alpha L(\mathbf{x}, \boldsymbol{\lambda}^1, \boldsymbol{\mu}^1) + (1 - \alpha)L(\mathbf{x}, \boldsymbol{\lambda}^2, \boldsymbol{\mu}^2)]$$

Since the minimum of a sum is greater than or equal to the sum of the minima,

$$h(\boldsymbol{\lambda}, \boldsymbol{\mu}) \geqslant \alpha h(\boldsymbol{\lambda}^1, \boldsymbol{\mu}^1) + (1 - \alpha)h(\boldsymbol{\lambda}^2, \boldsymbol{\mu}^2) \qquad 0 \leqslant \alpha \leqslant 1$$

which implies that the dual function is concave over D^* and the theorem is proved.

It is worthwhile considering several simple examples of these methods. Consider the example of Sec. 1-2-1. The Lagrangian may be written as

$$L = (x_1 - 1)^2 + (x_2 - 2)^2 + \lambda(x_2 - x_1 - 1) + \mu(x_2 + x_1 - 2)$$

which can be rearranged as

$$L = (x_1 - 1)^2 + x_1(\mu - \lambda) + (x_2 - 2)^2 + x_2(\lambda + \mu) - (\lambda + 2\mu)$$

The global minimum for the Lagrangian occurs at the point where its gradient with respect to $\mathbf{x}$ is zero in this example subject to the constraint that $\mathbf{x} \geqslant 0$. Thus

$$x_1{}^0 = 1 + \tfrac{1}{2}(\lambda - \mu) \qquad \lambda - \mu \geqslant -2$$

$$x_2{}^0 = 2 - \tfrac{1}{2}(\lambda + \mu) \qquad \lambda + \mu \leqslant 4$$

and the dual function is obtained by simply substituting $\mathbf{x}^0$ into the Lagrangian:

$$h(\lambda, \mu) = -\tfrac{1}{2}\mu^2 + \mu - \tfrac{1}{2}\lambda^2$$

$$D = \{\lambda, \mu \mid \lambda - \mu \geqslant -2, \lambda + \mu \leqslant 4, \mu \geqslant 0\}$$

The set D is diagrammed in Fig. 1-6. The maximum of the dual function over D is easily found to occur at

$$(\lambda, \mu) = (0, 1)$$

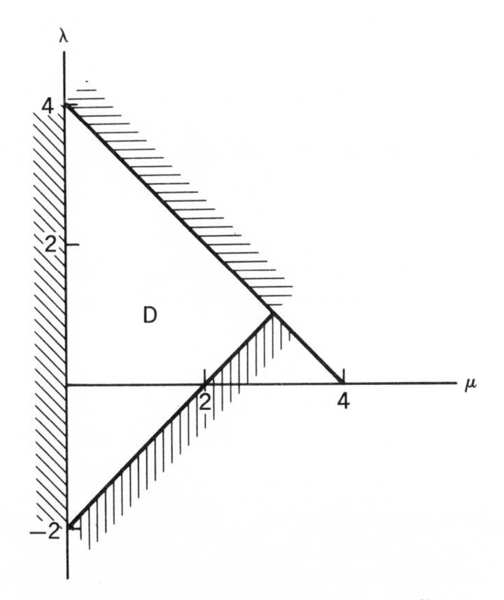

FIG. 1-6. The set of allowed multipliers.

and here $x_1{}^0 = \frac{1}{2}$ and $x_2{}^0 = \frac{3}{2}$, the same answers as obtained in Sec. 1-2-1. At this point, the dual function becomes

$$h(\lambda^0, \mu^0) = \tfrac{1}{2}$$

and the objective function

$$f(x^0) = \tfrac{1}{2}$$

Since these are equal, since the point (λ^0, μ^0) maximizes the dual function over D, and since the point $\mathbf{x}^0$ satisfies the primal problem, the point corresponds to a saddle point and is the desired solution to the optimization problem.

It is instructive to consider a purely linear programming problem from this point of view:

$$\min_{\mathbf{x}} f(\mathbf{x}) = x_1 - x_2$$

$$\text{subject to } g(\mathbf{x}) = x_1 + x_2 - 1 \leqslant 0$$

$$x_1 \geqslant 0 \qquad x_2 \geqslant 0$$

First, form the Lagrangian

$$\begin{aligned} L(\mathbf{x}, \boldsymbol{\mu}) &= x_1 - x_2 + \mu(x_1 + x_2 - 1) \\ &= x_1(1 + \mu) - x_2(1 - \mu) - \mu \end{aligned}$$

No minimum of L exists unless

$$\begin{aligned} \mu &\geqslant 0 \\ 1 + \mu &\geqslant 0 \\ 1 - \mu &\leqslant 0 \end{aligned}$$

and this determines the region D or $D = \{\mu \mid \mu \geqslant 1\}$. The optimum value of $\mathbf{x}$ is

$$\begin{aligned} x_1 &= 0 \qquad \text{for any} \quad \mu \in D \\ x_2 &= 0 \qquad \text{if} \quad \mu > 1 \end{aligned}$$

If $\mu = 1$, any x_2 is permitted.

The dual function becomes

$$h(\mu) = -\mu \qquad \mu \in D$$

The maximum clearly exists at $\mu = 1$ where $h(1) = -1$.

There remains the problem of finding an x_2 since it is not determined by the dual problem when $\mu = 1$. Substituting the known value $x_1{}^0 = 0$ into the constraint results in $x_2 \leqslant 1$ and clearly $f(\mathbf{x})$ is minimized if $x_2 = 1$ and here $f(\mathbf{x}^0) = -1 = h(\boldsymbol{\mu}^0)$. Thus an optimal solution has been found. In general, the linear programming problem may be formulated as

$$\min_{\mathbf{x}} f(\mathbf{x}) = \mathbf{c}'\mathbf{x}$$

$$\text{subject to } \mathbf{A}\mathbf{x} \leqslant \mathbf{b}$$

$$\mathbf{x} \geqslant 0$$

The Lagrangian for this problem is

$$L(\mathbf{x}, \boldsymbol{\mu}) = \mathbf{c}'\mathbf{x} + \boldsymbol{\mu}'(\mathbf{A}\mathbf{x} - \mathbf{b})$$
$$= (\mathbf{c}' + \boldsymbol{\mu}'\mathbf{A})\mathbf{x} - \boldsymbol{\mu}'\mathbf{b} \qquad \boldsymbol{\mu} \geqslant 0$$

and a minimum exists only if

$$\mathbf{c}' + \boldsymbol{\mu}'\mathbf{A} \geqslant 0$$

Under these conditions, the minimum is found by choosing $\mathbf{x}_i = 0$ if the ith component of $\mathbf{c} + \mathbf{A}'\boldsymbol{\mu}$ is nonzero. In any case,

$$(\mathbf{c}' + \mu'\mathbf{A})\mathbf{x} = 0$$

and the dual function is

$$h(\boldsymbol{\mu}) = -\mathbf{b}'\boldsymbol{\mu} \qquad \boldsymbol{\mu} \in D$$
$$D = \{\boldsymbol{\mu} \mid \mathbf{A}'\boldsymbol{\mu} + \mathbf{c} \geqslant 0, \boldsymbol{\mu} \geqslant 0\}$$

The dual problem is then

$$\max \; -\mathbf{b}'\boldsymbol{\mu}$$
$$\text{subject to } \mathbf{A}'\boldsymbol{\mu} + \mathbf{c} \geqslant 0$$
$$\boldsymbol{\mu} \geqslant 0$$

This is the classical dual of the linear programming problem and is useful in many circumstances.

1-2-5 Multilevel Solution of the Constrained Optimization Problem

The formulation of the equality and inequality constrained optimization problem in terms of primal and dual problems is very general, applying to finite as well as infinite constraint sets, differentiable and nondifferentiable objective and constraint functions, convex and nonconvex programming problems. Efficient solution of the problem is best handled by applying the multilevel principle discussed earlier, that is, converting the integrated constrained optimization problems into an interactive process carried out by a two-level structure. This may be done by assigning the integrated problem to the first-level unit, with some subset of the variables and multipliers called the *coordinating set* assumed fixed or known. The first-level unit solves this new optimization problem (which may or may not be constrained depending upon which coordinating set is chosen) assuming that the coordinating set is fixed and transmits the results of the optimization to the second-level unit whose function it is to determine the optimal values of the variables in the coordinating set and transmit them to the first-level unit.

The characteristics of the first-level problem vary considerably depending upon which variables are chosen to be in the coordinating set. For example, if the multipliers associated with all the constraints are selected to be the coordinating variables, the first-level problem is essentially that of finding the minimum of an unconstrained function (the Lagrangian). If the coordinating set is selected to be some subset of the $\mathbf{x}_i$, then the first-level unit has to solve a constrained optimization problem of lower dimensionality than the original. In general any combination of multipliers and variables can be used as the coordinating set as best suits the problem at hand.

The choice of the coordinating variables depends on two considerations:

1. The resulting second-level problem must be easy to solve. That is, either the space must be small enough to search or else gradients must exist so that efficient numerical optimizing schemes can be used.
2. The first-level problems must possess solutions and must be efficiently solved. This usually means that the first-level problem must be capable of decomposition into parallel, independent simpler problems.

The second condition is the more important one and usually provides the key to the choice of the coordinating variables. For example, optimization of a physical process involving several interconnected units is complex because of the interconnections. Choosing the multipliers associated with the interconnection constraints then results in independent first-level problems with one optimization problem associated with each physical unit. In such a situation, the existence of solutions is often ensured and methods of finding such solutions efficiently are often available.

The choice of a set of coordinating variables corresponding to a subset of the constraint multipliers is the most common case and consequently, it is important to investigate the first condition above, namely, how the second-level unit can determine the optimal multipliers to supply to the first-level unit.

The introduction of the dual problem in the previous section results in a natural two-level formulation of the optimization problem with the first level solving the primal problem and the second level solving the dual problem. However it is impractical in general for the first level to minimize the Lagrangian as an explicit function of the multipliers and therefore it is desirable to change the problem into a sequential search for the optimum. This may be done if the first level is assigned the task of minimizing the Lagrange function $L(\mathbf{x}, \boldsymbol{\lambda}^i, \boldsymbol{\mu}^i)$ where the multipliers $\boldsymbol{\lambda}^i$ and $\boldsymbol{\mu}^i$ are supplied by the second level. In turn, the first level supplies the second level with the $\mathbf{x}^i$ which minimizes that Lagrangian. The second level then determines new estimates for the multipliers, $\boldsymbol{\lambda}^{i+1}$ and $\boldsymbol{\mu}^{i+1}$, transmits them to the first level, and the process continues until the conditions of Table 1-3 are satisfied.

The second-level problem is well formed in the sense that $h(\boldsymbol{\lambda}, \boldsymbol{\mu})$ is concave over any convex subset of its domain D and hence if the gradient of $h(\boldsymbol{\lambda}, \boldsymbol{\mu})$ exists and can be found, any hill-climbing procedure can be used to find the optimal parameters.

Derivatives of the dual function with respect to the multipliers are not a trivial problem even though the dual function is equal to the Lagrangian (in which the multipliers enter linearly) because the values of $\mathbf{x}$ which minimize the Lagrangian over S_1 depend implicitly on the multipliers.[6]

Define right- and left-hand partial derivatives with respect to one of the μ_i as

RIGHT-HAND PARTIAL DERIVATIVE:

$$\left.\frac{\partial h(\boldsymbol{\lambda}, \boldsymbol{\mu})}{\partial \mu_i}\right]^+ = \lim_{\alpha \to 0+} \frac{h(\boldsymbol{\lambda}, \boldsymbol{\mu})|_{\mu_i \leftarrow \mu_i\ \alpha} - h(\boldsymbol{\lambda}, \boldsymbol{\mu})}{\alpha}$$

LEFT-HAND PARTIAL DERIVATIVE:

$$\left.\frac{\partial h(\boldsymbol{\lambda}, \boldsymbol{\mu})}{\partial \mu_i}\right]^- = \lim_{\alpha \to 0+} \frac{h(\boldsymbol{\lambda}, \boldsymbol{\mu}) - h(\boldsymbol{\lambda}, \boldsymbol{\mu})|_{\mu_i \leftarrow \mu_i - \alpha}}{\alpha}$$

If the set S_1 of allowable $\mathbf{x}$ is closed and bounded (and in almost any practical problem the set can be made to satisfy these conditions) and if the functions $f(\mathbf{x})$, $w_i(\mathbf{x})$, and $g_i(\mathbf{x})$ are continuous on S_1 then these partial derivatives may be shown to be[6]

$$\left.\frac{\partial \mathbf{h}(\boldsymbol{\lambda}, \boldsymbol{\mu})}{\partial \mu_i}\right]^+ = \min_{\mathbf{x} \in X(\boldsymbol{\lambda}, \boldsymbol{\mu})} g_i(\mathbf{x})$$

$$\left.\frac{\partial \mathbf{h}(\boldsymbol{\lambda}, \boldsymbol{\mu})}{\partial \mu_i}\right]^- = \max_{\mathbf{x} \in X(\boldsymbol{\lambda}, \boldsymbol{\mu})} g_i(\mathbf{x})$$

where $X(\boldsymbol{\lambda}, \boldsymbol{\mu})$ is the set of all $\mathbf{x} \in S_1$ which minimize the Lagrangian for a given $(\boldsymbol{\lambda}, \boldsymbol{\mu})$. That is, in minimizing the Lagrangian, a unique $\mathbf{x}$ may not be determined. In this case, the left- and right-hand partial derivatives of the dual function are not equal (the dual function is continuous but with discontinuous derivatives here). Clearly the partial derivatives of the dual function exist (are unique) only if the constraint functions $g_i(\mathbf{x})$ are constant over $X(\boldsymbol{\lambda}, \boldsymbol{\mu})$. That is equivalent of course to requiring the maximum and minimum to be the same. The above relations are written for the inequality constraints $g_i(\mathbf{x})$ but also hold for the equality constraints $w_i(\mathbf{x})$. Notice particularly that if the derivative exists it is equivalent to the derivative of the Lagrangian ignoring the dependence of the minimizing $\mathbf{x}$ on $\boldsymbol{\lambda}$ and $\boldsymbol{\mu}$, a convenient result. In order to apply these results to the linear programming example of the previous section, it is necessary to bound S_1, and so assume $x_1 \leqslant K$ and $x_2 \leqslant K$ where K is some large positive number.

Consider the first example of the previous section. There for $\mu > 1$ the Lagrangian was minimized by the unique point $x_1 = x_2 = 0$ so that $X(\boldsymbol{\mu})$ is a single point for $\mu > 1$. Then for $\mu > 1$, the derivative of the dual function exists and is equal to

$$\frac{\partial h(\mu)}{\partial \mu} = g(x) = -1 \qquad \text{for} \quad \mu > 1$$

On the other hand, for $\mu = 1$, the minimizing set is

$$X(\mu) = \{x \mid x_1 = 0, 0 \leqslant x_2 \leqslant K\}$$

and the constraint $g(x)$ is not constant over this set. Nonetheless, all the conditions are satisfied so that directional derivatives exist and we can calculate that

$$\left.\frac{\partial h(1)}{\partial \mu}\right]^+ = \min_X (x_2 - 1) = -1$$

$$\left.\frac{\partial h(1)}{\partial \mu}\right]^- = \max_X (x_2 - 1) = K - 1$$

In many practical problems, the directional derivatives do exist and may easily be calculated.

Hence the second-level optimization problem can be handled by the method of steepest ascent or any of the more efficient methods such as Fletcher-Powell,[14] conjugate gradients,[14] etc., if appropriate derivatives and directional derivatives are available. The term dual-feasible is used because the search at the second level is over the set D of allowed multipliers. If the optimum is on the boundary of D (as for example if inequality constraints are present and the multipliers associated with the inequality constraints are constrained to be positive), either a penalty function method or some gradient projection type of method[14] can be used. In any case, the concavity of the dual function makes most numerical optimization methods applicable.

Notice that the first-level unit solves a problem without constraints in the sense that the Lagrangian is minimized over S_1 and not over S (Table 1-3). Consequently, intermediate values for $\mathbf{x}^0$ produced for nonoptimal multipliers will not necessarily satisfy the equality or inequality constraints. Thus the solution must proceed until the optimum is reached since intermediate results are not approximate solutions generally. Examples of the application of the dual-feasible method to practical situations are developed in a later section.

When the coordinating variables involve some of the $\mathbf{x}$ variables in the original optimization problem, questions as to the existence of derivatives must be considered separately.[14]

1-3 APPLICATIONS OF MULTILEVEL TECHNIQUES

This section contains several examples of the application of multilevel techniques to the solution of typical problems, attempting to indicate how they may be applied in representative situations. It is important to bear in mind, however, that the main contribution of the method is not as a tool for the solution of problems which might be solved as integrated problems to begin with but rather as a tool for designing systems which by their very nature *must* be decomposed and in which coordination is therefore necessary. The industrial process control example discussed in Sec. 1-1 is an example in which each part of the system is independently designed but with the constraint that when the overall system is operating, it must be coordinated in order to achieve some overall system goal. The techniques of reformulating a problem into the multilevel structure and then decomposing it is useful in that circumstance for planning the individual tasks or subsystems in such a way that when completed the overall system may be coordinated.

The examples in the next three sections are simpler than that problem but are illustrative of the benefits and limitations of the technique.

1-3-1 Optimization of Coupled Subsystems via Multilevel Techniques

One of the most important applications of multilevel decomposition techniques is to the optimization of systems of coupled subsystems.[2,3,12,13,6] Consider a static system consisting of N coupled subsystems. Define the following variables

$\mathbf{M}^i$ = vector of manipulated variables of subsystem i

$\mathbf{Y}^i$ = vector of output variables of subsystem i which are not inputs to any subsystem

$\mathbf{X}^i$ = vector of output variables of subsystem i which are inputs to other subsystems

$\mathbf{Z}^i$ = vector of inputs to subsystem i which are outputs of other subsystems

Let the ith subsystem have the system relations (input-output relation)

$$\mathbf{F}_i(\mathbf{M}^i, \mathbf{Y}^i, \mathbf{X}^i, \mathbf{Z}^i) = 0$$

where $\mathbf{F}_i$ is a vector static relation. In general the operation of such a system is heavily constrained because of safety and economics. Suppose there are additional constraint equations associated with the operation of each subsystem, consisting of equality constraints

$$W_i(\mathbf{M}^i, \mathbf{Y}^i, \mathbf{X}^i, \mathbf{Z}^i) = 0$$

and inequality constraints

$$G_i(\mathbf{M}^i, \mathbf{Y}^i, \mathbf{X}^i, \mathbf{Z}^i) \leqslant 0$$

The only restriction on the constraints which can be handled in this formulation is that they involve only the variables associated with a single subsystem (this is so that decomposition can take place).

Because the subsystems are interconnected, the variables $\mathbf{Z}$ and $\mathbf{X}$ are actually the same. This may be expressed in the form of interconnection constraints

$$\mathbf{Z}^i = \sum_{j=i}^{N} \mathbf{C}_{ij}\mathbf{X}^j \qquad i = 1, 2, \ldots, N$$

Here the $\mathbf{C}_{ij}$ are constant matrices whose elements are 1 or 0 and the above relation merely indicates which element $\mathbf{z}$ corresponds to which element or elements $\mathbf{x}$, that is, which output of the other subsystems are the inputs to the ith subsystem.

Finally an objective function for the system must be assumed in the form

$$P = \sum_{i=1}^{N} P_i(\mathbf{M}^i, \mathbf{Y}^i, \mathbf{X}^i, \mathbf{Z}^i)$$

That is, it is assumed that the overall performance is the sum of performance functions for the individual subsystems. Again this is required if the resulting first-level problem is to be decomposed into one problem per subsystem. Relaxation of this assumption is discussed later. We assume here that P is to be *minimized.*

Consider first the conversion of this integrated problem into a multilevel problem through *goal coordination.* In this case, the interconnection variables are "cut," that is, allowed to be different on the two sides of the cut. In the above case, this means that the interconnection constraints need not be satisfied. Coordination is imposed by adding the interaction balance as a constraint and the Lagrange multipliers in the constraint are used as the coordinating variables. It is desirable to impose on the first level the task of ensuring that the resulting variables satisfy the system equations and the constraints associated with each subsystem. Therefore define the sets S_0^i as

$$S_0^i = \{(\mathbf{M}^i, \mathbf{Y}^i, \mathbf{X}^i, \mathbf{Z}^i) \mid \mathbf{F}_i = 0,\ W_i = 0, \text{ and } G_i \leqslant 0\} \qquad i = 1, 2, \ldots, N$$

Thus the sets S_0^i are the sets of feasible variables for a given subsystem, that is, the set of manipulated variables, input variables (from other subsystems), and output variables which satisfy the system equations and constraints. They do not satisfy in general the interconnection constraints.

The integrated problem together with the interconnection constraints may now be written as

$$\min P$$

$$\text{subject to } (\mathbf{M}^i, \mathbf{Y}^i, \mathbf{X}^i, \mathbf{Z}^i) \in S_0^i,\ \mathbf{Z}^i = \sum_{j=1}^{N} \mathbf{C}_{ij}\mathbf{X}^j \qquad i = 1, 2, \ldots, N$$

Assuming the existence of feasible solutions (nonempty sets S_0^i) and a minimum for the overall system, the objective is to find that optimum by searching for a saddle point of the constrained problem. Therefore, form the Lagrangian

$$L = \sum_{i=1}^{N} \left[P_i + \boldsymbol{\lambda}_i' \left(\mathbf{Z}^i - \sum_{j=1}^{N} \mathbf{C}_{ij}\mathbf{X}^j\right)\right]$$

The goal-coordination method now fixes the Lagrange multipliers and forms the first-level problem as follows:

FIRST-LEVEL PROBLEM:

$$\text{Determine } H(\boldsymbol{\lambda}) = \min L$$

$$(\mathbf{M}^i, \mathbf{Y}^i, \mathbf{X}^i, \mathbf{Z}^i) \in S_0^i \qquad i = 1, 2, \ldots, N$$

The first-level problem may immediately be decomposed into N independent optimization problems because of the separability of the performance function P. The individual subproblems are formed by grouping all terms in the Lagrangian involving $\mathbf{M}^i$, $\mathbf{Y}^i$, $\mathbf{X}^i$, and $\mathbf{Z}^i$. Thus for the

ith SUBPROBLEM:

$$\min \left[P_i(\mathbf{M}^i, \mathbf{Y}^i, \mathbf{X}^i, \mathbf{Z}^i) + \boldsymbol{\lambda}_i{}'\mathbf{Z}^i - \sum_{\substack{k=1 \\ k \neq i}}^{N} \boldsymbol{\lambda}_k{}'\mathbf{C}_{ki}\mathbf{X}^i \right]$$

$$\text{subject to } (\mathbf{M}^i, \mathbf{Y}^i, \mathbf{X}^i, \mathbf{Z}^i) \in S_0{}^i$$

Since each of these subproblems, after minimization, is a dual function depending only on $\boldsymbol{\lambda}$, the resulting minimum function can be written as $h_i(\boldsymbol{\lambda})$ and then it follows that the dual function for the integrated problem, the objective function of the second level, is simply

$$H(\boldsymbol{\lambda}) = \sum_{i=1}^{N} h_i(\boldsymbol{\lambda})$$

The second-level problem is of course the maximization of the dual function $H(\boldsymbol{\lambda})$ over its domain:

SECOND-LEVEL PROBLEM:

$$\max H(\boldsymbol{\lambda})$$

$$\text{subject to } \boldsymbol{\lambda} \in D$$

where the domain of the dual function is as before

$$D = \{\boldsymbol{\lambda} \mid \text{the first-level problem has a solution}\}$$

It was shown in Sec. 1-2-4 that the dual function $H(\boldsymbol{\lambda})$ is concave over any convex subset of its domain D and that furthermore, if the sets $S_0{}^i$ are closed and bounded (and they almost always are in practical problems) and if the functions P_i are continuous, then directional derivatives of $H(\boldsymbol{\lambda})$ exist and may easily be calculated. In the particular but usual case in which a unique set of manipulated variables exist for a given set of Lagrange multipliers, the derivatives of the dual function are simply the constraint functions. In this case

$$\nabla_{\boldsymbol{\lambda}_i} H(\boldsymbol{\lambda}) = Z_0{}^i - \sum_{j=1}^{N} \mathbf{C}_{ij}\mathbf{X}_0{}^j$$

where the subscript 0 indicates the optimal values of the variables as determined from the first-level minimizations for the given multipliers. Notice that the gradient of the objective function for the second-level coordinator is simply the *unbalance of the interaction variables*. From an economic point of view, the $\mathbf{Z}^i$ are the demand by the ith subsystem and the $\mathbf{X}^i$ are the supply from the ith subsystem and the second-level unit is attempting to match supply and demand when it attempts to satisfy the interaction-balance principle.

The Lagrange multipliers enter the individual first-level problem objective functions linearly and act like prices, adding to or subtracting from the performance function of each subproblem in direct proportion (with proper sign) to the amount of $\mathbf{Z}^i$ demanded and the amount of $\mathbf{X}^i$ produced. Thus the second-level goal coordination can be interpreted as modifying "prices" of the interacting variables in order to force the independent first-level problems to select the correct overall system optimum.

Since each of the first-level subproblems is itself a constrained optimization problem, there is no reason why these problems could not be converted to a multilevel form and then decomposed, which would result in a hierarchical form for the solution involving three levels. This can be continued indefinitely and might be useful in a situation where the overall problem involved two or more major subsystems with each of these major subsystems consisting of many similar but distinct subsystems. The resulting multilevel optimization would then involve individual optimization problems for each distinct unit at the first level together with coordination of similar units taking place in several second-level units and finally the overall coordination taking place at the third level. Assuming the existence of a saddle point for the overall optimization problem, the only restriction on the decomposition is that the individual subproblems at the first level actually possess solutions, so that the iterative process of choosing "prices" or Lagrange multipliers may succeed.

The other major approach to this coupled subsystem problem involves model coordination. Here, the integrated problem is converted to a multilevel problem by fixing the values of the interconnection variables at prescribed values, say $\mathbf{Z}^i$, $i = 1, 2, \ldots, N$. Thus in the first-level integrated problem, the $\mathbf{X}^i$ and $\mathbf{Z}^i$ are fixed and known. Define the set of permissible $\mathbf{M}^i$, $\mathbf{Y}^i$ to be $S_0{}^i$:

$$S_0{}^i = \left\{(\mathbf{M}^i, \mathbf{Y}^i) \mid F_i = 0, W_i = 0, G_i \leqslant 0, \mathbf{Z}^i = \sum_{j=1}^{N} \mathbf{C}_{ij}\mathbf{X}^j \text{ fixed}\right\}$$

Then the first-level problem is simply to find $H(\mathbf{Z})$:

FIRST-LEVEL PROBLEM:

$$\text{Determine } H(\mathbf{Z}) = \min P$$
$$\text{subject to } (\mathbf{M}^i, \mathbf{Y}^i) \in S_0{}^i, i = 1, 2, \ldots, N$$

Notice that the set of permissible $\mathbf{M}^i$ are those which produce the required $\mathbf{X}^i$ from each subsystem. Hence the sets $S_0{}^i$ may be empty in some problems if the dimensionality of the manipulated variables is not sufficient to permit any prescribed output to be obtained from a subproblem. Define the domain of $H(\mathbf{Z})$ to be D:

$$D = \{\mathbf{Z}^i, i = ,1\ 2,..., N \mid H(\mathbf{Z}) \text{ exists}\}$$

Clearly D contains the optimal values of the interaction variables and consequently the second-level objective is simply to minimize $H(\mathbf{Z})$:

SECOND-LEVEL PROBLEM:

$$\min H(\mathbf{Z})$$

$$\text{subject to } \mathbf{Z} \in D$$

Decomposition of the first-level problem into independent subproblems is again straightforward and results in the

*i*th SUBPROBLEM:

$$\text{Determine } h_i(\mathbf{Z}^i) = \min P_i(\mathbf{M}^i, \mathbf{Y}^i, \mathbf{Z}^i)$$

$$\text{subject to } (\mathbf{M}^i, \mathbf{Y}^i) \in S_0{}^i$$

Since the second level is required to minimize its objective function, some numerical algorithm is necessary. In general if gradients of $H(\mathbf{Z})$ exist, the minimization can be done very efficiently using any of the extended gradient methods.[14] To determine the existence of gradients of the second-level problem, write

$$P_i = P_i(\mathbf{M}^i, \mathbf{Y}^i, \mathbf{Z}^i)$$

since the interaction constraints are always satisfied. The minimization of P_i over the set $S_0{}^i$ produces $h_i(\mathbf{Z})$. Using the same theorems concerning the existence of derivatives of the dual functions, it follows that if the set over which the minimization is taking place $S_0{}^i$ is closed and bounded, and if the function P_i and the partial derivative of P_i with respect to any component of $\mathbf{Z}^i$ are continuous over $S_0{}^i$, then the gradient with respect to $\mathbf{Z}^i$ of P_i exists provided that P_i is constant over the minimizing set. This condition is true if a unique $(\mathbf{M}^i, \mathbf{Y}^i)$ results for a given $\mathbf{Z}^i$, which is the case in physical systems where the subsystems are well behaved. In this case, the gradient of P_i exists and is simply the gradient of P_i with respect to $\mathbf{Z}^i$ evaluated at the optimizing $(\mathbf{M}^i, \mathbf{Y}^i)$:

$$\nabla_{Z^i}[h_i(\mathbf{Z})] = \nabla_{Z^i}[P_i(\mathbf{M}_0{}^i, \mathbf{Y}_0{}^i, \mathbf{Z}^i)]$$

where the subscript 0 indicates evaluation at the minimizing values. With gradients of the second-level objective function available, an iterative solution is straightforward. In this model-coordinated approach, the first-level problems are again constrained optimization problems themselves, but of lower dimensionality since the interconnection or coordinating variables are fixed and known at each iteration.

Again since each subproblem is a constrained optimization problem, it follows that these subproblems can be converted to multilevel form and further decomposed, resulting in a hierarchical form for the solution.

There is no reason that in the formulation of a multilevel, decomposed solution to a coupled subsystem problem the same method of coordination be used throughout. For example, it is possible to separate some subsystems by the goal-coordination method and others by the model-coordination method, whichever is most advantageous. Moreover, if any of the first-level problems are difficult to solve because of inequality constraints, for example, the conversion to a multilevel form for that subproblem at least permits an efficient iterative scheme to be used for its solution.

1-3-2 The Multiitem Scheduling Problem

The multilevel decomposition technique can be useful in situations where the constraint sets are large but finite, a situation which is often difficult to handle in an integrated fashion. The multiitem scheduling problem is of that type and its decomposition is due to Lasdon.[6]

Consider a plant which makes I products or items using a number of machines each of which can make any of the products provided that it is properly set up (for which costs are involved). A schedule for the production of the ith item is denoted by

$$m_i = (m_{i1}, m_{i2}, \ldots, m_{iT})'$$

where m_{ik} is the number of machines to be used in time period k for the production of item i and the schedule is over T time periods. The problem is to choose m_i, $i = 1, 2, \ldots, I$ or schedules for the production of all items over this time period in such a way that setup costs (changing a machine from one item to another) and inventory costs are minimized.

Suppose setup cost is simply the number of setups. Then setup cost for the ith item is dependent on the schedule m_i used to produce this item and is given by $s_i(m_i)$:

$$s_i(m_i) = \sum_{t=1}^{T} \max \begin{cases} 0 \\ m_{it} - m_{i,t-1} \end{cases}$$

That is, the setup cost in any period is the number of changeovers provided that this number is nonnegative.

Let the inventory at the end of time period t of item i be h_{it}. The system equations for inventory then are simply

$$h_{i,t} = h_{i,t-1} + km_{i,t} - d_{it}$$

for $t = 1, 2,..., I$, and $i = 1, 2,..., T$ and $h_{i,0}$, the initial inventory. Here k is the rate of production on the machine so that km_{it} is the number of pieces produced in the time period and d_{it} is the demand during this time period (assumed known). In general both inventory and the number of machines used to produce an item are constrained:

$$m_{it} \leqslant K_i$$

$$H_i^{\min} \leqslant h_{it} \leqslant H_i^{\max}$$

That is, policy forbids inventory outside of a given range. All machines are not capable of producing all items simultaneously. Of course m_{it} must be an integer since it is the number of machines used to produce item i. With these restrictions on inventory, the costs associated with the inventory can often be taken to be proportional to the average inventory over an interval:

$$u_{it} = \frac{e(h_{i,t} + h_{i,t-1})}{2}$$

where e is the cost per unit. Total cost depends on the schedule used and is simply

$$u_i(m_i) = \sum_{t=1}^{T} u_{it}$$

Total cost for item i is then

$$c_i(m_i) = s_i(m_i) + u_i(m_i)$$

Notice that thus far there is no interaction between items, since all the above constraint equations involve only item i. Hence define

$$S_i = \{m_i \mid \text{item } i \text{ constraints are satisfied}\}$$

Thus any $M_i \in S_i$ is a *feasible* schedule for item i.

Then the problem of optimal scheduling would be to minimize $c_i(m_i)$ over S_i if there were no constraint on the maximum number of machines. But generally the total number of machines is limited and hence there is the *interconnection constraint*

$$\sum_{i=1}^{I} m_{it} \leqslant N \qquad t = 1, 2,..., T$$

It is desirable to use a multilevel scheme to permit the decomposition of this interacting optimization problem in order that at the first level, each item may be independently scheduled. To do this, form the Lagrangian

$$L(m, \boldsymbol{\lambda}) = \sum_{i=1}^{I} c_i(m_i) + \sum_{t=1}^{T} \lambda_t \left(\sum_{i=1}^{I} m_{it} - N \right)$$

Now form a multilevel problem by fixing the Lagrange multipliers (thus leading to *goal coordination*) so that the first-level problem becomes

FIRST-LEVEL PROBLEM:

$$\min L(\mathbf{m}, \boldsymbol{\lambda})$$

$$\text{subject to } m_i \in S_i \qquad i = 1, 2, \ldots, I$$

Since the multipliers are assumed fixed at this level, this problem decomposes immediately into I independent optimization problems:

*i*th SUBPROBLEM:

$$\text{Determine } h_i(\boldsymbol{\lambda}) = \min \left[c_i(m_i) + \sum_{t=1}^{T} \lambda_t m_{it} \right]$$

$$\text{subject to } m_i \in S_i$$

Denoting the minimum of the Lagrangian by $h(\boldsymbol{\lambda})$, it is clear that this is the dual function and consequently the second-level objective is to choose $\boldsymbol{\lambda}$ so as to maximize the dual function:

SECOND-LEVEL PROBLEM:

$$\max h(\boldsymbol{\lambda})$$

$$\text{subject to } \boldsymbol{\lambda} \in D$$

where D is the domain of $h(\boldsymbol{\lambda})$, that is, the set of all $\boldsymbol{\lambda}$ such that the first-level minimization exists.

Notice that since m_{it} must be an integer, the first-level constraint sets are finite but that this does not affect the problem formulation at all. Despite the finite set, the second-level problem is well behaved, being concave over any convex subset of D. Just as before, the gradient of the dual function with respect to the multipliers is the error in the fulfillment of the constraint:

$$\nabla_k h(\boldsymbol{\lambda}) = \left(\sum_{t=1}^{I} m_{it} \right) - N$$

This assumes of course that the minimizing set for the Lagrangian is unique and that the continuity assumptions hold. Lasdon[6] has shown that the use of a gradient procedure usually leads to good solutions to this problem. Moreover, there are many methods for the solution of the first-level problems (sometimes called the *single-item scheduling problems*), some of which are heuristic. Nonetheless, in this case, the decomposition leads to good, usable schedules.

1.3.3 Conversion of a Dynamic Optimization Problem to a Static One

Consider the now classical control problem of choosing an optimal trajectory for a dynamic system but with constraints on the state variables. Let the system be described by the state variable vector difference equation

$$\mathbf{x}_{k+1} = \mathbf{g}(\mathbf{x}_k, \mathbf{m}_k)$$

where $\mathbf{x}_k$ is the state variable and $\mathbf{m}_k$ is the manipulated variable at time k, and both are vectors. Let the constraints on the manipulated variables and state variables be of any form but denoted by

$$(\mathbf{x}_k, \mathbf{m}_k) \in S$$

Consider the problem of choosing a sequence of manipulated variables $\mathbf{m}_k$, $k = 0, 1, \ldots, N - 1$ such that the performance function

$$P = \sum_{k=0}^{N} f_k(\mathbf{m}_k, \mathbf{x}_k)$$

is minimized and the given boundary conditions $\mathbf{x}_0$ and $\mathbf{x}$ are satisfied. This is a conventional control problem and if the dimensionality of the state space is small enough it may be solved by dynamic programming. If there are no constraints on the state variables, many techniques are available for its solution.

This problem may be formulated as a static optimization problem by parameterizing time, that is, by considering the evolution of the system to be in space rather than time and using the goal-coordination method to decompose the problem into a manageable form.[12,13] Write the Lagrangian as

$$L(\mathbf{x}, \mathbf{m}, \boldsymbol{\lambda}) = \sum_{k=0}^{N} f_k(\mathbf{m}_k, \mathbf{x}_k) + \sum_{k=0}^{N-1} \boldsymbol{\lambda}_k'[\mathbf{x}_{k+1} - \mathbf{g}(\mathbf{x}_k, \mathbf{m}_k)]$$

Then fixing the Lagrange multipliers (to lead to goal coordination) yields the first-level problem

$$\min L(\mathbf{x}, \mathbf{m}, \boldsymbol{\lambda})$$

$$\text{subject to } (\mathbf{x}_k, \mathbf{m}_k) \in S \qquad k = 0, 1, 2, \ldots, N-1 \qquad \mathbf{x}_0 \text{ and } \mathbf{x}_N \text{ fixed}$$

This immediately decomposes into N independent subproblems each of the form

*k*th SUBPROBLEM ($k = 0, 1, \ldots, N-1$):

$$\min \left[f_k(\mathbf{m}_k, \mathbf{x}_k) + \boldsymbol{\lambda}_{k-1}\mathbf{x}_k - \boldsymbol{\lambda}_k \mathbf{g}(\mathbf{x}_k, \mathbf{m}_k) \right]$$

$$\text{subject to } \mathbf{x}_0 \text{ and } \mathbf{x}_N \text{ fixed}$$

$$(\mathbf{x}_k, \mathbf{m}_k) \in S$$

Denote the minimum of the *k*th subproblem objective function by $h_k(\boldsymbol{\lambda})$. The second-level problem is to maximize the dual function or

SECOND-LEVEL PROBLEM:

$$\max H(\boldsymbol{\lambda}) = \sum_{k=0}^{N-1} h_k(\boldsymbol{\lambda})$$

$$\text{subject to } \boldsymbol{\lambda} \in D$$

where again D is the domain of definition of the primal problem, that is, where the minima exist.

The classical control problem in which the system equation is linear and the performance function is quadratic and positive definite in both the state and manipulated variables clearly satisfies all the conditions under which a saddle point of the optimization problem exists and solves the problem. Consequently, the solution of the second-level problem is straightforward and can be done by any gradient scheme using as the elements of the gradient

$$\frac{\partial H(\boldsymbol{\lambda})}{\partial \boldsymbol{\lambda}_k} = \mathbf{x}_{k+1} - \mathbf{g}(\mathbf{x}_k, \mathbf{m}_k)$$

evaluated at the minimizing values of $\mathbf{x}_k$ and $\mathbf{m}_k$.

Notice also that the first-level problems can often be solved in closed form and that all first-level problems (except possibly the first and last) are identical in form so that the solutions can be done sequentially rather than in parallel. Solutions of simple optimal control problems can be effectively found in this manner and numerical examples may be found in Ref. 12.

REFERENCES:

1. Beltrami, E. J.: A Constructive Proof of the Kuhn-Tucker Multiplier Rule, *Proc. SIAM Natl. Meeting*, Toronto, 1968.
2. Brosilow, C. B., L. S. Lasdon, and J. D. Pearson: Feasible Optimization Methods for Interconnected Systems, *Proc. Joint Autom. Control Conf.*, 1965, pp. 79–84.
3. Brosilow, C. B., and L. S. Lasdon: A Two Level Optimization Technique for Recycle Processes, *Proc. A.I.Ch.E.-I. Chem. Eng. Symp.*, ser. 4, 1965.
4. Findeisen, W.: Parametric Optimization by Primal Method in Multi-level Systems, *Case Western Reserve Univ. Syst. Res. Cen. Rept.* SRC 113-C-67-46, 1967, and *Arch. Autom. Telemech.*, vol. 12, 1967.
5. Kuhn, H. W., and A. W. Tucker: Nonlinear Programming, *Proc. Symp. Math. Statist., Zd*, Berkeley, 1964, pp. 481–492.
6. Lasdon, L. S.: Duality and Decomposition in Mathematical Programming, *IEEE Trans. Syst. Sci. Cybernetics*, vol. SSC-4, no. 2, July, 1968.
7. Mesarovic, M. D.: Conceptual Framework for the Study of mlng Systems, *Case Western Reserve Univ. Syst. Res. Cen. Rept.* SRC 77-A-65-29, 1965.
8. Mesarovic, M. D., Macko, and Y. Takahara: Two Coordination Principles and Their Application in Large Scale Systems Control, *Proc. IFAC Congr.*, Warsaw, Poland, 1969.
9. Mesarovic, M. D., D. Macko, and Y. Takahara: "Theory of Multi-level Systems," Academic, 1970.
10. Pearson, J. D.: Multilevel Programming, *Case Western Reserve Univ. Syst. Res. Cen. Rept.* SRC 70-A-65-25, 1965.
11. Rosenbrock, H. H.: Discussion, Proc. A.I.Ch.E.-I. Chemical Engineering Symposium, London, 1965.
12. Schoeffler, J. D., and L. S. Lasdon: A Multi-level Technique for Optimization, *Proc. Joint Autom. Control Conf.*, 1965, pp. 85–91.
13. Schoeffler, J. D., and L. S. Lasdon: Decentralized Plant Control, *ISA Trans.*, vol. 5, no. 2, pp. 175–183, April, 1966.
14. Hadley, G.: "Nonlinear and Dynamic Programming," Addison-Wesley, 1964.
15. Karlin, S.: "Mathematic Methods and Theory in Games, Programming, and Economics," vol. 1, Addison-Wesley, 1959.

2

LARGE-SCALE LINEAR AND NONLINEAR PROGRAMMING

ARTHUR M. GEOFFRION[1]

2-1 INTRODUCTION

The two most striking features of the literature on large-scale mathematical programming are the following: (1) most of the problems arising in application possess distinctive structures that can be exploited, and (2) most of the solution methods proposed actually are based on a handful of exceedingly simple yet fundamental concepts that occur over and over again in various guises. This would suggest that an appropriate viewpoint from which to regard the literature is one that stresses similarities based on common problem structures and fundamental solution concepts. This is indeed the viewpoint of this chapter, with the emphasis on the fundamental concepts since these are quite obviously in greatest need of clarification.

The fundamental concepts are of two general types: *problem manipulations* and *solution strategies*. A problem manipulation is a device for restating a given problem in an alternative form that is essentially equivalent yet (hopefully) more amenable to solution. One familiar problem manipulation is dualization for linear programs. Another is the separation of a problem like (here $\mathbf{x}_k$ may be a vector)

$$\min_{\mathbf{x}_1,\ldots,\mathbf{x}_K} \sum_{k=1}^{K} f_k(\mathbf{x}_k)$$

$$\text{subject to } \mathbf{x}_k \in X_k \qquad k = 1,\ldots, K$$

[1] Support was provided by the National Science Foundation under Grant GP-8740.

into K independent smaller problems, each of the form

$$\min_{\mathbf{x}_k} f_k(\mathbf{x}_k)$$

$$\text{subject to } \mathbf{x}_k \in X_k$$

This chapter discusses and illustrates four types of problem manipulation in some detail: *projection* (Sec. 2-2), *inner linearization* (Sec. 2-4), *outer linearization* (Sec. 2-5), and *dualization* for general convex programs (Sec. 2-6). The end result of applying one or more of these devices to a given problem is what many authors call a "master" problem, to which some solution strategy must then be applied.

A solution strategy is a way of reducing an optimization problem to a *sequence* of related but simpler optimization problems. For example, the *feasible directions strategy*[23] can be viewed as reducing a problem to a sequence of one-dimensional optimizations along judiciously chosen directions. Another familiar example is the *penalty strategy*,[9] which reduces a constrained problem to a sequence of essentially unconstrained optimizations via penalty functions. In this chapter we take up two other solution strategies in detail: the *piecewise strategy* (Sec. 2-3) and *relaxation* (Sec. 2-5). (Lest the reader wonder what strategy the simplex method falls under, we point out that it is best viewed in terms of *restriction*,[11] a strategy complementary to relaxation that is not discussed here.) These and other solution strategies often lead to what are frequently called "subproblems" in the large-scale optimization literature; these subproblems in turn are usually amenable to solution by specialized algorithms.

Problem manipulations and solution strategies can be teamed up in various ways to clarify the derivation of many known algorithms and invent new ones. For instance, the pattern inner linearization/restriction leads to Dantzig-Wolfe decomposition (see Sec. 2-4); outer linearization/relaxation leads to cutting-plane algorithms;[4,15] projection/piecewise leads to Rosen's primal partition programming algorithm[19] (see Sec. 2-3); projection, outer linearization/relaxation leads to Benders' partitioning procedure;[1] and so on. By grouping computational proposals in terms of the manipulation/strategy pattern they can most naturally be viewed as using, one obtains a useful taxonomy of much of the large-scale mathematical programming literature. This has been done elsewhere by the author,[11] and will not be repeated here. One part of this taxonomy can be illustrated by noting that the following papers can all be viewed as applying the pattern dualization/feasible directions: Refs. 2 (sec. 3.2), 7, 8, 13, 14 (sec. III), 16, 17, 20, 21, 22.

The reader may consult Ref. 11 for a more detailed exposition and further illustrative applications of the fundamental concepts treated in this chapter.

2-2 PROJECTION

Projection, sometimes also called *partitioning*, is a simple problem manipulation which takes advantage in certain problems of the relative simplicity introduced by temporarily fixing the values of certain variables.

Consider the general problem

$$\max_{\substack{\mathbf{x}\in X \\ \mathbf{y}\in Y}} f(\mathbf{x}, \mathbf{y})$$

$$\text{subject to } \mathbf{g}(\mathbf{x}, \mathbf{y}) \geqslant \mathbf{0} \tag{1}$$

where $X \subseteq E^n$, $Y \subseteq E^m$, and $\mathbf{g}$ is a vector-valued function. The projection of Eq. (1) onto the space of the $\mathbf{y}$ variables alone is defined as

$$\max_{\mathbf{y}\in Y} \begin{bmatrix} \sup_{\mathbf{x}\in X} f(\mathbf{x}, \mathbf{y}) \\ \text{subject to } \mathbf{g}(\mathbf{x}, \mathbf{y}) \geqslant \mathbf{0} \end{bmatrix} \tag{2}$$

This is a problem in the $\mathbf{y}$ variables alone. Let $v(\mathbf{y})$ denote the maximand of Eq. (2)—the supremal value of the "inner" maximization problem within the brackets. If the inner maximization problem is infeasible for certain values of $\mathbf{y}$, v is defined to be $-\infty$ for those values. Obviously a point $\mathbf{y}^0 \in Y$ is useless in Eq. (2) if $v(\mathbf{y}^0) = -\infty$, and so the constraint $\mathbf{y} \in V$ may be added to Eq. (2) if desired, where

$$V \equiv \{\mathbf{y} \mid v(\mathbf{y}) > -\infty\} \equiv \{\mathbf{y} \mid \mathbf{g}(\mathbf{x}, \mathbf{y}) \geqslant \mathbf{0} \text{ for some } \mathbf{x} \in X\}$$

Thus Eq. (2) can be rewritten as

$$\max_{\mathbf{y}\in Y} v(\mathbf{y})$$

$$\text{subject to } \mathbf{y} \in V \tag{3}$$

The set V can be thought of as the projection of the constraints $\mathbf{x} \in X$ and $\mathbf{g}(\mathbf{x}, \mathbf{y}) \geqslant \mathbf{0}$ onto the space of the $\mathbf{y}$ variables alone.

It is straightforward to show that the relationship between Eqs. (1) and (3) is as follows [one may read (2) for (3) below, except that (2) can be feasible with value $-\infty$ when (1) is infeasible]. This result justifies the use of projection as a problem manipulation.

Theorem 2-1: Equation (1) is infeasible or has unbounded value if and only if the same is true of Eq. (3). If $(\mathbf{x}^0, \mathbf{y}^0)$ is optimal in Eq. (1), then $\mathbf{y}^0$ must be optimal in Eq. (3). If $\mathbf{y}^0$ is optimal in Eq. (3) and $\mathbf{x}^0$ achieves the supremum of $\mathbf{f}(\mathbf{x}, \mathbf{y}^0)$ subject to $\mathbf{x} \in X$ and $\mathbf{g}(\mathbf{x}, \mathbf{y}^0) \geqslant \mathbf{0}$, then $\mathbf{x}^0$ together with $\mathbf{y}^0$ is optimal in Eq. (1). If $\mathbf{y}^0$ is ϵ_1-optimal in Eq. (3) and $\mathbf{x}^0$ is within ϵ_2 of achieving $v(\mathbf{y}^0)$, then $(\mathbf{x}^0, \mathbf{y}^0)$ is $(\epsilon_1 + \epsilon_2)$-optimal in Eq. (1).

No special assumptions on X, Y, f, or $\mathbf{g}$ are required for Theorem 2-1 to hold. Under the convexity assumptions usually imposed on Eq. (1), however, it is reassuring that Eqs. (2) and (3) are concave programs. A proof of this result can be found, for example, in Ref. 11.

Theorem 2-2: Assume that X and Y are convex sets, and that f and $\mathbf{g}$ are concave on $X \times Y$. Then V is a convex set and v is a concave function on V.

The usefulness of projection as a problem manipulation depends on the projected problem being simpler than the original one. This, in turn, depends on the tractability of the supremal value function v. The key is usually to assign the variables of a given problem to the roles of $\mathbf{x}$ and $\mathbf{y}$ in Eq. (1) in such a way that the inner maximization problem is much easier than the given problem in its entirety. This ensures that at least the evaluation of v for any fixed $\mathbf{y}$ is relatively easy, although an explicit expression for the whole function v is not ordinarily available. If, then, the dimension of $\mathbf{y}$ is small, or good local information about v is inexpensively available as a by-product of its evaluation (such as the gradient if it exists), reasonably efficient methods for solving the projected problem can usually be contrived. The optimal solution of the given problem can then be found as indicated in Theorem 2-1.

Several examples of the use of projection will now be given.

Example 2-1

Consider the problem

$$\max_{\mathbf{x},\mathbf{y}} f(\mathbf{x}, \mathbf{y})$$

$$\text{subject to } \mathbf{h}(\mathbf{x}) \geqslant \mathbf{0}$$

$$\mathbf{x} - \mathbf{g}(\mathbf{y}) = \mathbf{0} \tag{4}$$

$$\mathbf{y} \in Y$$

where $\mathbf{h}$ and $\mathbf{g}$ are vector-valued functions. Projection onto $\mathbf{y}$ yields

$$\max_{\mathbf{y}\in Y} \begin{bmatrix} \sup_{\mathbf{x}} f(\mathbf{x}, \mathbf{y}) \\ \text{subject to } \mathbf{h}(\mathbf{x}) \geqslant \mathbf{0} \qquad \mathbf{x} = \mathbf{g}(\mathbf{y}) \end{bmatrix} \tag{5}$$

The inner supremum is trivial:

$$v(\mathbf{y}) = \begin{cases} f(\mathbf{g}(\mathbf{y}), \mathbf{y}) & \text{if } \mathbf{h}(\mathbf{g}(\mathbf{y})) \geqslant \mathbf{0} \\ -\infty & \text{if } \mathbf{h}(\mathbf{g}(\mathbf{y})) \ngeqslant \mathbf{0} \end{cases}$$

Thus $V = \{\mathbf{y} \mid \mathbf{h}(\mathbf{g}(\mathbf{y})) \geqslant \mathbf{0}\}$ in this case, and Eq. (5) becomes

$$\begin{aligned} &\max_{\mathbf{y} \in Y} f(\mathbf{g}(\mathbf{y}), \mathbf{y}) \\ &\text{subject to } \mathbf{h}(\mathbf{g}(\mathbf{y})) \geqslant \mathbf{0} \end{aligned} \tag{6}$$

Projecting on $\mathbf{y}$ merely amounts, in this special case, to substituting $\mathbf{x}$ out of the problem. Thus projection can be viewed as a generalization of the concept of substitution.

Example 2-2

One class of "multidivisional" problems can be represented as follows:

$$\begin{aligned} &\max_{\mathbf{x}, \mathbf{y}} \left[F(\mathbf{y}) + \sum_{k=1}^{K} f_k(\mathbf{x}_k)\right] \\ &\text{subject to } \mathbf{G}_k(\mathbf{y}) + \mathbf{g}_k(\mathbf{x}_k) \leqslant \mathbf{0} \qquad k = 1, \dots, K \\ &\qquad \mathbf{x}_k \in X_k \qquad k = 1, \dots, K \\ &\qquad \mathbf{y} \in Y \end{aligned} \tag{7}$$

where $\mathbf{x}_k$ is a vector of dimension n_k and $\mathbf{G}_k$ and $\mathbf{g}_k$ are vector-valued functions with the same number of components. One may interpret Eq. (7) in terms of an organization (or system or economy, etc.) with K divisions (or subsystems or sectors, etc.). Then $\mathbf{x}_k$ would be the activity vector, X_k the set of feasible activities, and f_k the payoff contribution associated with the kth division. The vector $\mathbf{y}$ would be the decision vector centrally controlled by the organization, rather than by any of the divisions, and $\mathbf{G}_k(\mathbf{y}) + \mathbf{g}_k(\mathbf{x}_k) \leqslant \mathbf{0}$ would reflect the effect of $\mathbf{y}$ on the feasibility of the decisions available to the divisions.

Equation (7) has coupling variables but no coupling constraints, since temporarily fixing $\mathbf{y}$ leads to separation into K independent problems. It is therefore natural to project on $\mathbf{y}$. The result is

$$\max_{\mathbf{y} \in Y} F(\mathbf{y}) + \sum_{k=1}^{K} v_k(\mathbf{y}) \tag{8}$$

where we define

$$\begin{aligned} v_k(\mathbf{y}) \equiv &\sup_{\mathbf{x}_k \in X_k} f_k(\mathbf{x}_k) \\ &\text{subject to } \mathbf{g}_k(\mathbf{x}_k) \leqslant -\mathbf{G}_k(\mathbf{y}) \end{aligned} \tag{9}$$

If $\mathbf{y}$ happens to be a scalar variable rather than a vector, then of course a simple direct-search approach for solving Eq. (8) is possible and quite likely to be efficient—especially if Eq. (7) is a concave program and parametric programming techniques are available to continually obtain the value of each function v_k as $\mathbf{y}$ varies. Otherwise, solving Eq. (8) requires a more sophisticated approach such as the piecewise strategy, the feasible directions strategy, or the pattern outer linearization/relaxation. The first of these approaches is applied to a special case of Eq. (8) in Sec. 2-3.

Example 2-3

Another important class of multidivisional problems are those with coupling constraints but no coupling variables:

$$\begin{aligned} &\max_{\mathbf{x}} \sum_{k=1}^{K} f_k(\mathbf{x}_k) \\ &\text{subject to } \mathbf{x}_k \in X_k \qquad k = 1,\ldots, K \\ &\sum_{k=1}^{K} \mathbf{g}_k(\mathbf{x}_k) \leqslant \mathbf{b} \end{aligned} \tag{10}$$

where $\mathbf{x}_k$ is a vector of dimension n_k and the $\mathbf{g}_k$ are all vector-valued functions of the dimension of $\mathbf{b}$. The symbols $\mathbf{x}_k$, X_k, and f_k may be interpreted as in the previous example. The vector $\mathbf{b}$ may be interpreted as the vector of resources available to the organization as a whole, with $\mathbf{g}_k(\mathbf{x}_k)$ being the amount of resources used up by the kth division on account of its activities.

A useful preliminary manipulation of this problem is to introduce the new vectors of variables $\mathbf{y}_1,\ldots, \mathbf{y}_K$, each with the dimension of $\mathbf{b}$, and rewrite Eq. (10) in the equivalent form

$$\begin{aligned} &\max_{\mathbf{x},\mathbf{y}} \sum_{k=1}^{K} f_k(\mathbf{x}_k) \\ &\text{subject to } \mathbf{x}_k \in X_k \qquad k = 1,\ldots, K \\ &\mathbf{g}_k(\mathbf{x}_k) \leqslant \mathbf{y}_k \qquad k = 1,\ldots, K \\ &\sum_{k=1}^{K} \mathbf{y}_k \leqslant \mathbf{b} \end{aligned} \tag{11}$$

The natural interpretation of $\mathbf{y}_k$ is that it represents the resources centrally allocated to the kth division. By this simple means the problem has been changed from one with coupling constraints and no coupling variables to one with coupling variables and no coupling constraints. The projection of Eq. (11) onto $\mathbf{y}$ now yields:

$$\max_{\mathbf{y}} \sum_{k=1}^{K} v_k(\mathbf{y}_k)$$

$$\text{subject to } \sum_{k=1}^{K} \mathbf{y}_k \leqslant \mathbf{b} \tag{12}$$

where

$$v_k(\mathbf{y}_k) \equiv \sup_{\mathbf{x}_k \in X_k} f_k(\mathbf{x}_k)$$

$$\text{subject to } \mathbf{g}_k(\mathbf{x}_k) \leqslant \mathbf{y}_k \tag{13}$$

One may think of Eq. (12) as the "projection of Eq. (10) onto the divisional resource space," and ascribe to it a very natural interpretation in terms of centrally allocating resources to the divisions. See Ref. 10 for a detailed development of three approaches for solving Eq. (12) when Eq. (10) is a concave program: the piecewise strategy, feasible directions strategy, and the pattern outer linearization/relaxation.

Example 2-4

A common class of discrete-time dynamic problems is as follows (here there are three periods):

$$\max_{\mathbf{x},\mathbf{y}} f_1(\mathbf{x}_1) + F_1(\mathbf{y}_1) + f_2(\mathbf{x}_2) + F_2(\mathbf{y}_2) + f_3(\mathbf{x}_3)$$

$$\begin{aligned} \text{subject to } \mathbf{g}_1(\mathbf{x}_1) + \mathbf{h}_1(\mathbf{y}_1) \qquad\qquad\qquad\qquad & \leqslant \mathbf{b}_1 \\ \mathbf{H}_1(\mathbf{y}_1) + \mathbf{g}_2(\mathbf{x}_2) + \mathbf{h}_2(\mathbf{y}_2) \qquad\quad & \leqslant \mathbf{b}_2 \\ \mathbf{H}_2(\mathbf{y}_2) + \mathbf{g}_3(\mathbf{x}_3) & \leqslant \mathbf{b}_3 \\ \mathbf{x}_t \in X_t \qquad t = 1, 2, 3 & \\ \mathbf{y}_t \in Y_t \qquad t = 1, 2 & \end{aligned} \tag{14}$$

where $\mathbf{x}_t$ is a vector of dimension n_t associated with time period t, $\mathbf{g}_1$ and $\mathbf{h}_1$ are vector-valued functions of the dimension of $\mathbf{b}_1$, and similarly for the constraints involving $\mathbf{b}_2$ and $\mathbf{b}_3$. The vector $\mathbf{y}_1$ couples periods one and two, while $\mathbf{y}_2$ couples periods two and three.

The point of this example is to show that projection can be applied more than once in a sequential fashion. The result is a reformulation which resembles dynamic programming.

First project Eq. (14) onto the variables $\mathbf{x}_1$, $\mathbf{y}_1$, $\mathbf{x}_2$, $\mathbf{y}_2$:

$$\max_{\mathbf{x}_1,\mathbf{x}_2,\mathbf{y}_1,\mathbf{y}_2} \{f_1(\mathbf{x}_1) + F_1(\mathbf{y}_1) + f_2(\mathbf{x}_2) + F_2(\mathbf{y}_2) + [\sup_{\mathbf{x}_3 \in X_3} f_3(\mathbf{x}_3) \quad \text{subject to } \mathbf{g}_3(\mathbf{x}_3) \leqslant \mathbf{b}_3 - \mathbf{H}_2(\mathbf{y}_2)]\}$$

$$\begin{aligned} \text{subject to } & \mathbf{g}_1(\mathbf{x}_1) + \mathbf{h}_1(\mathbf{y}_1) \leqslant \mathbf{b}_1 \\ & \mathbf{H}_1(\mathbf{y}_1) + \mathbf{g}_2(\mathbf{x}_2) + \mathbf{h}_2(\mathbf{y}_2) \leqslant \mathbf{b}_2 \\ & \mathbf{x}_1 \in X_1 \\ & \mathbf{x}_2 \in X_2 \\ & \mathbf{y}_1 \in Y_1 \\ & \mathbf{y}_2 \in Y_2 \end{aligned} \tag{15}$$

Now project Eq. (15) onto $(\mathbf{x}_1, \mathbf{y}_1)$. The result is

$$\max_{\substack{\mathbf{x}_1 \in X_1 \\ \mathbf{y}_1 \in Y_1}} [f_1(\mathbf{x}_1) + F_1(\mathbf{y}_1) + \phi_1(\mathbf{y}_1)] \quad \text{subject to } \mathbf{g}_1(\mathbf{x}_1) + \mathbf{h}_1(\mathbf{y}_1) \leqslant \mathbf{b}_1 \tag{16}$$

where

$$\phi_1(\mathbf{y}_1) \equiv \sup_{\mathbf{x}_2,\mathbf{y}_2} [f_2(\mathbf{x}_2) + F_2(\mathbf{y}_2) + \phi_2(\mathbf{y}_2)]$$

$$\begin{aligned} \text{subject to } & \mathbf{g}_2(\mathbf{x}_2) + \mathbf{h}_2(\mathbf{y}_2) \leqslant \mathbf{b}_2 - \mathbf{H}_1(\mathbf{y}_1) \\ & \mathbf{x}_2 \in X_2 \\ & \mathbf{y}_2 \in Y_2 \end{aligned} \tag{17}$$

and

$$\phi_2(\mathbf{y}_2) \equiv \sup_{\mathbf{x}_3 \in X_3} f_3(\mathbf{x}_3) \quad \text{subject to } \mathbf{g}_3(\mathbf{x}_3) \leqslant \mathbf{b}_3 - \mathbf{H}_2(\mathbf{y}_2) \tag{18}$$

Probably the most obvious way of approaching Eq. (16) computationally is by a direct "inside-out" attack in the spirit of dynamic programming. Namely, first explicitly compute the function ϕ_2 over the relevant range of values for $\mathbf{y}_2$; then explicitly compute ϕ_1; and finally solve Eq. (16) itself. This yields optimal values for $\mathbf{x}_1$ and $\mathbf{y}_1$, from which the optimal values of $\mathbf{x}_2$ and $\mathbf{y}_2$ and finally $\mathbf{x}_3$ can be found. This is likely to be a fairly efficient solution procedure if $\mathbf{y}_1$ and $\mathbf{y}_2$ are both scalar variables, especially if parametric programming procedures are available to efficiently compute ϕ_2 and ϕ_1 (Ref. 5, p. 61). Otherwise, however, this direct procedure could be onerous and might better be replaced by a more sophisticated approach that does not require the explicit computation of ϕ_2 and ϕ_1.

2-3 PIECEWISE STRATEGY

The piecewise solution strategy applies to problems in which some of the functions are significantly simpler when temporarily restricted to certain regions of their domain. The domain is implicitly subdivided into such regions, and the problem is solved by considering these regions one at a time. Usually it is necessary to explicitly consider only a small fraction of them.

To be more specific, consider the problem

$$\max_{\mathbf{y} \in Y} v(\mathbf{y}) \tag{19}$$

where Y is a convex set and v is a "piecewise-simple" concave function in the sense that there are regions or pieces $P^j (j = 1, \ldots, p)$ of its domain on which it coincides with a relatively tractable function v^j. The piecewise strategy would then be as follows:

Step 1. Let a point $\hat{\mathbf{y}} \in Y$ be given. Determine the piece $\hat{P}$ containing $\hat{\mathbf{y}}$ and the corresponding function $\hat{v}$. Go to step 2.

Step 2. Maximize $\hat{v}(\mathbf{y})$ subject to $\mathbf{y} \in \hat{P} \cap Y$. Let $\bar{\mathbf{y}}$ be an optimal solution of this problem (assume that an optimal solution exists unless the optimal value is infinite, in which case termination obtains). Go to step 3.

Step 3. Determine a piece $\bar{P}$ adjacent to $\hat{P}$ at $\bar{\mathbf{y}}$ such that $v(\mathbf{y}) > v(\bar{\mathbf{y}})$ for some $\mathbf{y} \in \bar{P} \cap Y$ [if none exists, $\bar{\mathbf{y}}$ is optimal in Eq. (19)]. Determine the corresponding function $\bar{v}$ and return to step 2 with $\bar{P}$, $\bar{v}$, $\bar{\mathbf{y}}$ in place of $\hat{P}$, $\hat{v}$, $\hat{\mathbf{y}}$.

Example 2-5

Constrained games and military applications frequently lead to max-min problems of the form

$$\max_{\mathbf{y} \in Y} \left[\min_{\mathbf{x} \in X} f(\mathbf{x}, \mathbf{y})\right] \tag{20}$$

where f is convex-concave, that is, f is convex on the convex set $X \subseteq E^n$ for each fixed $\mathbf{y}$ in Y, and concave on the convex set $Y \subseteq E^m$ for each fixed $\mathbf{x}$ in X. As a simple case, let us take $n = m$ and

$$X \equiv \{\mathbf{x} \in E^n \mid \mathbf{x} \geqslant 0\}$$

$$f(\mathbf{x}, \mathbf{y}) \equiv \sum_{i=1}^{n} f_i(x_i, y_i) = \sum_{i=1}^{n} x_i^2 a_{1i} + 2x_i y_i a_{2i} + y_i^2 a_{3i}$$

with $a_{1i} > 0$ and $a_{3i} \leqslant 0$ for all i (it is easy to verify that this choice of f is convex-concave). Since f is separable with respect to each component of $\mathbf{x}$ for fixed $\mathbf{y}$, we may write the maximand of Eq. (20) as

$$v(\mathbf{y}) = \sum_{i=1}^{n} \min_{x_i \geqslant 0} f_i(x_i, y_i) = \sum_{i=1}^{n} \begin{cases} y_i^2 a_{3i} & \text{if } y_i a_{2i} \geqslant 0 \\ y_i^2 \left(a_{3i} - \dfrac{a_{2i}^2}{a_{1i}} \right) & \text{if } y_i a_{2i} \leqslant 0 \end{cases} \tag{21}$$

The very simple structure of f_i makes it easy to derive the explicit expression given for its minimum (for y fixed) subject to $x_i \geqslant 0$.

It is obvious that v is piecewise-simple (actually, piecewise-quadratic) in this case, with each "piece" being the points in E^n such that $y_i a_{2i}$ has a particular sign for all i. If $n = 2$, $a_{21} > 0$, and $a_{22} < 0$, for example, there would be four pieces and v would be a quadratic function on each as indicated in Fig. 2-1.

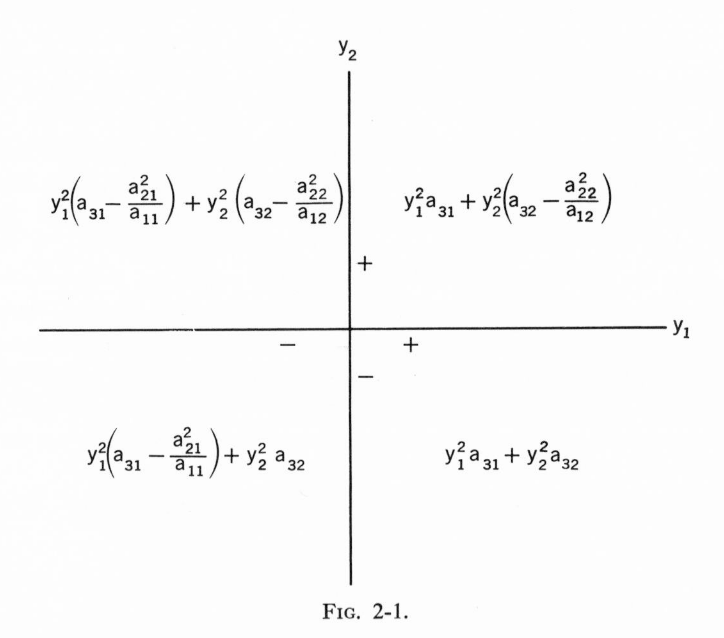

FIG. 2-1.

How to carry out step 1 of the piecewise strategy should be evident. Step 2 requires an algorithm capable of maximizing a quadratic function subject to $\mathbf{y} \in Y$ and extra constraints of the form $y_i \leqslant 0$ or $y_i \geqslant 0$ for each i. Step 3 could be done by trial and error over each of the pieces on whose boundary $\bar{\mathbf{y}}$ lies; that is, one could try all possible combinations of sign changes in the extra constraints corresponding to $\bar{y}_i = 0$. No doubt further analysis of Eq. (21) would reveal ways of reducing the number of necessary trials well below the maximum possible number $2^q - 1$ when q is large, where q is the number of components for which $\bar{y}_i = 0$.

Example 2-2 (Continued)

Rosen[19] has shown how the piecewise strategy can be applied to Eq. (8) in the completely linear case. Assume that Eq. (7) is feasible and has finite optimal value, and that $Y \equiv E^m$, $X_k \equiv E^{n_k}$, $F(\mathbf{y}) \equiv \mathbf{b}_0'\mathbf{y}$, $f_k(\mathbf{x}_k) \equiv \mathbf{b}_k'\mathbf{x}_k$, $\mathbf{G}_k(\mathbf{y}) \equiv \mathbf{y}'\mathbf{D}_k - \mathbf{c}_k'$, $\mathbf{g}_k(\mathbf{x}_k) \equiv \mathbf{x}_k'\mathbf{A}_k$, where $\mathbf{b}_0'$, $\mathbf{b}_k'$, $\mathbf{c}_k'$ are row vectors and $\mathbf{D}_k$, $\mathbf{A}_k$ matrices of appropriate dimension. Then Eq. (8) becomes

$$\max_{\mathbf{y}} \left[\mathbf{b}_0'\mathbf{y} + \sum_{k=1}^{K} v_k(\mathbf{y})\right] \tag{22}$$

where $v_k(\mathbf{y})$ is defined as the supremal value of the linear program

$$\max_{\mathbf{x}_k} \mathbf{b}_k'\mathbf{x}_k$$

$$\text{subject to } \mathbf{x}_k'\mathbf{A}_k \leqslant \mathbf{c}_k' - \mathbf{y}'\mathbf{D}_k \tag{23}$$

The appropriateness of the piecewise strategy for Eq. (22) derives from the fact that each function v_k is piecewise-linear as well as concave. For, if $\hat{\mathbf{y}}$ is any point such that Eq. (23) is feasible for all k, then by hypothesis and the dual theorem of linear programming it follows (see Ref. 11, sec. 4.2, or Ref. 19) that $v_k(\mathbf{y}) = (\mathbf{c}_k' - \mathbf{y}'\mathbf{D}_k)\,\hat{\mathbf{u}}_k$ so long as $\mathbf{y}$ satisfies certain linear "feasibility" constraints, where $\hat{\mathbf{u}}_k$ is the vector of optimal dual variables for Eq. (23) with $\mathbf{y} = \hat{\mathbf{y}}$. The feasibility constraints can easily be written down explicitly from the optimal tableau of Eq. (23). Thus step 1 of the piecewise strategy is easily done, and the problem to be solved at step 2 is simply the linear program

$$\max_{\mathbf{y}} \left[\mathbf{b}_0'\mathbf{y} + \sum_{k=1}^{K} (\mathbf{c}_k' - \mathbf{y}'\mathbf{D}_k)\hat{\mathbf{u}}_k\right]$$

$$\text{subject to linear feasibility constraints on } \mathbf{y} \text{ for each } k \tag{24}$$

See Ref. 19 for a prescription of how to carry out step 3.

2-4 INNER LINEARIZATION AND DANTZIG-WOLFE DECOMPOSITION

The so-called Dantzig-Wolfe decomposition method for structured linear programs can be derived in several different ways, but the derivation in the original paper[6] is probably the most intuitively appealing. It employs the problem manipulation inner linearization, followed by the simplex method with a column generation feature. It will be appropriate to discuss inner linearization in general before considering its specific application in the Dantzig-Wolfe context.

Probably the simplest instance of the problem manipulation inner linearization is the approximation of a function of one variable by linear interpolation between selected points. Figure 2-2 shows a convex

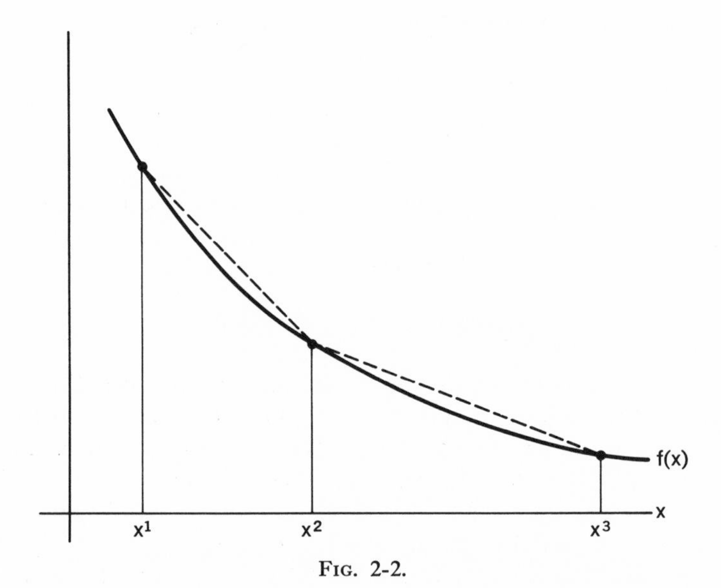

FIG. 2-2.

function (solid line) and an inner linearization of it (broken line) based on evaluations of the function at the points $\mathbf{x}^1$, $\mathbf{x}^2$, $\mathbf{x}^3$. These points are called the *base* of the inner linearization. The approximation can, of course, be made as accurate as desired by making the base sufficiently dense in the domain of the function. The use of such approximations to linearize nonlinear problems has been common at least since the early days of linear programming.[3]

In general, the inner linearization $\hat{f}$ of a convex function f on E^n with respect to the base $\{\mathbf{x}^1, \ldots, \mathbf{x}^p\}$ is as follows:

$$\hat{f}(\mathbf{x}) = \inf_{\boldsymbol{\alpha} \geq \mathbf{0}} \sum_{j=1}^{p} \alpha_j f(\mathbf{x}^j)$$

$$\text{subject to } \sum_{j=1}^{p} \alpha_j = 1 \qquad \sum_{j=1}^{p} \alpha_j \mathbf{x}^j = \mathbf{x} \tag{25}$$

To see the validity of this expression, one must distinguish two cases: either a given $\mathbf{x}$ is in the convex hull of the base points, or it is not. If not, then there will be no $\boldsymbol{\alpha}$ satisfying the constraints and the infimum is taken to be $+\infty$. This is the usual convention for defining an infimum over an empty set, and is the correct one to employ here as no interpolation is possible at such a point. If $\mathbf{x}$ does lie in the convex hull of $\{\mathbf{x}^j\}$, there will be at least one vector $\boldsymbol{\alpha}$ satisfying the constraints and $f(\mathbf{x}) \leqslant \sum \alpha_j f(\mathbf{x}^j)$ for any such $\boldsymbol{\alpha}$ by the convexity of f. The desired approximation to $f(\mathbf{x})$ is obviously the smallest value of $\sum \alpha_j f(\mathbf{x}^j)$ over such vectors $\boldsymbol{\alpha}$ (it is evident that the infimum is achieved). Note that $f(\mathbf{x}) \leqslant \hat{f}(\mathbf{x})$ for all $\mathbf{x}$; that is, inner linearization never underestimates the value of a convex function.

The use of this approximation is not nearly so cumbersome as might seem at first glance, for the minimization in Eq. (25) actually need not be done explicitly. This remark is clarified in the following example.

Example 2-6

Consider the problem

$$\min_{\mathbf{x}} f(\mathbf{x})$$

$$\text{subject to } \mathbf{g}(\mathbf{x}) \leqslant \mathbf{0}$$

$$h(\mathbf{x}) \leqslant 0 \tag{26}$$

where f, $\mathbf{g}$, and h are convex functions defined on E^n ($\mathbf{g}$ may be a vector-valued function). If h is inner linearized with respect to the base $\{\mathbf{x}^j\}$, then Eq. (26) becomes

$$\min_{\mathbf{x}} f(\mathbf{x})$$

$$\text{subject to } \mathbf{g}(\mathbf{x}) \leqslant 0 \text{ and } \begin{bmatrix} \inf\limits_{\boldsymbol{\alpha} \geqslant \mathbf{0}} \sum\limits_j \alpha_j h(\mathbf{x}^j) \\ \text{subject to } \sum\limits_j \alpha_j = 1 \\ \sum\limits_j \alpha_j \mathbf{x}^j = \mathbf{x} \end{bmatrix} \leqslant 0 \tag{27}$$

It is easy to see that the constraint involving h is equivalent to the following simpler one:

$$\sum_j \alpha_j h(\mathbf{x}^j) \leqslant 0 \qquad \text{for some } \boldsymbol{\alpha} \geqslant \mathbf{0} \text{ such that } \sum_j \alpha_j = 1 \qquad \sum_j \alpha_j \mathbf{x}^j = \mathbf{x} \tag{28}$$

Consequently, Eq. (27) is equivalent to

$$\begin{aligned} &\min_{\mathbf{x},\boldsymbol{\alpha}} f(\mathbf{x}) \\ &\text{subject to } \mathbf{g}(\mathbf{x}) \leqq \mathbf{0} \\ &\qquad \sum_j \alpha_j h(\mathbf{x}^j) \leqslant 0 \\ &\qquad \sum_j \alpha_j = 1 \\ &\qquad \sum_j \alpha_j \mathbf{x}^j = \mathbf{x} \\ &\qquad \boldsymbol{\alpha} \geqslant \mathbf{0} \end{aligned} \tag{29}$$

Projecting this onto the $\boldsymbol{\alpha}$ variables and reasoning as in Example 2-1, one obtains

$$\begin{aligned} &\min_{\boldsymbol{\alpha}\geqslant\mathbf{0}} f\left(\sum_j \alpha_j \mathbf{x}^j\right) \\ &\text{subject to } \mathbf{g}\left(\sum_j \alpha_j \mathbf{x}^j\right) \leqslant \mathbf{0} \\ &\qquad \sum_j \alpha_j h(\mathbf{x}^j) \leqslant 0 \\ &\qquad \sum_j \alpha_j = 1 \end{aligned} \tag{30}$$

This is the resulting problem normally referred to when one speaks of inner linearizing h in Eq. (26) with respect to $\{\mathbf{x}^j\}$. Note that the $\mathbf{x}$ variables are replaced by the normalized weighting variables $\boldsymbol{\alpha}$.

If, on the other hand, the objective function of Eq. (26) rather than a constraint is to be inner linearized with respect to the base $\{\mathbf{x}^j\}$, one would proceed as follows. The counterpart of Eq. (27) would be

$$\begin{aligned} &\min_{\mathbf{x}} \left[\inf_{\boldsymbol{\alpha}\geqslant\mathbf{0}} \sum_j \alpha_j f(\mathbf{x}^j) \text{ subject to } \begin{array}{l} \sum_j \alpha_j = 1 \\ \sum_j \alpha_j \mathbf{x}^j = \mathbf{x} \end{array} \right] \\ &\text{subject to } \mathbf{g}(\mathbf{x}) \leqslant \mathbf{0} \\ &\qquad\qquad h(\mathbf{x}) \leqslant 0 \end{aligned} \tag{31}$$

This can be viewed as the projection of the following problem onto $\mathbf{x}$:

$$\begin{aligned} \min_{\mathbf{x},\boldsymbol{\alpha}} \sum_j \alpha_j f(\mathbf{x}^j) & \\ \text{subject to } \mathbf{g}(\mathbf{x}) &\leqslant \mathbf{0} \\ h(\mathbf{x}) &\leqslant 0 \\ \sum_j \alpha_j &= 1 \\ \sum_j \alpha_j \mathbf{x}^j &= \mathbf{x} \\ \boldsymbol{\alpha} &\geqq \mathbf{0} \end{aligned} \tag{32}$$

If Eq. (32) is projected onto $\boldsymbol{\alpha}$ (cf. Example 2-1), the result is

$$\begin{aligned} \min_{\boldsymbol{\alpha}\geqslant\mathbf{0}} \sum_j \alpha_j f(\mathbf{x}^j) & \\ \text{subject to } \mathbf{g}\left(\sum_j \alpha_j \mathbf{x}^j\right) &\leqslant \mathbf{0} \\ h\left(\sum_j \alpha_j \mathbf{x}^j\right) &\leqslant 0 \\ \sum_j \alpha_j &= 1 \end{aligned} \tag{33}$$

This is what is ordinarily referred to when one speaks of inner linearizing f in Eq. (26) with respect to $\{\mathbf{x}^j\}$.

It is clear from this example that one can freely choose which functions are to be inner linearized and which are not. Successful application of this problem manipulation usually requires that this choice be made judiciously.

Not only can convex or concave functions be inner linearized, but so can convex sets. Figure 2-3 illustrates how the convex set $X \subseteq E^2$ can be approximated by an inner convex polytope (broken line) the vertices of which are the points $\{\mathbf{x}^1,..., \mathbf{x}^4\}$ in X. These points are again called the "base" of the inner linearization. In general, the inner linearization of a convex set X with respect to the finite collection of points $\{\mathbf{x}^j\}$ in X is simply the convex hull $\{\mathbf{x} \mid \mathbf{x} = \sum \alpha_j \mathbf{x}^j$ for some $\boldsymbol{\alpha} \geqslant \mathbf{0}$ such that $\sum \alpha_j = 1\}$.

Since the points lying on or above the graph of a convex function constitute a convex set, the inner linearization of a convex function can be considered as the inner linearization of a special class of convex sets.

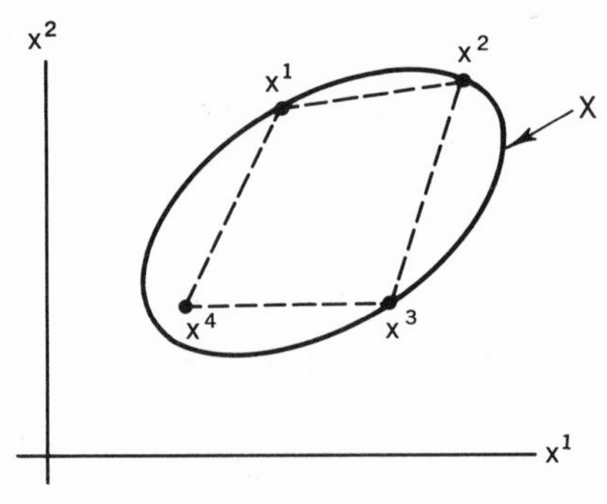

FIG. 2-3.

Example 2-7

Consider the linear program

$$\max_{\mathbf{x}\geqslant\mathbf{0}} \mathbf{c}'\mathbf{x}$$

$$\text{subject to } \mathbf{A}\mathbf{x} = \mathbf{b} \tag{34}$$

$$\bar{\mathbf{A}}\mathbf{x} = \bar{\mathbf{b}}$$

where the constraints $\bar{\mathbf{A}}\mathbf{x} = \bar{\mathbf{b}}$ are "complicating" in the sense that the problem would be much easier if they were not present. For instance, $\mathbf{A}$ might be block diagonal, so that the problem would separate into several independent problems were it not for the coupling constraints $\bar{\mathbf{A}}\mathbf{x} = \bar{\mathbf{b}}$; this structure, a linear case of Example 2-3, is the one usually associated with Dantzig-Wolfe decomposition.[6] Define the convex polyhedron $P \equiv \{\mathbf{x} \geqslant 0 \mid \mathbf{A}\mathbf{x} = \mathbf{b}\}$, so that the problem can be written as

$$\max_{\mathbf{x}\in P} \mathbf{c}'\mathbf{x}$$

$$\text{subject to } \bar{\mathbf{A}}\mathbf{x} = \bar{\mathbf{b}} \tag{35}$$

Assuming that P is nonempty and bounded, it is well known that it admits an exact inner linearization with respect to the base consisting of all extreme points $\{\mathbf{x}^1,\ldots,\mathbf{x}^p\}$ of P. Thus Eq. (35) is equivalent to the linear program

$$\max_{\alpha\geqslant\mathbf{0}} \mathbf{c}'\left(\sum_{j=1}^{p} \alpha_j \mathbf{x}^j\right)$$

$$\text{subject to } \bar{\mathbf{A}}\left(\sum_{j=1}^{p} \alpha_j \mathbf{x}^j\right) = \bar{\mathbf{b}} \tag{36}$$

$$\sum_{j=1}^{p} \alpha_j = 1$$

This is called the "master" problem in the terminology of Dantzig-Wolfe decomposition, and it can be solved by applying the revised simplex method. Of course, determining all extreme points of P in advance is out of the question, and so provision must be made to generate them as needed in the course of carrying out the revised simplex method. This is known as *column generation*. The idea can be sketched as follows (see also Chapter 3 of this book). Suppose that an initial basic feasible solution of Eq. (36) is available and that the corresponding simplex multipliers are $\mathbf{u}$ (associated with the $\bar{\mathbf{b}}$ constraints) and σ (associated with the normalization constraint). The optimality condition is that the "reduced costs" be nonnegative:

$$\sigma + \mathbf{u}'\bar{\mathbf{A}}\mathbf{x}^j - \mathbf{c}'\mathbf{x}^j \geqslant 0 \qquad j = 1,\ldots, p$$

which is equivalent to

$$\sigma + \min_{1 \leqslant i \leqslant p} (\mathbf{u}'\bar{\mathbf{A}} - \mathbf{c}')\mathbf{x}^j \geqslant 0$$

or, since the minimum of a linear function over P occurs at an extreme point,

$$\sigma + \min_{\mathbf{x} \in P} (\mathbf{u}'\bar{\mathbf{A}} - \mathbf{c}')\mathbf{x} \geqslant 0 \tag{37}$$

Thus the optimality test in the revised simplex method can be done by solving the auxiliary linear program

$$\begin{gathered} \min_{\mathbf{x} \geqslant \mathbf{0}} (\mathbf{u}'\bar{\mathbf{A}} - \mathbf{c}')\mathbf{x} \\ \text{subject to } \mathbf{A}\mathbf{x} = \mathbf{b} \end{gathered} \tag{38}$$

and checking the inequality (37). This procedure replaces the naïve one of exhaustively checking the signs of the reduced costs of each column of Eq. (36). If Eq. (37) does not hold, then Eq. (38) yields a profitable nonbasic variable to bring into the basis, namely α_{j_0}, where $\mathbf{x}^{j_0}$ is an optimal basic feasible solution of Eq. (38). The "generated column" in this event is

$$\begin{bmatrix} \mathbf{c}'\mathbf{x}^{j_0} \\ \bar{\mathbf{A}}\mathbf{x}^{j_0} \\ 1 \end{bmatrix}$$

This shows that the revised simplex method for Eq. (36) can be carried out without having to specify all the extreme points of P in advance.

This is the Dantzig–Wolfe decomposition approach for Eq. (34). It seems appealing when the auxiliary linear programs (38) are much simpler than the original problem itself; that is why the constraints $\bar{\mathbf{A}}\mathbf{x} = \bar{\mathbf{b}}$ were assumed to be "complicating."

See Ref. 11, sec. 4.3, for a discussion of generalizations that permit nonlinearities and great flexibility as to which sets and functions are inner linearized. Some of these generalizations require a solution strategy called ***restriction***, which can be viewed as a nonlinear generalization of the strategy on which the simplex method is based (Ref. 11, sec. 3.2). It would not be appropriate to go into details here.

2-5 OUTER LINEARIZATION/RELAXATION

The use of projection and dualization as problem manipulations in large-scale optimization usually leads to functions and sets that are not explicitly available, yet can be built up by tangential approximation. The pattern outer linearization/relaxation provides one approach by which this fact can be exploited. The result is often an algorithm similar to those commonly referred to as *cutting-plane* methods.[4,15] Outer linearization and relaxation are also of independent interest, and will be discussed separately before being juxtaposed.

Outer linearization is the complement of inner linearization in that it approximates convex functions tangentially rather than interpolatively, and convex sets from without rather than from within. Figure 2-4 shows a

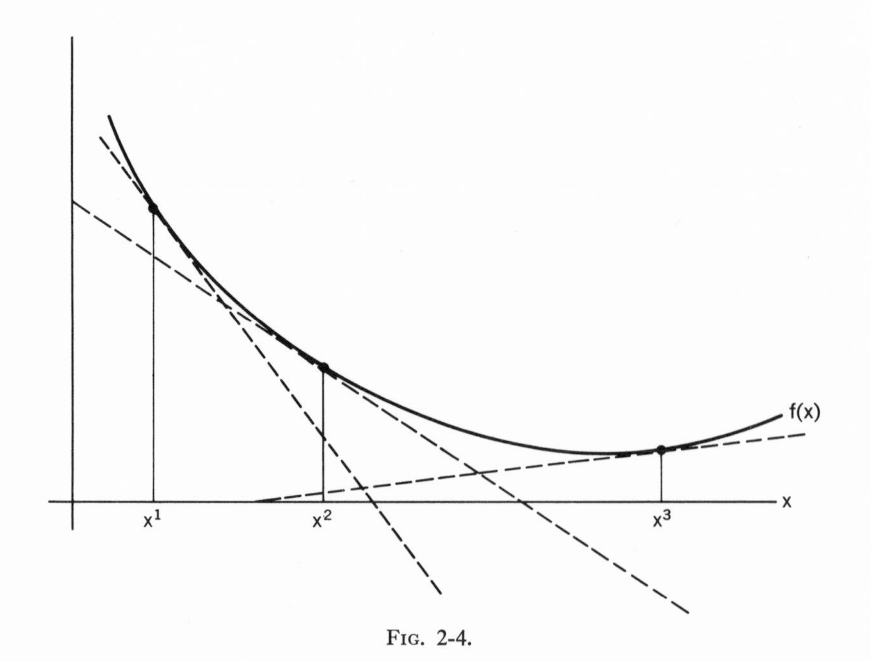

Fig. 2-4.

function with tangents constructed at the points $\{\mathbf{x}^1, \mathbf{x}^2, \mathbf{x}^3\}$; the corresponding outer-linearized approximation is the upper envelope of these tangents. Compare this with Fig. 2-2. In general, an approximation of a convex function f based on outer linearization is the pointwise maximum (upper envelope) of a collection of linear supports to f at various points. A linear support at a point $\bar{\mathbf{x}}$ is defined as any linear function that nowhere exceeds f in value, and equals f in value at $\bar{\mathbf{x}}$. [If f is differentiable then of course the gradient $\nabla \mathbf{f}(\bar{\mathbf{x}})$ at $\bar{\mathbf{x}}$ yields a linear support: $f(\bar{\mathbf{x}}) + \nabla \mathbf{f}(\bar{\mathbf{x}})(\mathbf{x} - \bar{\mathbf{x}})$.] Outer linearization never overestimates the value of a convex function.

It is easy to see that the pointwise maximum can be dealt with at the expense of increasing the number of constraints. Suppose, for example, that $f(\mathbf{x})$ is approximated by $\max \{s_1(\mathbf{x}),\ldots, s_p(\mathbf{x})\}$, where each function s_i is a linear support to f at some point. If f appears as a constraint, say $f(\mathbf{x}) \leqslant b$, then the approximation

$$\max \{s_1(\mathbf{x}),\ldots, s_p(\mathbf{x})\} \leqslant b$$

is equivalent to the p constraints $s_i(\mathbf{x}) \leqslant b$, $i = 1,\ldots, p$. If f is the objective function, as in $\min f(\mathbf{x})$ subject to $\mathbf{x} \in X$, then the outer-linearized approximation to the problem is

$$\min_{\mathbf{x},\sigma} \sigma$$

$$\text{subject to } \sigma \geqslant s_i(\mathbf{x}) \qquad i = 1,\ldots, p \tag{39}$$

where σ is a scalar variable.

When applied to a convex set, outer linearization refers to the intersection of a number of half spaces which contain the set—especially supporting half spaces. The result is a convex polytope that contains the given convex set.

The main obstacle faced by outer linearization is that an excessive number of supports may be required for an adequate approximation. The usual way around this difficulty in most applications is to generate the supports sequentially "as needed," rather than all in advance. The solution strategy permitting this is relaxation.

Relaxation is appropriate for convex programs with a large number of inequality constraints, some of which may not be explicitly available in advance. The basic idea is this: Solve a relaxed version of the given problem ignoring some (perhaps most) of the inequality constraints; if the resulting solution does not satisfy all the ignored constraints, then generate and include at least one violated constraint in the relaxed problem and re-solve it; continue in this fashion until a relaxed problem solution satisfies all the ignored constraints, at which time an optimal solution of the given problem has been found. If the number of inequality constraints which are candidates for being relaxed is finite, it can be shown that the amply satisfied constraints among them in a relaxed problem can be dropped from the next one so long as the successive optimal values of the relaxed problems are strictly decreasing (assuming that the original problem is stated as a maximization). This modification, which tends to keep the number of constraints in the relaxed problems small, will not interfere with the inherent finiteness of the procedure.

The following example illustrates the application of relaxation.

Example 2-8

An airline has m types of containerized air cargo to move from one airport to another, and must do it by making as many flights as necessary with a certain type of aircraft (the only type available). Define

d_i = containers of cargo type i to be moved

a_{ij} = containers of cargo type i carried on one flight with the plane loaded in manner j

s_i = space consumed by one container of cargo type i

S = total cargo space available on a single flight

C = cost to the airline of a single flight (independent of the manner of loading)

The possible ways of loading an aircraft for a flight are many in number, but they can be characterized: a vector $[y_1, \ldots, y_m]$ of nonnegative integers equals $(a_{1j}, \ldots, a_{mj})$ for some j if and only if $\sum_{i=1}^{m} y_i s_i \leqslant S$.

Suppose that a railroad wishes to contract the job of moving the cargo on the basis of an undetermined price π_i dollars per container of cargo type i. If $\boldsymbol{\pi}$ must satisfy certain convex governmental regulatory constraints $\mathbf{h}(\boldsymbol{\pi}) \leqslant \mathbf{0}$ and also be such that the airline must find it economical to sign the contract, then the determination of $\boldsymbol{\pi}$ so as to maximize revenue subject to these constraints leads to the problem:

$$\max_{\boldsymbol{\pi} \geqslant 0} \sum_{i=1}^{m} d_i \pi_i$$

$$\text{subject to } \sum_{i=1}^{m} \pi_i a_{ij} \leqslant C \qquad \text{for all possible loading patterns } j \tag{40}$$

$$\mathbf{h}(\boldsymbol{\pi}) \leqslant \mathbf{0}$$

Note that $\boldsymbol{\pi}$ must be such that no allowable loading pattern will permit the airline to move a load of cargo for less than what the railroad would move it for. It will be assumed that this problem possesses a feasible solution.

The relaxation strategy will be used to cope with the great number of loading pattern constraints in Eq. (40). This approach seems reasonable since it would be expected that only a handful of the more efficient loading patterns would contribute constraints binding at the optimum. In stating this version of the strategy, $(40)^J$ *will denote the relaxed version of Eq.* (40) *in which the loading constraints whose indices are not in the index set* J *are omitted.*

Step 1. Let J_0 be any index set of loading patterns with the property that, for each i, $a_{ij} > 0$ for some $j \in J_0$. Put $J = J_0$ and go to step 2.

Step 2. Solve $(40)^J$ for an optimal solution $\boldsymbol{\pi}^J$. If $\sum \pi_i^J a_{ij} \leqslant C$ for all $j \notin J$, then $\boldsymbol{\pi}^J$ is an optimal solution of Eq. (40); otherwise, go to step 3.

Step 3. Determine the most violated constraint, say j^*. Add j^* to J, and delete any index j in J for which $\sum \pi_i^J a_{ij} < C$ provided that $\sum d_i \pi_i^J$ is smaller than the corresponding value at the previous execution of step 2. Return to step 2.

The choice of J_0 at step 1 is designed to guarantee that $(40)^{J_0}$ will have a bounded feasible region. Thus if $\mathbf{h}$ is continuous, the initial relaxed version of Eq. (40)—and, consequently, all succeeding relaxed versions—will possess an optimal solution as required by step 2. In general, the "better" the loading patterns in J_0, the fewer the number of required executions of step 2. The choice of an algorithm capable of solving $(40)^J$ at step 2 is a tactical matter deliberately left outside the scope of the relaxation strategy.

The determination at step 2 of whether a violated constraint exists can, of course, actually be accomplished at step 3. Seeking a *most* violated constraint to append to the current relaxed problem may appear to be a somewhat arbitrary criterion, since in theory any violated constraint will do, but it is a quite plausible criterion if one reasons geometrically in terms of obtaining as "deep a cut" as possible. At any rate, finding the most violated constraint can be stated as the problem

$$\max_j \sum \pi_i^J a_{ij} - C$$

or, equivalently,

$$\max_{\mathbf{y} \geqslant \mathbf{0}} \sum \pi_i^J y_i - C$$

$$\text{subject to } \sum y_i s_i \leqslant S \qquad (41)$$

$$\mathbf{y} \text{ integer}$$

This is a problem of the "knapsack" variety for which efficient methods are available. It has a very natural interpretation in terms of finding the allowable loading pattern that moves the greatest total value of cargo measured at the trial price vector $\boldsymbol{\pi}^J$.

The stage is now set for an example using outer linearization and relaxation together.

Example 2-9

Consider the following min-max problem, which might arise in a competitive situation:

$$\min_{\mathbf{y}\in Y}\begin{bmatrix}\max_{\mathbf{x}\geqslant 0}\mathbf{x}'(\mathbf{C}\mathbf{y}+\mathbf{d})\\ \text{subject to }\mathbf{A}\mathbf{x}=\mathbf{b}\end{bmatrix} \tag{42}$$

This is a special case of Eq. (20) in Example 2-5, but here the outer linearization/relaxation pattern rather than the piecewise strategy will be applied.

It will be assumed that Y is a convex set, and that $\{\mathbf{x}\geqslant 0 \mid \mathbf{A}\mathbf{x}=\mathbf{b}\}$ is nonempty and bounded. Let the extreme points of the latter set be $\{\mathbf{x}^1,\dots,\mathbf{x}^p\}$. Since for each given $\mathbf{y}$ the inner maximum of Eq. (42) must be achieved at one of these points, it follows that this quantity can be expressed as

$$\max_{1\leqslant j\leqslant p}(\mathbf{x}^j)'(\mathbf{C}\mathbf{y}+\mathbf{d})$$

In other words, the minimand of Eq. (42) can be outer linearized without error as the pointwise maximum of at most p linear functions. Thus Eq. (42) can be expressed [cf. Eq. (39)] as

$$\begin{aligned}&\min_{\mathbf{y},\sigma}\ \sigma\\ &\text{subject to } \sigma\geqslant(\mathbf{x}^j)'(\mathbf{C}\mathbf{y}+\mathbf{d}) \qquad j=1,\dots,p\\ &\mathbf{y}\in Y\end{aligned} \tag{43}$$

Relaxation can be applied to Eq. (43) with the p constraints as the candidates for being relaxed. In view of the previous example, only one aspect needs to be discussed, namely, how the analog of step 3 will find a (most) violated constraint if one exists at the current solution $(\sigma^J, \mathbf{y}^J)$ to Eq. (43) with the constraints not in J relaxed ($J\subseteq\{1,\dots,p\}$). The answer resides in the subsidiary problem

$$\max_{1\leqslant j\leqslant p}(\mathbf{x}^j)'(\mathbf{C}\mathbf{y}^J+\mathbf{d})-\sigma^J$$

which is obviously equivalent [cf. Eq. (37)] to the linear program

$$\max_{\mathbf{x}\geqslant 0} \mathbf{x}'(\mathbf{C}\mathbf{y}^J + \mathbf{d}) - \sigma^J$$

$$\text{subject to } \mathbf{A}\mathbf{x} = \mathbf{b} \tag{44}$$

If the optimal value of Eq. (44) is not greater than zero, then $(\sigma^J, \mathbf{y}^J)$ is optimal in Eq. (43) and consequently $\mathbf{y}^J$ is optimal in Eq. (42). If the optimal value is greater than zero, the most violated constraint of Eq. (43) at $(\sigma^J, \mathbf{y}^J)$ is given by $\sigma \geqslant (\mathbf{x}^J)'(\mathbf{C}\mathbf{y} + \mathbf{d})$, where $\mathbf{x}^J$ is an optimal basic feasible solution of Eq. (44).

It is perhaps worth noting that σ^J provides a nondecreasing sequence of lower bounds on the optimal value of Eq. (42), and that $(\mathbf{x}^J)'(\mathbf{C}\mathbf{y}^J + \mathbf{d})$ provides a sequence (not necessarily nonincreasing) of upper bounds on the optimal value of Eq. (42).

2-6 DUALIZATION AND LAGRANGE DECOMPOSITION

Taking the dual of a convex program is a problem manipulation of considerable utility in large-scale mathematical programming. To obtain the full benefit, it is essential to realize that the dual can be taken with respect to *any* subset of the constraints. Consider the problem

$$\min_{\mathbf{x}\in X} f(\mathbf{x})$$

$$\text{subject to } g_j(\mathbf{x}) \leqslant 0 \qquad j = 1,\ldots, m \tag{45}$$

where f and each g_j is convex on the convex set $X \subseteq E^n$. The dual with respect to $g_1 ,\ldots, g_{m_1}$ $(1 \leqslant m_1 \leqslant m)$, say, is

$$\max_{\mathbf{u}\geqslant 0} \begin{bmatrix} \inf_{\mathbf{x}\in X} f(\mathbf{x}) + \sum_{j=1}^{m_1} u_j g_j(\mathbf{x}) \\ \text{subject to } g_j(\mathbf{x}) \leqslant 0, j = m_1 + 1,\ldots, m \end{bmatrix} \tag{46}$$

where $\mathbf{u}$ is an m_1-vector of dual variables. The maximand of Eq. (46) is evidently a concave function, for it is the pointwise infimum of a collection (indexed by $\mathbf{x}$) of functions linear in $\mathbf{u}$. Note that dual variables have been introduced only for those constraints with respect to which the dual was taken. In most cases one should dualize only with respect to constraints that are "complicating" (cf. Example 2-7). Several examples of this will be given, but first the nature of the relationship between a primal problem and its dual will be reviewed (see Ref. 12 or 18 for a more detailed review).

The key hypothesis required to establish the relationship between Eqs. (45) and (46) is called *stability*. Equation (45) is said to be stable with respect to the first m_1 constraints if its infimal value is finite and does not decrease infinitely steeply in any perturbation direction as the right-hand sides of the first m_1 constraints vary. Several equivalent mathematical renderings of this property could be given, but for present purposes it is enough to know that it is implied by every so-called constraint qualification assumption ever used to prove the existence of optimal multipliers (associated with the first m_1 constraints) when Eq. (45) is known to have an optimal solution. In fact, stability is necessary as well as sufficient for the existence of such multipliers. Thus stability is a quite weak property that will usually hold for convex problems of practical interest. The relationship between Eqs. (45) and (46) can now be stated (a proof is given in Ref. 12).

Strong Duality Theorem: If Eq. (45) is stable, then

1. Equation (46) has an optimal solution.
2. The optimal values of Eqs. (45) and (46) are equal.
3. $\mathbf{u}^*$ is an optimal solution of Eq. (46) if and only if $-\mathbf{u}^*$ is the normal of a linear support at the origin of the infimal value of Eq. (45) as a function of perturbations in the right-hand side of the first m_1 constraints.
4. Every optimal solution $\mathbf{u}^*$ of Eq. (46) characterizes the set of all optimal solutions (if any) of Eq. (45) as the minimizers of $f(\mathbf{x}) + \sum_{j=1}^{m_1} u_j{}^*g_j(\mathbf{x})$ over X which also satisfy $g_j(\mathbf{x}) \leqslant 0, j = 1,\ldots,m_1$ and the complementary slackness condition $\sum_{j=1}^{m_1} u_j{}^*g_j(\mathbf{x}) = 0$.

Note that, according to statements 2 and 3, an optimal solution of the dual tells a good deal about the primal problem even before an optimal solution to it is recovered.

The trouble with the dual problem, so far as obtaining a numerical solution is concerned, is the same as the trouble with a problem that has been projected: the objective function is not explicitly available, and even evaluating it requires solving an "inner" Lagrange optimization problem [cf. Eq. (2)]. It would seem absurd to incur this difficulty unless there were a way around it, as indeed there is in many specific cases. In most if not all of these, the trick is to dualize only with respect to those constraints which are "complicating," so that the resulting Lagrange optimization problem is much simpler than the original problem itself. This ensures that evaluating the maximand of Eq. (46), at least, is less difficult than solving Eq. (45) itself.

For instance, suppose that $m = 1$ and that Eq. (45) is a transportation problem with one additional linear constraint $g(\mathbf{x}) \leqslant 0$. Dualizing with respect to this constraint leads to a Lagrange optimization problem that is again a transportation problem (for fixed $\mathbf{u}$). Since Eq. (46) in this case involves maximizing a concave function of but a single variable, an efficient solution technique can be based on a parametric search over the positive values of $\mathbf{u}$. Another instance, and one that will be taken up in more detail below, is provided by the multidivisional problem with coupling constraints of Example 2-3. Dualizing with respect to the coupling constraints alone yields

$$\min_{\mathbf{u} \geqslant 0} \left[\mathbf{u}'\mathbf{b} + \sum_{k=1}^{K} \sup_{\mathbf{x}_k \in X_k} f_k(\mathbf{x}_k) - \mathbf{u}'\mathbf{g}_k(\mathbf{x}_k) \right] \tag{47}$$

where we have taken advantage of the separation of the Lagrange optimization problem into K independent smaller optimization problems. Examples such as these, in which the Lagrange optimization problem separates or reduces to a problem substantially simpler than the given problem, are sometimes referred to as instances of *Lagrange decomposition.*

Assuming that dualization has been judiciously applied so that the Lagrange optimization problem of Eq. (46) is indeed simpler than Eq. (45), the other half of the story concerns methods of optimizing with respect to the dual variables. Only when m_1 equals 1 or 2 is a brute force search over $\mathbf{u}$ likely to be satisfactory. The rest of the time the most likely ways to go about solving Eq. (46) involve outer linearization/relaxation, the feasible directions strategy, and (when applicable) the piecewise strategy. The following two examples illustrate, respectively, the first two of these approaches.

Example 2-10

Consider again, as in Example 2-7, the linear program

$$\begin{aligned} &\max_{\mathbf{x} \geqslant 0} \mathbf{c}'\mathbf{x} \\ &\text{subject to } \mathbf{A}\mathbf{x} = \mathbf{b} \\ &\qquad\qquad \bar{\mathbf{A}}\mathbf{x} = \bar{\mathbf{b}} \end{aligned} \tag{48}$$

where the constraints $\bar{\mathbf{A}}\mathbf{x} = \bar{\mathbf{b}}$ are complicating. The dual with respect to the complicating constraints is (this requires a trivial modification of the duality results reviewed above to accommodate linear equality constraints):

$$\min_{\mathbf{u}} \left[\begin{aligned} &\sup_{\mathbf{x}} \mathbf{c}'\mathbf{x} + \mathbf{u}'(\bar{\mathbf{b}} - \bar{\mathbf{A}}\mathbf{x}) \\ &\text{subject to } \mathbf{A}\mathbf{x} = \mathbf{b} \\ &\qquad\qquad \mathbf{x} \geqslant 0 \end{aligned} \right] \tag{49}$$

Solving Eq. (49) by outer linearization/relaxation is accomplished exactly as in Example 2-9, assuming that $\{\mathbf{x} \geqslant 0 \mid \mathbf{Ax} = \mathbf{b}\}$ is nonempty and bounded. The exact outer linearization of Eq. (49) is

$$\min_{\mathbf{u},\sigma} \mathbf{u}'\bar{\mathbf{b}} + \sigma$$

$$\text{subject to } \sigma \geqslant (\mathbf{c}' - \mathbf{u}'\bar{\mathbf{A}})\mathbf{x}^j \qquad j = 1,\dots,p \tag{50}$$

with the $\mathbf{x}^j$'s being the extreme points of the aforementioned set. When relaxation is applied to Eq. (50), a most violated constraint is generated (if one exists) at a typical iteration by the subsidiary linear program

$$\max_{\mathbf{x} \geqslant 0} (\mathbf{c}' - \mathbf{u}'\bar{\mathbf{A}})\mathbf{x} - \sigma$$

$$\text{subject to } \mathbf{Ax} = \mathbf{b} \tag{51}$$

where $\mathbf{u}$ and σ are fixed at an optimal solution of the current relaxed version of Eq. (50). See Example 2-9 for further details.

Besides illustrating the pattern dualization, outer linearization/relaxation, this example also demonstrates an instance in which this pattern is essentially equivalent to the Dantzig–Wolfe decomposition idea illustrated in Example 2-7. This follows from the fact that Eq. (50) is the linear dual of Eq. (36), and Eq. (51) is equivalent to Eq. (38).

Example 2-11

In this example the method of Uzawa[21] will be applied to Example 2-3 with $X_k = E^{n_k}$, namely,

$$\max_{\mathbf{x}} \sum_{k=1}^{K} f_k(\mathbf{x}_k)$$

$$\text{subject to } \sum_{k=1}^{K} \mathbf{g}_k(\mathbf{x}_k) \leqslant \mathbf{b} \tag{52}$$

as an illustration of the pattern dualization/feasible directions (a number of references also employing this pattern were mentioned at the end of Sec. 2-1). In terms of Eq. (52), Uzawa's assumptions are these: f_k is strictly concave and $\mathbf{g}_k$ is convex on E^{n_k}, Eq. (52) has an optimal solution and satisfies Slater's qualification that there exists a point $\mathbf{x}^0$ such that every component of $\sum \mathbf{g}_k(\mathbf{x}_k{}^0)$ is strictly less than the corresponding component of $\mathbf{b}$, and the Lagrange function $f_k(\mathbf{x}_k) - \mathbf{u}'\mathbf{g}_k(\mathbf{x}_k)$ must have for each k a finite maximum $\bar{\mathbf{x}}_k(\mathbf{u})$ for any $\mathbf{u} \geqslant 0$ ($\mathbf{u}$, of course, is a vector whose dimension matches $\mathbf{b}$).

The dual of Eq. (52) with respect to all the constraints is [cf. Eq. (47)]:

$$\min_{\mathbf{u}\geqslant 0}\left[\mathbf{u}'\mathbf{b} + \sum_{k=1}^{K} \max_{\mathbf{x}_k} f_k(\mathbf{x}_k) - \mathbf{u}'\mathbf{g}_k(\mathbf{x}_k)\right] \tag{53}$$

The assumption of Slater's qualification implies that Eqs. (52) and (53) are related according to the strong duality theorem. Thus an optimal solution $\mathbf{u}^*$ of Eq. (53) exists, and would yield each component $\mathbf{x}_k{}^*$ of the optimal solution $\mathbf{x}^*$ of Eq. (52) as $\mathbf{x}_k{}^* = \bar{\mathbf{x}}_k(\mathbf{u}^*)$.

Uzawa's proposal can be viewed as a short-step feasible directions method for solving Eq. (53). Beginning with any initial $\mathbf{u} \geqslant 0$, each trial value $\mathbf{u}^1$ is determined from the previous trial value $\mathbf{u}^0$ according to the rule

$$\mathbf{u}^1 = \max\{0, \mathbf{u}^0 - \rho\gamma_{\mathbf{u}^0}\} \tag{54}$$

where the maximum is taken component by component, ρ is a predetermined small positive "step size" parameter, and $\gamma_{\mathbf{u}^0}$ is defined according to

$$\gamma_{\mathbf{u}} \equiv \mathbf{b} - \sum_{k=1}^{K} \mathbf{g}_k(\bar{\mathbf{x}}_k(\mathbf{u})) \tag{55}$$

If the minimand of Eq. (53) is differentiable at the point $\mathbf{u}^0$, then $\gamma_{\mathbf{u}^0}$ is its gradient at $\mathbf{u}^0$; in general, it is easy to show that $\gamma_{\mathbf{u}^0}$ is always a subgradient of the minimand at $\mathbf{u}^0$; that is, it is the normal to a linear support of the minimand at $\mathbf{u}^0$. This fact renders the use of $\gamma_{\mathbf{u}^0}$ as a feasible direction in Eq. (54) plausible, and indeed Uzawa has proved convergence. The component-by-component maximum in Eq. (54), of course, keeps the trial $\mathbf{u}$'s nonnegative.

REFERENCES:

1. Benders, J. F.: Partitioning Procedures for Solving Mixed-variables Programming Problems, *Numer. Math.*, vol. 4, pp. 238–252, 1962.
2. Bradley, S. P.: Decomposition Programming and Economic Planning, *Univ. Calif. Operations Res. Cen. Rept.* 67-20, Berkeley, June, 1967.
3. Charnes, A., and C. Lemke: Minimization of Nonlinear Separable Convex Functionals, *Naval Res. Log. Quart.*, vol. 1, pp. 301–312, 1954.
4. Cheney, E. W., and A. A. Goldstein: Newton's Method for Convex Programming and Tchebycheff Approximation, *Numer. Math.*, vol. 1, pp. 253–268, 1959.
5. Dantzig, G. B.: On the Status of Multistage Linear Programming Problems, *Management Sci.*, vol. 6, no. 1, pp. 53–72, October, 1959.
6. Dantzig, G. B., and P. Wolfe: Decomposition Principle for Linear Programs, *Operations Res.*, vol. 8, no. 1, pp. 101–111, January-February, 1960. See also *Econometrica*, vol. 29, no. 4, pp. 767–778.
7. Falk, J.: An Algorithm for Separable Convex Programming under Linear Equality Constraints, *Res. Anal. Corp. Tech. Paper* 148, McLean, Va., March, 1965.

8. Falk, J.: Lagrange Multipliers and Nonlinear Programming, *J. Math. Anal. Appl.*, vol. 19, no. 1, pp. 141–159, 1967.
9. Fiacco, A. V., and G. P. McCormick: "Nonlinear Programming," Wiley, 1968.
10. Geoffrion, A. M.: Primal Resource-Directive Approaches for Optimizing Nonlinear Decomposable Systems, *Operations Res.*, vol. 18, no. 3, pp. 375–403, May-June, 1970.
11. Geoffrion, A. M.: Elements of Large-scale Mathematical Programming, *Management Sci.*, vol. 16, no. 11, pp. 652–691, July, 1970.
12. Geoffrion, A. M.: Duality in Nonlinear Programming: A Simplified Applications-oriented Development, *Western Management Sci. Inst. Working Paper* 150, University of California, Los Angeles, August, 1969. To appear in *SIAM Review.*
13. Golshtein, E. G.: A General Approach to the Linear Programming of Block Structures, *Soviet Physics—Doklady*, vol. 11, no. 2, pp. 100–103, August, 1966.
14. Grinold, R. C.: Steepest Ascent for Large-scale Linear Programs, *Univ. Calif. Cen. Res. Management Sci. Working Paper* 282, Berkeley, September, 1969.
15. Kelley, J. E.: The Cutting-plane Method for Solving Convex Programs, *SIAM J.*, vol. 8, no. 4, pp. 703–712, 1960.
16. Lasdon, L. S.: Duality and Decomposition in Mathematical Programming, *IEEE Trans. Syst. Sci. Cybernetics*, vol. SSC-4, no. 2, pp. 86–100, July, 1968.
17. Pearson, J. D.: Decomposition, Coordination, and Multi-level Systems, *IEEE Trans. Syst. Sci. Cybernetics*, vol. SSC-2, no. 1, pp. 36–40, August, 1966.
18. Rockafellar, R. T.: Duality in Nonlinear Programming, in G. B. Dantzig and A. F. Veinott, Jr. (eds), "Mathematics of the Decision Sciences," pt. 1, American Mathematical Society, 1968.
19. Rosen, J. B.: Primal Partition Programming for Block Diagonal Matrices, *Numer. Math.*, vol. 6, pp. 250–260, 1964.
20. Takahashi, I.: Variable Separation Principle for Mathematical Programming, *J. Operations Res. Soc. Japan*, vol. 6, no. 1, pp. 82–105, February, 1964.
21. Uzawa, H.: Iterative Methods for Concave Programming, in K. Arrow, L. Hurwicz, and H. Uzawa (eds.), "Studies in Linear and Nonlinear Programming," chap. 10, Stanford University Press, 1958.
22. Wilson, R.: Computation of Optimal Controls, *J. Math. Anal. Appl.*, vol. 14, no. 1, pp. 77–82, April, 1966.
23. Zoutendijk, G.: "Methods of Feasible Directions," Elsevier, 1960.

3

GENERALIZED LINEAR PROGRAMMING

GEORGE B. DANTZIG[1]
and
RICHARD M. VAN SLYKE[2]

3-1 INTRODUCTION

Many of the fundamental notions in the theory of mathematical programming have direct analogs to the various aspects of optimal control theory. Moreover, the well-developed and very general computational methods of mathematical programming frequently offer efficient solution methods for large classes of optimal control problems.

Our aim here in the application of mathematical programming techniques to optimal control theory is not to attempt to formulate all control problems as mathematical programs but rather to apply mathematical programming to those classes of control problems for which the algorithms and concepts of mathematical programming are particularly useful. We will be especially interested in devices which allow us to transform apparently nonlinear problems into problems which if not linear at least may be attacked by linear methods.

The algorithms suggested here have the additional advantage that two-point boundary problems need not be solved and the question of normality never arises.

[1] The research of the first author on which this chapter was based was partially supported by the Office of Naval Research under Contracts ONR-N-00014-67-A-0112-0011 to 0016, the U. S. Atomic Energy Commission under Contract AT(04-3)-326 PA#18, the National Science Foundation under Grant GP-6431, and the U. S. Army Research Office under Contract DAHC04-67-C-0028.

[2] The research of the second author on which this chapter was based was partially supported by the Office of Naval Research under Contract Nonr-222(83) and the National Science Foundation under Grant GP-8695 with the Operations Research Center, University of California, Berkeley.

The basic problem we keep in mind for this analysis is

$$\begin{aligned} &\min_{\mathbf{u}} x_0(T) \\ &\text{subject to } \frac{d}{dt}\mathbf{x}(t) = \mathbf{A}(t)\mathbf{x}(t) + \mathbf{B}(t)\mathbf{u}(t) \\ &\qquad \mathbf{u}(t) \in \mathbf{U}(t) \qquad\qquad t \in [0, T] \\ &\qquad \mathbf{x}(0) = \mathbf{x}^0 \\ &\qquad x_i(T) = x_i \qquad\qquad = 1, \ldots, n \end{aligned} \tag{1}$$

where $\mathbf{x}(t) = (x_0(t), \ldots, x_n(t))$ is an $n + 1$ component vector function of time, $\mathbf{u}(t) = (u_1(t), \ldots, u_r(t))$ is an r-component vector function of time, $\mathbf{A}(t)$ and $\mathbf{B}(t)$ are respectively $(n + 1) \times (n + 1)$ and $(n + 1) \times r$ matrix functions. Further $\mathbf{U}(t)$ will be a closed and convex subset of E^r, r dimensional Euclidean space, for each t in $[0, T]$.

We will also be interested in the following discrete analog of Eq. (1) which can either arise from the intrinsic nature of the problem or as an approximation to Eq. (1).

$$\begin{aligned} &\min x_0^K \\ &\text{subject to } \mathbf{x}^{k+1} = \mathbf{A}^k\mathbf{x}^k + \mathbf{B}^k\mathbf{u}^k \qquad k = 0, 1, \ldots, K-1 \\ &\qquad \mathbf{x}^0 = \mathbf{v}^0 \\ &\qquad x_i^K = x_i^T \qquad\qquad i = 1, \ldots, n \\ &\qquad \mathbf{u}^k \in \mathbf{U}^k \qquad\qquad k = 0, 1, \ldots, K-1 \end{aligned} \tag{2}$$

where the vectors and matrices have the same dimensions as in Eq. (1) and each $\mathbf{U}^k$ is a closed, convex set in E^r.

The basic problem (1) can be generalized to

$$\begin{aligned} &\min \int_0^T c(\mathbf{x}, \mathbf{u}, \mathbf{t})\, dt \\ &\text{subject to } \frac{d}{dt}\mathbf{x}(t) = \mathbf{A}(t)\mathbf{x}(t) + \mathbf{B}(t)\mathbf{u}(t) \\ &\qquad \mathbf{u}(t) \in \mathbf{U}(t) \qquad\qquad t \in [0, T] \\ &\qquad \mathbf{x}(t) \in \mathbf{X}(t) \qquad\qquad t \in [0, T] \\ &\qquad \mathbf{x}(0) \in \mathbf{X}^0 \\ &\qquad \mathbf{x}(T) \in \mathbf{X}^T \end{aligned} \tag{3}$$

where now $\mathbf{x}(t) = (x_1(t), \ldots, x_n(t))$, $\mathbf{A}(t)$ is an $n \times n$ matrix, $\mathbf{B}(t)$ is an $n \times r$ matrix, $\mathbf{U}(t)$, $\mathbf{X}(t)$, $\mathbf{X}^0$, $\mathbf{X}$ are closed convex subsets, and $\mathbf{c}(\mathbf{x}, \mathbf{u}, t)$ is a convex function of $\mathbf{x}$ and $\mathbf{u}$ for each t. The system dynamics are allowed to be time varying so that the more general problems of nonlinear optimal control can be handled iteratively by linear approximation (Rosen[35,36]).

In order to solve Eq. (1) or (3) it is often necessary to consider the auxiliary problem

$$\begin{aligned} &\min \| \mathbf{X}^T - \mathbf{x}(T) \| \\ &\text{subject to } \frac{d}{dt}\mathbf{x}(t) = \mathbf{A}(t)\mathbf{x}(t) + \mathbf{B}(t)\mathbf{u}(t) \\ &\qquad \mathbf{u}(t) \in \mathbf{U}(t) \qquad\qquad t \in [0, T] \\ &\qquad \mathbf{x}(0) \in \mathbf{X}^0 \end{aligned} \tag{4}$$

where in this problem we consider a state vector of only n components ignoring the x_0 component. The problem then is to minimize the distance from the closed convex set $\mathbf{X}^T$ at time T starting from $\mathbf{X}^0$ at time zero. Clearly Eq. (3) has no feasible trajectory unless the minimum value for Eq. (4) is zero. The formulation (4) is also used in solving minimum time problems. The idea is to start with $T = 0$ and to increase T until an optimal value of zero is obtained for Eq. (4). The norm chosen in the objective is of crucial importance. The $l_2[\| \mathbf{x} \|_2 = (\sum x_i^2)^{1/2}]$ norm leads to quadratic programming and the $l_1(\| \mathbf{x} \|_1 = \sum | x_i |)$ and l_∞ $(\| \mathbf{x} \|_\infty = \max_i | x_i |)$ norms lead to linear programs.

In the next section, we give a brief review of some topics in mathematical programming which we will need. These include the simplex method, the simplex method for problems with upper-bounded variables, generalized linear programming, the decomposition methods of Dantzig-Wolfe and Benders, and a method of convex programming.

It is assumed that the reader has some knowledge of linear programming and therefore the material on the better-known algorithms is designed to refresh the reader's memory and more important to interpret these algorithms in a framework appropriate to the optimal control problems considered in Secs. 3-3 and 3-4. Most of the topics considered are discussed in more detail in Dantzig.[11] In all cases, references will be given to the literature for the complete details of each algorithm.

In Sec. 3-3, we describe a direct method for solving continuous time linear optimal control problems[12] as distinguished from the methods involving discrete time which are described in Sec. 3-4. The direct method is based on the generalized linear programming algorithm (see Dantzig,[11] chap. 22).

Finally in the last section we give a summary of methods of applying mathematical programming to the solution of optimal control problems and some comments on the validity of approximating continuous control problems by discrete ones.

3-2 MATHEMATICAL PROGRAMMING ALGORITHMS

3-2-1 The Simplex Method

Let $\mathbf{A}$ be an $m \times n$ matrix, $\mathbf{A}^j$ be the jth column of $\mathbf{A}$, $\mathbf{b}$ be a column vector with m components, and $\mathbf{c}$ a row vector with n elements. We consider the linear programming problem in standard form[11] of finding the n-vector which solves

$$\begin{aligned} \min_{\mathbf{x}} z &= \mathbf{cx} \\ \text{subject to } \mathbf{Ax} &= \mathbf{b} \\ \mathbf{x} &\geqslant \mathbf{0} \end{aligned} \tag{5}$$

The most common solution method used for this problem is the simplex method and we will outline its essential features. We assume for the time being that the equations $\mathbf{Ax} = \mathbf{b}$ are neither redundant nor inconsistent, that is, that the rows of $\mathbf{A}$ are linearly independent. In this situation a *basis* $\mathbf{B}$ is a set of m linearly independent columns of $\mathbf{A}$, $\mathbf{B} = [\mathbf{B}^1, \ldots, \mathbf{B}^m]$, which we will assume for notational convenience to be the first m columns of $\mathbf{A}$. Thus $\mathbf{A} = [\mathbf{B} \mid \mathbf{C}]$ where $\mathbf{C}$ denotes the remaining $n - m$ columns. We partition $\mathbf{x} = (\mathbf{x}_\mathbf{B}, \mathbf{x}_\mathbf{C})$ and $\mathbf{c} = (\mathbf{c}_\mathbf{B}, \mathbf{c}_\mathbf{C})$ accordingly, where $\mathbf{x}_\mathbf{B}$ is called the vector of *basic variables* and $\mathbf{x}_\mathbf{C}$ the vector of *nonbasic variables*. The basis $\mathbf{B}$ is said to be a *feasible basis* if $\mathbf{x}_\mathbf{B}{}^0 = \mathbf{B}^{-1}\mathbf{b} \geqslant \mathbf{0}$, because in this case $\mathbf{x}_\mathbf{B} = \mathbf{x}_\mathbf{B}{}^0$, $\mathbf{x}_\mathbf{C} = \mathbf{0}$ (which we call the *basic solution* corresponding to $\mathbf{B}$) satisfies the constraints on $\mathbf{x}$ in Eq. (5). The second temporary assumption we make is that we have a feasible basis at hand. We will remove these two assumptions later. In the simplex algorithm a sequence of changes of basis is made where each feasible basis is obtained from the previous feasible basis by removing one basic variable and replacing it by a formerly nonbasic variable. The value of z decreases for each of the corresponding basic solutions (this last assertion requires mild qualification as we shall see later). Geometrically, basic solutions correspond to extreme points of the convex polyhedral set $K = \{\mathbf{x} \mid \mathbf{Ax} = \mathbf{b}, \mathbf{x} \geqslant \mathbf{0}\}$, and the algorithm can be interpreted as traversing a sequence of distinct adjacent extreme points with the corresponding values of z always decreasing. The fact that K has a finite number of extreme points and that the minimum value of a linear function occurs at an extreme point provides a proof of the effectiveness and finiteness of the algorithm. To completely describe the algorithm we must specify a column choice rule which determines the column to enter the basis, a rule for choosing the column which leaves the basis, and a stopping rule.

To get started we transform Eq. (5) into the *canonical form* associated with our basis $\mathbf{B}$. This is done by premultiplying $\mathbf{Ax} = \mathbf{b}$ by $\mathbf{B}^{-1}$, the inverse of $\mathbf{B}$, and then subtracting $\pi[\mathbf{Ax} - \mathbf{b}]$ from $\mathbf{cx}$ where $\pi = \mathbf{c_B B^{-1}}$. This leads to the form

$$
\begin{array}{llcl}
\min z & & & \\
\text{subject to } x_1 & \bar{a}_{1m+1}x_{m+1} + \cdots + \bar{a}_{1s}x_s + \cdots + \bar{a}_{1n}x_n & = & \bar{b}_1 \\
\quad\ddots & \vdots \qquad\qquad \vdots \qquad\qquad \vdots & & \vdots \\
\qquad x_r & \bar{a}_{rm+1}x_{m+1} + \cdots + \bar{a}_{rs}x_s + \cdots + \bar{a}_{rn}x_n & = & \bar{b}_r \\
\qquad\quad\ddots & \vdots \qquad\qquad \vdots \qquad\qquad \vdots & & \vdots \\
\qquad\qquad x_m + & \bar{a}_{mm+1}x_{m+1} + \cdots + \bar{a}_{ms}x_s + \cdots + \bar{a}_{mn}x_n & = & \bar{b}_n \\
 & \bar{c}_{m+1}\ x_{m+1} + \cdots + \bar{c}_s\ x_s + \cdots + \bar{c}_n\ x_n & = & z - z_0
\end{array}
\tag{6}
$$

where we have retreated momentarily from matrix notation. The bars over the coefficients indicate that they differ from those of Eq. (5). The fact that the columns of coefficients for the basic variables $\mathbf{x}_1, \ldots, \mathbf{x}_m$ are unit vectors follows from the fact that the coefficients of the basic variables in Eq. (5) make up $\mathbf{B}$ and to get Eq. (6) we premultiplied by $\mathbf{B}^{-1}$. The missing terms corresponding to the basic variables in the equation for z result from the subtraction of $\pi(\mathbf{Ax} = \mathbf{b}) = \mathbf{c_B B}^{-1}(\mathbf{Ax} = \mathbf{b})$ from $\mathbf{cx} = z$. We remember that if $\mathbf{W}$, $\mathbf{U}$, $\mathbf{V}$ are matrices such that $\mathbf{W} = \mathbf{UV}$, then $\mathbf{W}_i = \mathbf{U}_i\mathbf{V}$, $\mathbf{W}^j = \mathbf{UV}^j$, and $w_{ij} = \mathbf{W}_i^j = \mathbf{U}_i\mathbf{V}^j$ where $\mathbf{W}_i$ is the ith row of $\mathbf{W}$, $\mathbf{W}^j$ the jth column, and $\mathbf{W}_i^j = w_{ij}$, the element in row i, column j. Then making use of these relations we have

$$
\bar{\mathbf{A}}^j = \begin{pmatrix} \bar{a}_{1j} \\ \vdots \\ \bar{a}_{mj} \end{pmatrix} = \mathbf{B}^{-1}\mathbf{A}^j \qquad \bar{\mathbf{b}} = \mathbf{B}^{-1}\mathbf{b} \qquad \mathbf{z}_0 = \pi\mathbf{b} \tag{7}
$$

and $\bar{\mathbf{c}} = (0, \ldots, 0, \bar{c}_{m+1}, \ldots, \bar{c}_n) = \mathbf{c} - \pi\mathbf{A}$ where $\pi = \mathbf{c_B B}^{-1}$. In particular $\bar{c}_j = c_j - \pi\mathbf{A}^j$.

Our optimality rule follows immediately from Eq. (6). If $\bar{c}_j \geqslant 0$, $j = m + 1, \ldots, n$, then the basic solution $x_i = \bar{b}_i$ for $i = 1, \ldots, m$ and $\mathbf{x}_j = 0$ for $j = m + 1, \ldots, n$ is optimal. This is true because $z = z_0$ is a lower bound for z since the terms $\bar{c}_j x_j \geqslant 0$ and the value $z = z_0$ is attained by the basic solution.

Optimality Criterion: If $\bar{c}_j \geqslant 0$ for $j = m + 1, \ldots, n$ then the basic solution $\mathbf{x}_B = \mathbf{B}^{-1}\mathbf{b} = \bar{\mathbf{b}}$, $\mathbf{x}_C = \mathbf{0}$ is optimal and the value for z is $z = z_0 = \pi\mathbf{b}$.

Suppose now that not all the $\bar{c}_j \geqslant 0$; in particular suppose that $\bar{c}_s < 0$. We will now increase x_s and maintain feasibility by adjusting the values of the basic variables. That is if we parameterize on the value θ of x_s we have

$$\begin{aligned} x_i(\theta) &= \bar{b}_i - \bar{a}_{is}\theta \qquad i = 1,\dots, m \\ x_s(\theta) &= \theta \\ x_j(\theta) &= 0 \qquad j = m+1,\dots, s-1, s+1,\dots, n \end{aligned} \tag{8}$$

This corresponds to moving along an edge of the polyhedral set $K = \{\mathbf{x} \mid \mathbf{Ax} = \mathbf{b}, \mathbf{x} \geqslant \mathbf{0}\}$. The change in z is $\bar{c}_s$ per unit increase in x_s and since $\bar{c}_s < 0$ we obtain an improvement in the value of z. Theoretically any choice for s works as long as $\bar{c}_s < 0$; however, from the interpretation of $\bar{c}_s$ it makes sense to use the following:

Column Test: Introduce into the basis $\mathbf{A}^s$ where $\bar{c}_s = \min \bar{c}_j < 0$.

This gives us the greatest decrease in z per unit increase of the nonbasic variable. By Eq. (7) the column test can also be written $\min_j (c_j - \boldsymbol{\pi}\mathbf{A}^j)$. From the viewpoint of economics $\boldsymbol{\pi}$ can be interpreted as a price vector, hence the column test is often called *pricing out*.

The column to leave the basis is determined by the relations (8). Since all the basic variables must be nonnegative we must have

$$x_i(\theta) = \bar{b}_i - \bar{a}_{is}\theta_{\max} \geqslant 0 \qquad \text{for} \quad i = 1,\dots, m$$

which implies

$$\theta_{\max} = \min_i \left\{ \frac{\bar{b}_i}{\bar{a}_{is}} \,\middle|\, \bar{a}_{is} > 0 \right\}$$

If

$$\frac{\bar{b}_s}{\bar{a}_{rs}} = \theta_{\max}$$

then

$$x_r(\theta_{\max}) = 0$$

and we remove $\mathbf{A}^r$ from the basis. This completes one iteration. As long as $\theta_{\max} > 0$, z decreases so that no basic solution can be repeated. Eventually we run out of basic solutions so that the algorithm must terminate. Cases where $\theta_{\max} = 0$ must be handled separately. The modifications required in this case are discussed in Dantzig,[11] chap. 10. Note that if $\bar{a}_{is} \leqslant 0$ for all $i = 1,\dots, m$ the value of z is unbounded below since we can increase x_s arbitrarily decreasing z by $\bar{c}_s < 0$ per unit increase of x_s.

Row Test: Delete from the basis a column $\mathbf{A}^r$, with $\bar{a}_{rs} > 0$, and

$$\frac{\bar{b}_s}{\bar{a}_{rs}} = \min_i \left\{ \frac{\bar{b}_i}{\bar{a}_{is}} \,\middle|\, \bar{a}_{is} > 0 \right\}$$

If $\bar{a}_{is} \leqslant 0$ for $i = 1,\dots, m$ terminate the problem, the class of solutions $x_i = \bar{b}_i - \theta\bar{a}_{is}$ for $i = 1,\dots, m$, $x_s = \theta$, $x_j = 0$ for $j = m + 1,\dots, s - 1, s + 1,\dots, n$ yields a class of feasible solutions with $z \rightarrow -\infty$ as $\theta \rightarrow +\infty$.

The remaining detail is how to get the canonical form corresponding to the new basis. This can be done in several ways depending on how information about the problem is stored. In standard simplex method an image of Eq. (6) is maintained in the computer and is transformed for each successive basis. In the revised simplex method the original data (5) plus $\mathbf{B}^{-1}$ and $\bar{\mathbf{b}}$ are stored, and then the barred coefficients $\bar{c}_j$ for $j = 1,\dots, n$ and $\bar{a}_{is}$ for $i = 1,\dots, m$ are generated as needed by means of Eq. (8). For our purposes it will be convenient to use the latter approach. A discussion of the relative merits of the two approaches is given in Dantzig,[11] pp. 216–217. In this case the new inverse $\mathbf{B}^{-1}$ is computed from the old one $\mathbf{B}^{-1}$ by

$$\mathbf{B}^{-1} = \mathbf{P}\mathbf{B}^{-1}$$

where the *pivot matrix* $\mathbf{P}$ is given by

$$\mathbf{P} = \begin{bmatrix} 1 & & & \dfrac{-\bar{a}_{1s}}{\bar{a}_{rs}} & & & \\ & \ddots & & \vdots & & & \\ & & 1 & \dfrac{-\bar{a}_{r-1,s}}{\bar{a}_{rs}} & & & \\ & & & \dfrac{1}{\bar{a}_{rs}} & & & \\ & & & \dfrac{-\bar{a}_{r+1,s}}{\bar{a}_{rs}} & 1 & & \\ & & & \vdots & & \ddots & \\ & & & \dfrac{-\bar{a}_{ms}}{\bar{a}_{rs}} & & & 1 \end{bmatrix} \tag{9}$$

The new $\bar{\mathbf{b}}$—let us denote it $\bar{\bar{\mathbf{b}}}$—is given by $\bar{\bar{\mathbf{b}}} = \mathbf{P}\bar{\mathbf{b}}$. This is called a *pivot step*.

The last point connected with the basic simplex method is the removal of the temporary assumptions we made at the beginning of the discussion. Let $\mathbf{A}$ be an arbitrary $m \times n$ matrix and $\mathbf{b}$ an arbitrary nonnegative vector of compatible size. We now define the *phase* 1 *problem*

$$\begin{aligned} &\min_{\mathbf{x},\mathbf{y}} \mathbf{w} = \mathbf{e}\mathbf{y} \\ &\text{subject to } \mathbf{I}\mathbf{y} + \mathbf{A}\mathbf{x} = \mathbf{b} \\ &\mathbf{x} \geqslant \mathbf{0} \qquad \mathbf{y} \geqslant \mathbf{0} \end{aligned} \tag{10}$$

where $\mathbf{e} = (1,\ldots, 1)$, $\mathbf{I}$ is an m-dimensional identity matrix, and the m-dimensional vector $\mathbf{y}$ is called the vector of artificial variables. The purpose of solving this problem is to

1. Determine if $\mathbf{Ax} = \mathbf{b}$, $\mathbf{x} \geqslant \mathbf{0}$ has any solutions
2. Determine if $\mathbf{Ax} = \mathbf{b}$ has redundant equations
3. Obtain a starting feasible basis for Eq. (5)

Subtracting each of the equations from the objective yields a linear program in canonical from (6) to which the simplex algorithm can be applied. Requiring $\mathbf{b}$ to be nonnegative is no restriction since if $b_i < 0$ then multiplying the ith row of $\mathbf{Ax} = \mathbf{b}$ by -1 yields an equivalent system with the corresponding $\mathbf{b}$ element positive. $\mathbf{Ax} = \mathbf{b}$, $\mathbf{x} \geqslant \mathbf{0}$ has solutions if and only if min $\mathbf{w} = 0$. The system is redundant if it has solutions and there is an artificial variable which cannot be removed from the basis. In this case the equation corresponding to the offending artificial variables are vacuous in $\mathbf{x}$ and if removed leave a nonredundant equivalent system (Dantzig,[11] chap. 5). The final result for feasible problems is an equivalent canonical form (6) in which all the artificial variables have zero values in the corresponding basic feasible solution. This leaves us with a feasible basis for the original problem, and we can now apply the revised simplex method as previously described.

REMARKS:

One consequence of the simplex algorithm which we will make use of in the development of the decomposition algorithm is

The Unbounded Solution Theorem: If the linear program (5) has feasible solutions, then $\mathbf{cx}$ is unbounded below on $\mathbf{K} = \{\mathbf{x} \mid \mathbf{Ax} = \mathbf{b}, \mathbf{x} \geqslant \mathbf{0}\}$ if and only if there exists a vector $\mathbf{y} \geqslant \mathbf{0}$ such that $\mathbf{Ay} = \mathbf{0}$ and $\mathbf{cy} < 0$.

Proof: Suppose $\mathbf{y} \geqslant \mathbf{0}$, $\mathbf{Ay} = \mathbf{0}$, and $\mathbf{cy} < 0$ and let $\mathbf{x} \geqslant \mathbf{0}$ be any feasible point for Eq. (5). Then $\mathbf{x} + \lambda\mathbf{y} \geqslant \mathbf{0}$, $\mathbf{A}(\mathbf{x} + \lambda\mathbf{y}) = \mathbf{Ax} + \lambda\mathbf{Ay} = \mathbf{b} + \mathbf{0}$, and $\mathbf{c}(\mathbf{x} + \lambda\mathbf{y})$ decreases linearly in λ for $\lambda > 0$. Thus $\mathbf{cx}$ is unbounded below on K.

On the other hand, if $\mathbf{cx}$ is unbounded below on K, at some point in the simplex method there is a feasible basis $\mathbf{B}$ and s, $1 \leqslant s \leqslant n$, such that $\bar{c}_s = \mathbf{c}_s - \pi\mathbf{A}^s < 0$, and $\bar{\mathbf{A}}^s = \mathbf{B}^{-1}\mathbf{A}^s \leqslant 0$, where $\pi = \mathbf{c}_\mathbf{B}\mathbf{B}^{-1}$. Then $\mathbf{y} = (-a_{1s},\ldots, -a_{ms}, 0,\ldots, 0, 1, 0,\ldots, 0)$ satisfies $\mathbf{0} = \mathbf{Ay} = -\mathbf{B}\bar{\mathbf{A}}^s + \mathbf{A}^s$, and $\mathbf{cy} = \mathbf{c}_s - \mathbf{c}_\mathbf{B}\mathbf{B}^{-1}\mathbf{A}^s = \mathbf{c}_s - \pi\mathbf{A}^s < 0$.

3-2-2 Duality

Associated with every linear program (5) is a dual problem:

$$\max_{\pi} v = \pi \mathbf{b}$$
$$\text{subject to } \pi \mathbf{A} \leqslant \mathbf{c} \tag{11}$$

The first observation we make is that if $\mathbf{Ax} = \mathbf{b}$, $\mathbf{x} \geqslant \mathbf{0}$ and if $\pi \mathbf{A} \leqslant \mathbf{c}$ then $\pi \mathbf{b} \leqslant \mathbf{cx}$. To see this we simply observe that $(\pi \mathbf{A})\mathbf{x} \leqslant \mathbf{cx}$, and $\pi(\mathbf{Ax}) = \pi \mathbf{b}$. This allows several immediate observations. If Eq. (5) has solutions for which $z = \mathbf{cx}$ is unbounded below then Eq. (11) has no solutions. Conversely if $\pi \mathbf{b}$ is unbounded above then Eq. (5) has no solution. Finally, if we find π^* such that $\pi^* \mathbf{A} \leqslant \mathbf{c}$, and $\mathbf{x}^*$ such that $\mathbf{Ax}^* = \mathbf{b}$, $\mathbf{x}^* \geqslant \mathbf{0}$ with the additional property that $\pi^* \mathbf{b} = \mathbf{cx}^*$ then $\mathbf{x}^*$ and π^* are respectively optimal for Eqs. (5) and (11). The simplex method tells us that if Eq. (5) has an optimal solution there exists a vector of simplex multipliers π associated with the optimal basis which satisfies

$$\mathbf{c} = \mathbf{c} - \pi \mathbf{A} \geqslant \mathbf{0}$$
$$z_0 = \min z = \pi \mathbf{b}$$

This follows from Eq. (7) and the optimality criterion. Thus we have proved the

Duality Theorem: Exactly one of the following statements holds:

1. Equations (5) and (11) have optimal solutions and $\max v = \min z$.
2. Equation (5) is unbounded below in z and Eq. (11) is infeasible.
3. Equation (11) is unbounded above in v and Eq. (5) is infeasible.
4. Both Eqs. (5) and (11) are infeasible.

We have also proved the

Complementary Slackness Theorem: $\mathbf{x}^*$ optimizes Eq. (5) and π^* optimizes Eq. (11) if and only if $\mathbf{x}^*$ and π^* are feasible for their respective problems and $(\mathbf{c} - \pi^* \mathbf{A})\, \mathbf{x}^* = \mathbf{0}$.

Proof (of Complementary Slackness): $(\mathbf{c} - \pi^* \mathbf{A})\, \mathbf{x}^* = \mathbf{cx}^* - \pi^* \mathbf{Ax}^* = \mathbf{cx}^* - \pi^* \mathbf{b} = \mathbf{0}$ if and only if $\mathbf{x}^*$, π^* are optimal.

There are other forms of linear programs than Eq. (5). A common formulation is called the symmetric formulation because of the similarity in the appearance of the primal problem and its dual. The primal is

$$\begin{aligned} &\min_{\mathbf{x}} \mathbf{cx} \\ &\text{subject to } \mathbf{Ax} \geqslant \mathbf{b} \\ &\mathbf{x} \geqslant \mathbf{0} \end{aligned}$$

Its dual is

$$\begin{aligned} &\max_{\mathbf{y}} \mathbf{yb} \\ &\text{subject to } \mathbf{yA} \leqslant \mathbf{c} \\ &\mathbf{y} \geqslant \mathbf{0} \end{aligned}$$

For the general mixed primal problem,

$$\begin{aligned} &\min_{\mathbf{x}_1,\mathbf{x}_2} \mathbf{z} = \mathbf{c}^1\mathbf{x}^1 + \mathbf{c}^2\mathbf{x}^2 \\ &\text{subject to } \mathbf{Dx}^1 + \mathbf{Ex}^2 \geqslant \mathbf{b}^1 \\ &\mathbf{Fx}^1 + \mathbf{Gx}^2 = \mathbf{b}^2 \\ &\mathbf{x}^1 \geqslant \mathbf{0} \qquad \mathbf{x}^2 \text{ unrestricted in sign} \end{aligned} \tag{12}$$

We have (Dantzig,[11] chap. 6) the dual

$$\begin{aligned} &\max_{\mathbf{y}_1,\mathbf{y}_2} \mathbf{v} = \mathbf{y}^1\mathbf{b}^1 + \mathbf{y}^2\mathbf{b}^2 \\ &\text{subject to } \mathbf{y}^1\mathbf{D} + \mathbf{y}^2\mathbf{F} \leqslant \mathbf{c}^1 \\ &\mathbf{y}^1\mathbf{E} + \mathbf{y}^2\mathbf{G} = \mathbf{c}^2 \\ &\mathbf{y}^1 \geqslant \mathbf{0} \qquad \mathbf{y}^2 \text{ unrestricted in sign} \end{aligned} \tag{13}$$

We will also have occasion to consider a slightly different from of the standard linear program obtained by substituting in Eq. (5)

$$\hat{\mathbf{A}}^j = \begin{pmatrix} \mathbf{c}_j \\ \mathbf{A}^j \end{pmatrix} \qquad \hat{\mathbf{b}} = \begin{pmatrix} 0 \\ \mathbf{b} \end{pmatrix} \qquad \mathbf{U}_0 = \begin{pmatrix} 1 \\ 0 \\ \vdots \\ 0 \end{pmatrix} \qquad \text{and} \qquad \mathbf{x}_0 = -\mathbf{z}$$

Eq. (5) then becomes

$$\begin{aligned} &\max \mathbf{x}_0 \\ &\text{subject to } \mathbf{U}_0\mathbf{x}_0 + \hat{\mathbf{A}}\mathbf{x} = \hat{\mathbf{b}} \\ &\mathbf{x} \geqslant 0 \end{aligned} \tag{14}$$

The application of the simplex method to Eq. (14) is fairly obvious. The only remark we make concerns the column test procedure; namely, if we let $\hat{\pi} = (1, -\pi)$, where π is the simplex multiplier vector in terms of Eq. (5), then the new column rule is

Column Test [for Eq. (7)]: Pick as the new basic variable x_s where $\hat{\pi}\hat{\mathbf{A}}^s = \min \hat{\pi}\hat{\mathbf{A}}^s < 0$.

3-2-3 Linear Programs with Upper-bounded Variables

We now turn to some generalizations of the basic simplex algorithm. The first deals with problems where the components of $\mathbf{x}$ have upper and lower bounds. If x_j is restricted to the interval $l_j \leqslant x_j \leqslant u_j$ the transformation of variables

$$\hat{x}_j = \frac{x_j - l_j}{u_j - l_j}$$

yields a variable restricted to $0 \leqslant \hat{x}_j \leqslant 1$. Thus we consider the problem

$$\begin{aligned} &\min \mathbf{cx} \\ &\text{subject to } \mathbf{Ax} = \mathbf{b} \\ &0 \leqslant x_j \leqslant 1 \qquad j = 1,\dots, n \end{aligned} \tag{15}$$

The obvious modifications required when only some of the variables have upper and lower variables will be pointed out after the algorithm for Eq. (15) is given. The fundamental idea of the algorithm is that the nonbasic variables will be allowed to take on the value one as well as zero, that is, the nonbasic variables are divided into two subclasses: those *nonbasic at lower bound* and those *nonbasic at upper bound.* Let $\mathbf{U}$ be the subset of $\{1,\dots, n\}$ corresponding to the nonbasic variables at upper bound and $\mathbf{L}$ the subset of nonbasic variables at lower bound. These sets will change from iteration to iteration. We assume that we have obtained an initial basis B leading to a system of the form (6) and a set U such that the corresponding basic solution is feasible, i.e.,

$$\begin{aligned} x_i = \hat{b}_i = \bar{b}_i - \sum_{j\in U} \bar{a}_{ij}x_j = \bar{b}_i - \sum_{j\in U} \bar{a}_{ij} \qquad i = 1,\dots, m \\ x_j = 0 \quad j \in L \\ x_j = 1 \quad j \in U \\ z = z_0 + \sum_{j\in U} \bar{c}_j x_j + \sum_{j\in L} \bar{c}_j x_j = z_0 + \sum_{j\in U} \bar{c}_j \end{aligned} \tag{16}$$

The optimality criterion follows.

Optimality Criterion: If $\bar{c}_j \geqslant 0$ for $j \in L$, and $\bar{c}_j \leqslant 0$ for $j \in U$, then the basic solution (16) is optimal.

To see this we note, from Eq. (16), that $z_0 + \sum_{j \in U} \bar{c}_j$ is a lower bound for z. Moreover the bound is achieved for the basic feasible solution (16).

If the optimality criterion is not satisfied then we can decrease the value of z by either increasing a nonbasic variable currently at lower bound with $\bar{c}_s < 0$ or decreasing a nonbasic variable currently at upper bound with $\bar{c}_s > 0$ predicated of course on the fact that $0 < \hat{b}_i = \bar{b}_i - \sum_{j \in U} \bar{a}_{ij} < 1$. The decrease in any case is $| \bar{c}_s |$ per unit change in the nonbasic variable. This leads to the

Column Test: Change the status of the column A^s which minimizes

$$\min[\{\bar{c}_j |_{j \in L}\} \cup \{-\bar{c}_i |_{i \in U}\}]$$

Note we do not say that $\mathbf{A}^s$ will necessarily be introduced into the basis; it may simply go from lower bound to upper or conversely. The determination of the variable to leave (if any) is similar to the previous method but there are more cases to consider. The test is based on the restraint

$$0 \leqslant x_i = \hat{b}_i = \bar{b}_i - \sum_{j \in U} \bar{a}_{ij} \leqslant 1$$

Let θ be the *change* in x_s; that is, $x_s = 1 - \theta$ if x_s was originally at upper bound and $x_s = \theta$ if it was originally at lower bound. We illustrate by going through one of the cases and then state the general results for the other cases.

Suppose the variable x_s coming into the basis was originally at upper bound, and we decrease its value by θ, adjusting the basic variables to maintain feasibility and leaving the other nonbasic variables unchanged. That is, we consider the solution

$$\begin{aligned} x_i(\theta) &= \bar{b}_i - \sum_{\substack{j \in \mathbf{U} \\ j \neq s}} \bar{a}_{ij} - (1 - \theta)\bar{a}_{is} = \bar{\mathbf{b}}_i - \sum_{j \in \mathbf{U}} \bar{a}_{ij} + \theta \bar{a}_{is} \\ &= \hat{b}_i + \theta \bar{a}_{is} \qquad \text{for} \quad i = 1, \ldots, m \\ x_s(\theta) &= \theta \\ x_j(\theta) &= 1 \qquad j \neq s \;.\; j \in \mathbf{U} \\ x_j(\theta) &= 0 \qquad j \in \mathbf{L} \end{aligned} \tag{17}$$

Since $0 \leqslant x_i(\theta) \leqslant 1$ must be satisfied we must have (we drop the super $^-$'s for clarity)

$$\theta \leqslant \frac{\hat{b}_i}{-a_{is}} \qquad \text{if} \quad a_{is} < 0$$

$$\theta \leqslant \frac{1 - \hat{b}_i}{a_{is}} \qquad \text{if} \quad a_{is} > 0$$

and finally $\theta \leqslant 1$ since $x_s(\theta) \geqslant 0$ must be satisfied. Thus if $s \in U$

$$\theta = \min(\theta_1, \theta_2, 1) \tag{18a}$$

where

$$\theta_1 = \min\left\{\frac{\hat{b}_i}{-a_{is}} \,\middle|\, a_{is} < 0\right\} = \frac{\hat{b}_{r_1}}{-a_{r_1 s}} \tag{18b}$$

$$\theta_2 = \min\left\{\frac{1 - \hat{b}_i}{a_{is}} \,\middle|\, a_{is} > 0\right\} = \frac{1 - \hat{b}_{r_2}}{a_{r_2 s}} \tag{18c}$$

If θ equals one then x_s is changed from nonbasic at upper bound to nonbasic at lower bound, the values of $\hat{b}_i$ are corrected, and the next iteration is performed. If $\theta = \theta_1$, let $r = r_1$, and then pivot to remove x_r from the basis and make it nonbasic at lower bound; if $\theta = \theta_2$, then let $r = r_2$, and pivot and make x_r nonbasic at upper bound.

In the case where the entering variable x_s is not lower bound ($s \in \mathbf{L}$) let

$$\theta = \min(\theta_1, \theta_2, 1) \tag{19a}$$

where

$$\theta_1 = \min\left\{\frac{\hat{b}_i}{a_{is}} \,\middle|\, a_{is} > 0\right\} = \frac{\hat{b}_{r_1}}{a_{r_1 s}} \tag{19b}$$

$$\theta_2 = \min\left\{\frac{1 - \hat{b}_i}{-a_{is}} \,\middle|\, a_{is} < 0\right\} = \frac{\hat{b}_{r_2}}{-a_{r_2 s}} \tag{19c}$$

If θ equals one change x_s from nonbasic at lower bound to nonbasic at upper bound; if $\theta = \theta_1$, x_s enters the basis, and x_{r_1} becomes nonbasic at upper bound. Finally if $\theta = \theta_2$, x_{r_2} becomes nonbasic at lower bound. The algorithm then proceeds as before; at each iteration one must have access to the original data, $\mathbf{b}$, $\mathbf{A}$, and the inverse of the current basis (for the revised simplex method).

Using the relation in Eq. (7), $\bar{\mathbf{c}} = \mathbf{c} - \boldsymbol{\pi}\mathbf{A}$ (which holds for canonical form and the optimality criterion for linear programs with upper bounds), one obtains for the optimal solution $\mathbf{x}^*$

$$\begin{aligned} 0 < x_j^* < 1 &\text{ implies } \bar{c}_j = c_j - \boldsymbol{\pi}\mathbf{A}^j = 0 \\ x_j^* = 0 &\text{ implies } \bar{c}_j \geqslant 0 \\ x_j^* = 1 &\text{ implies } \bar{c}_j \leqslant 0 \end{aligned} \tag{20}$$

If some of the variables are not sign restricted the corresponding tests of Eqs. (18) and (19) are omitted for these variables. For example if $s \in \mathbf{L}$ and x_i has no upper bound,

$$\frac{1 - \hat{b}_i}{-a_{is}}$$

is not considered in the test for θ_2 even if $a_{is} < 0$. The other modifications are obvious. The phase 1 procedure similar to the one described before is used to get an initial basis. The artificial variables can be upper bounded if desired simply by multiplying the corresponding rows of $\mathbf{AX} = \mathbf{b}$ by scalars so that $0 \leqslant \hat{b}_i \leqslant 1$. More details of the algorithm can be found in Dantzig,[11] chap. 9.

3-2-4 Generalized Linear Programming and Convex Sets

The next extension of the basic algorithm is to allow an infinite number of columns for $\mathbf{A}$. To motivate the approach we consider the following problem

$$\begin{aligned} &\max x_0 \\ &\text{subject to } \mathbf{U}_0\mathbf{x}_0 + \mathbf{y} = \hat{\mathbf{b}} \\ &\mathbf{y} \in \mathbf{Y} \end{aligned} \tag{21}$$

where

$$\mathbf{Y} = \{\mathbf{y} \mid \mathbf{y} = \hat{\mathbf{A}}\mathbf{x}, \mathbf{x} \geqslant \mathbf{0}\}$$

and

$$\mathbf{U}_0 = (1, 0, \ldots, 0)$$

This problem is clearly equivalent to Eq. (14). Another equivalent problem is obtained if we partition $\hat{\mathbf{A}} = \left[\begin{smallmatrix}\mathbf{c}\\ \mathbf{A}\end{smallmatrix}\right]$ into two submatrices ${}^1\hat{\mathbf{A}}$ consisting of $\mathbf{c}$ and the first m_1 rows of $\mathbf{A}$ and ${}^2\mathbf{A}$ consisting of the $m_2 = m - m_1$ remaining rows. We similarly partition $\hat{\mathbf{b}} = \left(\begin{smallmatrix}0\\ \mathbf{b}\end{smallmatrix}\right)$ into $\hat{\mathbf{b}}^1$ and $\hat{\mathbf{b}}^2$. Then if $\mathbf{Y} = \{y \mid y = {}^1\hat{\mathbf{A}}\mathbf{x}$ for some $\mathbf{x} \geqslant 0$, and ${}^2\mathbf{A}\mathbf{x} = \mathbf{b}^2\}$, Eq. (14) is equivalent to

$$\begin{aligned} &\max \mathbf{x}_0 \\ &\text{subject to } \mathbf{U}_0\mathbf{x}_0 + \mathbf{y} = \bar{\mathbf{b}}^1 \\ &\mathbf{y} \in Y \end{aligned} \tag{22}$$

In Eq. (21) $\mathbf{Y}$ is a convex cone and in Eq. (22) Y is a convex polyhedral set. In general we allow $\mathbf{Y}$ to be an arbitrary closed convex set which is "compact enough." In much of what follows $\mathbf{Y}$ will in fact be compact. That is we consider

$$\begin{aligned} &\max x_0 \\ &\text{subject to } \mathbf{U}_0\mathbf{x}_0 + \mathbf{P} = 0 \\ &\mathbf{P} \in \mathscr{K} \end{aligned} \tag{23}$$

where $\mathscr{K} \subset \mathbf{E}^n$, n-dimensional Euclidean space, is closed, convex, and "compact enough" (defined below). This problem we will call the *generalized linear program.* See Dantzig, Blattner, and Rao[13] for applications of generalized linear programming to ship routing and Gomory[17] for applications to trim and network problems. Before we proceed let us define our terminology more carefully.

Compact (closed and bounded) convex sets have two essentially different types of characterizations, one as the intersection of half spaces, the other as the set of all weighted averages of its corners or more accurately its extreme points. Generalized linear programming and decomposition algorithms are based on switching from one characterization to the other. Let us illustrate the two characterizations in the case of the unit square $\mathbf{S}$ in the plane. One representation is as the intersection of half spaces $\mathbf{S} = \{(\mathbf{x}, \mathbf{y}) \mid x \leqslant 1, x \geqslant -1, y \leqslant 1, y \geqslant -1\}$; the other in terms of extreme points is

$$\mathbf{S} = \{(x, y) \mid (x, y) = \lambda_1(1, 1) + \lambda_2(1, -1) + \lambda_3(-1, 1) + \lambda_4(-1, -1),$$
$$\lambda_1 + \lambda_2 + \lambda_3 + \lambda_4 = 1, \lambda_1 \geqslant 0, \lambda_2 \geqslant 0, \lambda_3 \geqslant 0, \lambda_4 \geqslant 0\}$$

This is illustrated in Fig. 3-1.

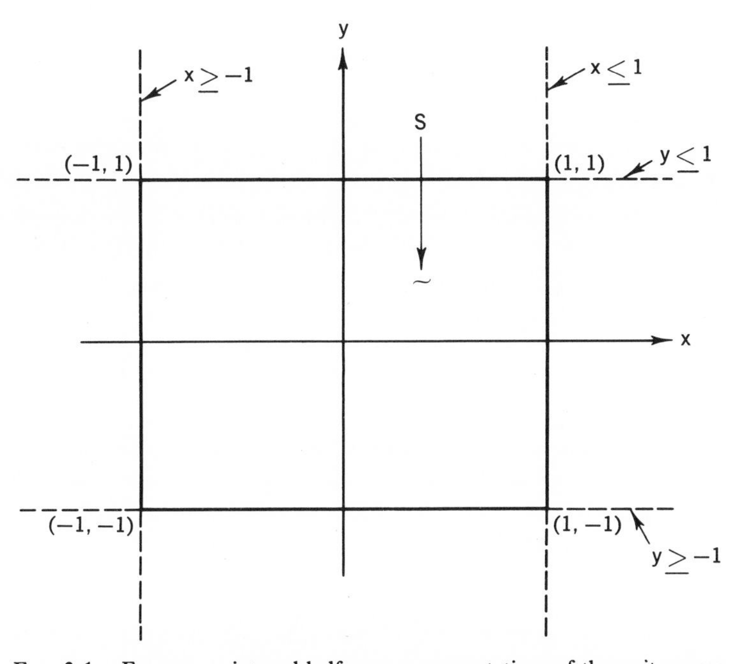

FIG. 3-1. Extreme point and half-space representations of the unit square.

Normally as in Eq. (5) the feasible $\mathbf{x}$'s are characterized as the intersection of half spaces. (Note that the solution of the equation $\sum a_j x_j = b$ is the intersection of the two half spaces determined by $\sum a_j x_j \leqslant b$, $\sum a_j x_j \geqslant b$.) In the methods we are about to discuss the representation in terms of extreme points is used.

Unfortunately, in general, the number of extreme points for an arbitrary closed and bounded convex set is infinite. For example every point on the circumference of a disc in the plane is an extreme point of the disc. It is only in the case where the set in question is a convex polyhedron that the number of extreme points is finite. Fortunately, the important case of linear programs falls in this category. Moreover, if the set is not bounded it is not the set of weighted averages of its extreme points. In these cases a more general analysis must be carried out. We state without proof results from Hirsch and Hoffman[23] which completely characterized the situation.

Let $\mathscr{K}$ be a closed convex set in $\mathbf{E}^n$ and assume for convenience $\mathbf{0} \in \mathscr{K}$. We wish to consider the cross section of $\mathscr{K}$ if it contains a translate of a linear subspace. To this end let $\mathscr{L}$ be the largest linear subspace contained in $\mathscr{K}$. Then $\mathbf{x} + \mathscr{L} = \{\mathbf{x} + \mathbf{y} \mid \mathbf{y} \in \mathscr{L}\}$ is the largest affine subspace (translate of a linear subspace) containing $\mathbf{x}$ in $\mathscr{L}$ for all $\mathbf{x} \in \mathscr{K}$. Let $\mathscr{L}^\perp$ be the set of all vectors which are perpendicular to every vector in $\mathscr{L}$; then the cross section is given by $\mathscr{K} \cap \mathscr{L}^\perp$. Now let $\mathscr{C}$ be the largest cone with vertex 0 in $\mathscr{K} \cap \mathscr{L}^\perp$, then it is shown in Hirsch and Hoffman[23] that $\mathbf{x} + \mathscr{C}$ is the largest cone with vertex $\mathbf{x}$ contained in $\mathscr{K} \cap \mathscr{L}^\perp$ for any $\mathbf{x}$ in $\mathscr{K} \cap \mathscr{L}^\perp$. Moreover $\mathscr{C}$ is a closed pointed convex cone (a cone $\mathscr{C}$ is pointed if $\mathbf{y} \in \mathscr{C}$ and $-\mathbf{y} \in \mathscr{C}$ implies $\mathbf{y} = \mathbf{0}$). Finally, let $\mathscr{H}$ be the convex hull of the extreme points of $\mathscr{K} \cap \mathscr{L}^\perp$, the cross section of $\mathscr{K}$. It can be shown that $\mathscr{H}$ is nonempty. If $\mathscr{H}$ is compact, the set $\mathscr{K}$ is said to be *compact enough.*[40] Convex polyhedral sets are compact enough. The set $\mathscr{K} = \{(\mathbf{x}, \mathbf{y}) \mid \mathbf{xy} \geqslant 1, \mathbf{x} \geqslant 0, \mathbf{y} \geqslant 0\}$ is not compact enough because each point on the curve $\mathbf{xy} = 1$ is an extreme point (see Fig. 3-2*a* and *b*). Finally

$$\mathscr{K} = \mathscr{L} + \mathscr{C} + \mathscr{H} = \{\mathbf{x} \mid l + \mathbf{c} + \mathbf{h}, l \in \mathscr{L}, \mathbf{c} \in \mathscr{C}, \mathbf{h} \in \mathscr{H}\}$$

If we let $\mathbf{L} = [L^1,\ldots, L^r]$ be a basis for $\mathscr{L}$, $\mathscr{C}$ be generated by $\mathbf{C} = \{\mathbf{C}^j \mid j \in J\}$ (that is, $\mathbf{y} \in \mathscr{C}$ if and only if $\mathbf{y} = \sum \mu_j \mathbf{C}^j$, $\mu_j \geqslant 0$), and $\mathscr{H}$ be the convex hull of the extreme points $\mathbf{E} = \{\mathbf{E}^i \mid i \in \mathbf{I}\}$ then $\mathbf{P} \in \mathscr{K}$ if and only if

$$\mathbf{P} = \sum \sigma_i \mathbf{L}^i + \sum \mu_j \mathbf{C}^j + \sum \lambda_k \mathbf{E}^k$$

where $\lambda_k \geqslant 0$, $\mu_j \geqslant 0$, $\sum \lambda_k = 1$; moreover, the number of nonzero coefficients can be taken to be $n + 1$ or less. An example of the decomposition is given in Fig. 3-3.

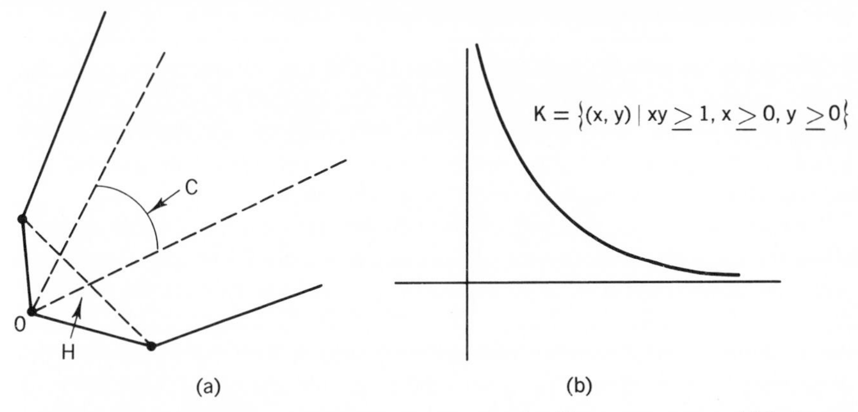

FIG. 3-2. (*a*) Polyhedral sets are compact enough; (*b*) a convex set which is not compact enough.

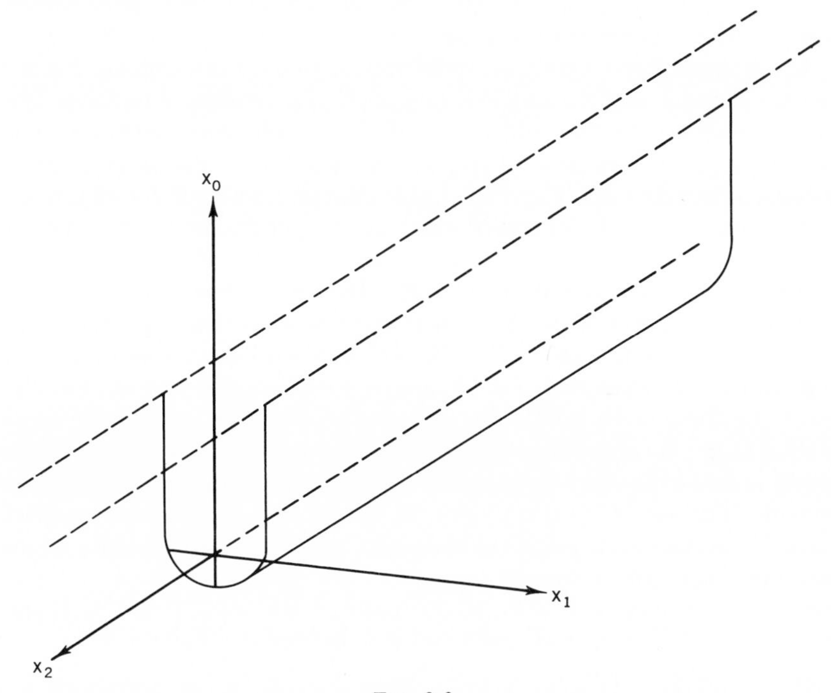

FIG. 3-3.

$$\mathscr{K} = \{(x_0, x_1, x_2) \mid -1 \leqslant x_1 \leqslant 1, x_0 \geqslant -(1 - x_1^2), x_2 \text{ arbitrary}\}$$
$$\mathscr{L} = \{(x_0, x_1, x_2) \mid x_1 = x_0 = 0\},$$
$$\mathscr{C} = \{(x_0, x_1, x_2) \mid x_1 = x_2 = 0, x_0 \geqslant 0\}$$
$$\mathscr{L}^\perp = \{(x_0, x_1, x_2) \mid x_2 = 0\},$$
$$\mathscr{H} = \{(x_0, x_1, x_2) \mid 0 \geqslant x_0 \geqslant -(1 - x_1^2), -1 \leqslant x_1 \leqslant 1, x_2 = 0\}$$
$$L = \{(0, 1, 0)\},$$
$$C = \{(1, 0, 0)\},$$
$$E = \{(x_0, x_1, x_2) \mid x_2 = 0, x_0^2 + x_1^2 = 1, x_0 \leqslant 0\}$$

3-2-5 Decomposition—Polyhedral Case

Now we are prepared to discuss the *decomposition method.* Consider the problem

$$\begin{aligned} \min z &= \mathbf{cx} \\ \text{subject to } \mathbf{Ax} &= \mathbf{b} \\ \mathbf{x} &\in \mathbf{K} \end{aligned} \tag{24}$$

where $\mathbf{K}$ is a general closed convex set. If $\mathbf{A} = \mathbf{I}$, the identity matrix, we obtain the generalized linear program as a special case. From the previous remarks $\mathbf{K} = \mathscr{L} + \mathscr{C} + \mathscr{H}$. Then an equivalent problem to Eq. (24) is

$$\begin{aligned} \min_{\lambda,\mu,\sigma} \mathbf{z} &= \sum_{i\in\mathbf{I}'} \lambda_i\alpha_i + \sum_{j\in\mathbf{J}'} \mu_j\beta_j + \sum_{k=1}^{r} \sigma_k\gamma_k \\ \text{subject to} \quad & \sum_{i\in\mathbf{I}'} \lambda_i\mathbf{P}^i + \sum_{j\in\mathbf{J}'} \mu_j\mathbf{R}^j + \sum_{k=1}^{r} \sigma_k\mathbf{S}^k = \mathbf{b} \\ & \sum \lambda_i = 1 \\ & \lambda_i \geqslant 0 \qquad \mu_j \geqslant 0 \end{aligned} \tag{25}$$

and $\mathbf{I}'$ and $\mathbf{J}'$ arbitrary finite sets contained in $\mathbf{I}$ and $\mathbf{J}$ respectively where

$$\begin{aligned} \alpha_i &= \mathbf{cE}^i & \beta_j &= \mathbf{cC}^j & \gamma_k &= \mathbf{cL}^k \\ \mathbf{P}^i &= \mathbf{AE}^i & \mathbf{R}^j &= \mathbf{AC}^j & \mathbf{S}^k &= \mathbf{AL}^k \end{aligned} \tag{26}$$

The equivalence of Eqs. (24) and (25) follows because a point $\mathbf{x}$ is feasible if and only if (1) $\mathbf{x} \in K$; that is,

$$\mathbf{x} = \sum \lambda_i\mathbf{E}^i + \sum \mu_j\mathbf{C}^j + \sum \sigma_k\mathbf{L}^k$$

for some finite sets $\mathbf{I}'$ and $\mathbf{J}'$ and $\boldsymbol{\lambda}_i$, $\boldsymbol{\mu}_j$, $\boldsymbol{\sigma}_k$ such that $\mathbf{I}' \subset \mathbf{I}$, $\mathbf{J}' \subset \mathbf{J}$, and $\sum \lambda_i = 1$, $\lambda_i \geqslant 0$, $\mu_j \geqslant 0$, and (2) $\mathbf{Ax} = \mathbf{b}$; that is,

$$\begin{aligned} \mathbf{Ax} &= \mathbf{A}\left(\sum_{i\in\mathbf{I}'} \lambda_i\mathbf{E}^i + \sum_{j\in\mathbf{J}'} \mu_j\mathbf{C}^j + \sum_{k=1}^{r} \sigma_k\mathbf{L}^k\right) = \mathbf{b} \\ &= \sum_{i\in\mathbf{I}'} \lambda_i\mathbf{AE}^i + \sum_{j\in\mathbf{J}'} \mu_j\mathbf{AC}^j + \sum_{k=1}^{r} \sigma_k\mathbf{AL}^k = \mathbf{b} \\ &= \sum_{i\in\mathbf{I}'} \lambda_i\mathbf{P}^i + \sum_{j\in\mathbf{J}'} \mu_j\mathbf{R}^j + \sum_{k=1}^{r} \sigma_k\mathbf{S}^k = \mathbf{b} \end{aligned}$$

Finally

$$z = \mathbf{cx} = \sum_{i\in \mathbf{I}'} \lambda_i(\mathbf{cE}^i) + \sum_{j\in \mathbf{J}'} \mu_j(\mathbf{cC}^j) + \sum_{k=1}^{r} \sigma_k(\mathbf{cL}^k)$$

$$= \sum_{i\in \mathbf{I}'} \lambda_i\alpha_i + \sum_{j\in \mathbf{J}'} \mu_j\beta_j + \sum_{k=1}^{r} \sigma_k\gamma_k$$

This transformation is the basis of the decomposition method. Equation (25) is called the *master program.* Thus, if we had a characterization of K in terms of **L**, **C**, and **E** we could use Eq. (25) to solve the problem. The real economy of the decomposition method comes by generating the $\mathbf{L}^k$, $\mathbf{C}^j$, and $\mathbf{E}^i$ only as needed since in even the simplest cases the sets **I** and **J** can be very large (Balinskii[2]).

Before we treat the general problem we consider the important special case which is

$$\begin{aligned} &\min_{\mathbf{x}} \mathbf{cx} \\ &\text{subject to } \mathbf{Ax} = \mathbf{b} \\ &\qquad \mathbf{Fx} = \mathbf{f} \\ &\qquad \mathbf{x} \geqslant 0 \end{aligned} \tag{27}$$

where $\mathbf{x}$ is a n-vector, $\mathbf{A}$ is an $m \times n$ matrix, $\mathbf{F}$ is an $l \times n$ matrix, and **c**, **b**, and **f** are vectors of appropriate dimension. For this problem $\mathscr{K}$ is the closed, convex, polyhedral set $\{\mathbf{x} \mid \mathbf{Fx} = \mathbf{f}, \mathbf{x} \geqslant 0\}$. Since $\mathbf{x} \geqslant 0$ there are no nontrivial affine subspaces of $\mathscr{K}$; thus $\mathscr{K} = \mathscr{C} + \mathscr{H}$ where $\mathscr{C}$ is a polyhedral cone characterized by its extreme rays $\mathbf{C}^1,\ldots, \mathbf{C}^r$ which are finite in number. $\mathscr{C} = \{\mathbf{x} \mid \mathbf{x} = \sum \mu_j C^j, \mu_j \geqslant 0\}$. Similarly $\mathscr{H}$ is the convex hull of the finite number of extreme points $\mathbf{E}^1,\ldots, \mathbf{E}^h$; thus

$$\mathscr{H} = \left\{\mathbf{x} \,\middle|\, \mathbf{x} = \sum_{i=1}^{h} \lambda_i \mathbf{E}^j, \sum_{i=1}^{h} \lambda_i = 1, \lambda_i \geqslant 0\right\}$$

The master problem corresponding to Eq. (27) is

$$\begin{aligned} &\min_{\lambda,\mu} \sum_{i=1}^{h} \lambda_i\alpha_i + \sum_{j=1}^{r} \mu_j\beta_j \\ &\text{subject to } \sum_{i=1}^{h} \lambda_i\mathbf{P}^i + \sum_{j=1}^{r} \mu_j\mathbf{R}^j = \mathbf{b} \\ &\qquad \sum_{i=1}^{h} \lambda_i = 1 \\ &\qquad \lambda_i \geqslant 0 \qquad i = 1,\ldots, h \\ &\qquad \mu_j \geqslant 0 \qquad j = 1,\ldots, r \end{aligned} \tag{28}$$

where

$$\alpha_i = \mathbf{cE}^i \qquad \mathbf{P}^i = \mathbf{AE}^i \qquad i = 1,\ldots, h$$

$$\beta_j = \mathbf{cC}^j \qquad \mathbf{R}^j = \mathbf{AC}^j \qquad j = 1,\ldots, r$$

As we mentioned before, h and r in general will be very large, and so one does not wish to catalog all the $\mathbf{E}^i$'s and the $\mathbf{C}^j$'s but rather to generate them as needed. We assume that at some iteration t we have a subset $\mathbf{I}^t \subset \mathbf{I} = \{1,\ldots, r\}$ and a subset $\mathbf{J}^t \subset \mathbf{J} = \{1,\ldots, h\}$ such that

$$\begin{aligned} &\min_{\boldsymbol{\lambda},\boldsymbol{\mu}} \sum_{i\in\mathbf{I}^t} \lambda_i\alpha_i + \sum_{j\in\mathbf{J}^t} \mu_j\beta_j \\ &\text{subject to } \sum_{i\in\mathbf{I}^t} \lambda_i\mathbf{P}^i + \sum_{j\in\mathbf{J}^t} \mu_j\mathbf{R}^j = \mathbf{b} \\ &\sum_{i\in\mathbf{I}^t} \lambda_i = 1 \\ &\lambda_i \geqslant 0 \;\; i\in\mathbf{I}^t \qquad \mu_j \geqslant 0 \;\; j\in\mathbf{J}^t \end{aligned} \tag{29}$$

has a feasible basis. That is, there are $m + 1$ columns which are linearly independent and the corresponding solution for the λ_i's and the μ_j's is nonnegative. In such a case Eq. (29) is called a ***restricted master*** since it is of the same form as the master problem (28) but contains only a subset of its columns.

Let $\mathbf{I}^{t,0} \subset \mathbf{I}^t$ and $\mathbf{J}^{t,0} \subset \mathbf{J}^t$ define an optimal basis for the restricted master (28), and let $\hat{\boldsymbol{\pi}}^t = (\boldsymbol{\pi}^t, \delta_t)$ be the simplex multipliers determined by

$$\begin{aligned} \boldsymbol{\pi}^t\mathbf{P}^i + \delta_t &= \alpha_i \qquad i\in\mathbf{I}^{t,0} \\ \boldsymbol{\pi}\mathbf{R}^j &= \beta_j \qquad j\in\mathbf{J}^{t,0} \end{aligned} \tag{30}$$

Now we wish to add another column to the restricted master which "prices out" negatively, by which we mean a $\mathbf{P}^i$, $i\in\mathbf{I} - \mathbf{I}^t$ such that $\boldsymbol{\pi}^t\mathbf{P}^i + \delta_t > \alpha_i$ or an $\mathbf{R}^j$, $j\in\mathbf{J} - \mathbf{J}^t$ such that $\boldsymbol{\pi}^t\mathbf{R}^j > \beta_j$. In terms of the set $\mathbf{K} = \{\mathbf{x} \mid \mathbf{Ex} = \mathbf{f}, \mathbf{x} \geqslant 0\}$ we seek $\mathbf{E}^i$ such that $(\boldsymbol{\pi}^t\mathbf{A})\,\mathbf{E}^i + \delta_t > \mathbf{cE}^i$ or a $\mathbf{C}^j$ such that $(\boldsymbol{\pi}^t\mathbf{A})\,\mathbf{C}^j > \mathbf{cC}^j$. To do this we solve the *subproblem*:

$$\begin{aligned} &\min_{\mathbf{x}} (\mathbf{c} - \boldsymbol{\pi}^t\mathbf{A})\mathbf{x} + \delta_t \\ &\text{subject to } \mathbf{x}\in K \end{aligned} \tag{31}$$

or equivalently

$$\begin{aligned} &\text{subject to } \mathbf{Fx} = \mathbf{f} \\ &\mathbf{x} \geqslant 0 \end{aligned}$$

As we have seen in Sec. 3-2-2 if we apply the simplex algorithm to Eq. (31) one of three things can happen: (1) Equation (31) is infeasible, (2) Eq. (31) has an optimal solution which is achieved at an extreme $\mathbf{E}^t$, or (3) $(\mathbf{c} - \pi^t\mathbf{A})\mathbf{x} + \delta_t$ is unbounded below for $\mathbf{x} \in \mathbf{K}$. If (1) occurs the original problem (27) is infeasible and we can stop. If (2) occurs $\mathbf{P}^t = \mathbf{AE}^t$ is added to the restricted master (29) which is then reoptimized with t increased to $t + 1$. Finally if (3) occurs there is by the unbounded solutions theorem in Sec. 3-2-1 a ray $\mathbf{C}^t \geqslant 0$ such that $\mathbf{EC}^t = 0$, and $(\pi\mathbf{A})\,\mathbf{C}^t < 0$; in fact the ray constructed in the proof of the theorem is an extreme ray. Then we add $\mathbf{R}^t = \mathbf{AC}^t$ to Eq. (29) and reoptimize in the next, $t + 1$st, iteration. This procedure is continued until either the restricted master (29) has an unbounded solution, in which case the original problem (20) had an unbounded objective function, or until in the subproblem (30) the minimum of $(\pi^t\mathbf{A})\,\mathbf{x} + \delta$ is nonnegative, at which point

$$\mathbf{x} = \sum_{i \in \mathbf{I}^{t,0}} \lambda_i \mathbf{E}^i + \sum_{j \in \mathbf{J}^{t,0}} \mu_j \mathbf{R}^j \tag{32}$$

is optimal. The algorithm is clearly finite since it is equivalent to the simplex method being applied to the linear program (28).

The decomposition algorithm applied to Eq. (27) is particularly effective when $\mathbf{A}$ has few rows and $\mathbf{F}$ has simple structure. Two well-known examples are, first, where $\mathbf{E}$ is block diagonal in which case the subproblem (31) decouples and breaks into independent smaller problems, and, second, when $\mathbf{F}$ represents the constraints of a network flow problem for which methods exist which are much more efficient than the simplex method.[16] There are few computational results published in the open literature. Results have been reported in Hellerman[22] and in the comprehensive study by Broise, Huard, and Sentenac.[4] In the latter report the possibility of multilevel decomposition is considered in some detail. It seems from what little knowledge there is available that in the case where $\mathbf{E}$ is block diagonal quite large problems can be solved if $\mathbf{A}$ has very few rows. However, if $\mathbf{A}$ has even as many as 100 rows, computation time is often excessive. In Kunzi, Tzschach, and Zehnder,[31] Fortran and Algol listings of a decomposition computer program are given.

Now we return to the consideration of Eq. (24) where $\mathbf{K}$ is a general convex set rather than a polyhedron. Since, in general, $\mathbf{K}$ will have an infinite number of extreme points we cannot expect a finite algorithm as we obtained for Eq. (27).

3-2-6 Decomposition—General Case

As before we assume at iteration t we have sets $\mathbf{I}^{t,0} \subset \mathbf{I}^t \subset \mathbf{I}$, $\mathbf{J}^{t,0} \subset \mathbf{J}^t \subset \mathbf{J}$, and $\mathbf{K}^{t,0} \subset \{1,\ldots, r\}$ such that the restricted master (33) corresponding to the master problem (25) has linearly independent rows and such that $\mathbf{I}^{t,0}$, $\mathbf{J}^{t,0}$, and $\mathbf{K}^{t,0}$ define a feasible basis for Eq. (33).

$$
\begin{aligned}
\min_{\boldsymbol{\lambda},\boldsymbol{\mu},\boldsymbol{\sigma}} z &= \sum_{i\in \mathbf{I}^t} \lambda_i \alpha_i + \sum_{j\in \mathbf{J}^t} \mu_j \beta_j + \sum_{k \in K^t} \sigma_k \gamma_k \\
\text{subject to} & \sum_{i\in \mathbf{I}^t} \lambda_i \mathbf{P}^i + \sum_{j\in \mathbf{J}^t} \mu_j \mathbf{R}^j + \sum_{k\in K^t} \sigma_k \mathbf{S}^k = \mathbf{b} \\
& \sum_{i\in \mathbf{I}^t} \lambda_i = 1 \\
& \lambda_i \geqslant 0 \quad i \in \mathbf{I}^t \qquad \mu_j \geqslant 0 \quad j \in \mathbf{J}^t
\end{aligned}
\tag{33}
$$

The simplex multipliers $\hat{\boldsymbol{\pi}}^t = (\boldsymbol{\pi}^t, \delta_t)$ are defined by

$$\pi^t \mathbf{P}^i + \delta_t = \alpha_i \qquad i \in \mathbf{I}^{t,0} \tag{34a}$$

$$\pi^t \mathbf{R}^j = \beta_j \qquad j \in \mathbf{J}^{t,0} \tag{34b}$$

$$\pi^t \mathbf{S}^k = \gamma_k \qquad k \in \mathbf{K}^{t,0} \tag{34c}$$

Again we look for a column which prices out negative. That is, we seek a $\mathbf{P}^i$ such that $\boldsymbol{\pi}^t\mathbf{P}^i + \delta_t > \alpha_i$, an $\mathbf{R}^j$ such that $\boldsymbol{\pi}^t\mathbf{R}^j > \beta_j$, or a $\mathbf{S}^k$ such that $\boldsymbol{\pi}\mathbf{S}^k \neq \sigma_k$. Equivalently using the relations (26) we seek

$$\mathbf{E}^i \in \mathscr{H} \qquad \text{such that} \quad (\boldsymbol{\pi}^t\mathbf{A})\mathbf{E}^i + \delta_t > (\boldsymbol{\pi}^t\mathbf{c})\mathbf{E}^i \tag{35a}$$

$$\mathbf{C}^j \in \mathscr{C} \qquad \text{such that} \quad (\boldsymbol{\pi}^t\mathbf{A})\mathbf{C}^j > (\boldsymbol{\pi}^t\mathbf{c})\mathbf{C}^j \tag{35b}$$

or

$$\mathbf{L}^k \in \mathscr{L} \qquad \text{such that} \quad (\boldsymbol{\pi}^t\mathbf{A})\mathbf{L}^k \neq (\boldsymbol{\pi}\mathbf{c})\mathbf{L}^k \tag{35c}$$

The not-equal relation in Eq. (35*c*) occurs because the σ_i's are not sign restricted. In most applications the way we do this is to solve the *subproblem*:

$$
\begin{aligned}
&\min_{\mathbf{x}} \; (\mathbf{c} - \boldsymbol{\pi}^t\mathbf{A})\mathbf{x} + \delta_t \\
&\text{subject to } \mathbf{x} \in \mathbf{K}
\end{aligned}
\tag{36}
$$

The efficiency or even the feasibility of the decomposition method depends on the ease with which Eq. (36) can be solved. If Eq. (36) is not feasible (that is, $\mathbf{K} = \phi$) then of course neither is the original problem. If Eq. (36) has a finite optimum value it is clear that Eq. (35*b*) or Eq. (35*c*) cannot be satisfied for any $\mathbf{C}^j \in \mathscr{C}$ or $\mathbf{L}^k \in \mathscr{L}$. For if

$(\mathbf{c} - \boldsymbol{\pi}^t\mathbf{A})\,\mathbf{L}^k < 0$ then $\mathbf{x} = \mathbf{x}^0 + \sigma\mathbf{L}^k \in \mathbf{K}$ for any scalar σ and any $\mathbf{x}^0 \in \mathbf{K}$ and $\lim_{\sigma\to\infty}(\mathbf{c} - \boldsymbol{\pi}^t\mathbf{A})\,\mathbf{x} + \delta_t = -\infty$ if $(\mathbf{c} - \boldsymbol{\pi}^t\mathbf{A})\,\mathbf{L}^k > 0$ then $\lim_{\sigma\to-\infty}(\mathbf{c} - \boldsymbol{\pi}^t\mathbf{A})\,\mathbf{x} + \delta_t = -\infty$ and similarly if Eq. (36) has a finite minimum $(\mathbf{c} - \boldsymbol{\pi}^t\mathbf{A})\,\mathbf{C}^j \geqslant 0$ must hold for all $j \in \mathbf{J}$. On the other hand, if $(\mathbf{c} - \boldsymbol{\pi}^t\mathbf{A})\,\mathbf{L}^k = 0$ for all $k \in \{1,\ldots, r\}$ and $(\mathbf{c} - \boldsymbol{\pi}^t\mathbf{A})\,\mathbf{C}^j \geqslant 0$ for $j \in \mathbf{J}$ then for any $\mathbf{x} = \sum \lambda_i\mathbf{E}^i + \sum \mu_j\mathbf{C}^j + \sum \sigma_k\mathbf{S}^k$ we have

$$(\mathbf{c} - \boldsymbol{\pi}^t\mathbf{A})\mathbf{x} = \sum \lambda_i(\mathbf{c} - \boldsymbol{\pi}^t\mathbf{A})\mathbf{E}^i + \sum \pi_j(\mathbf{c} - \boldsymbol{\pi}^t\mathbf{A})\mathbf{C}^j + \sum \sigma_k(\mathbf{c} - \boldsymbol{\pi}^t\mathbf{A})\mathbf{S}^k$$

$$\geqslant \left(\sum \lambda_i\right) \min_{i\in\mathbf{I}} (\mathbf{c} - \boldsymbol{\pi}^t\mathbf{A})\mathbf{E}^i = \min_{i\in\mathbf{I}} (\mathbf{c} - \boldsymbol{\pi}^t\mathbf{A})\mathbf{E}^i$$

Since we have assumed that $\mathbf{K}$ is compact enough $\mathscr{H}$ is compact and therefore $\min\{(\mathbf{c} - \boldsymbol{\pi}^t\mathbf{A})\mathbf{E}^i \mid i \in \mathbf{I}\}$ exists. Thus there are only three possibilities: either Eq. (36) has a finite optimum achieved at an extreme point $\mathbf{E}^i$ or Eq. (35*b*) or Eq. (35*c*) is satisfied. An important case that can occur when $\mathbf{K}$ is not compact enough is that Eq. (36) could have a finite infimum which is not achieved. This would occur, for example, if one tried to minimize the x coordinate of $K = \{(x, y) \mid xy \geqslant 1,\ x \geqslant 0,\ y \geqslant 0\}$ (see Fig. 3-2*b*). If Eq. (36) has a finite optimum value which is greater than or equal to zero this indicates that the current optimal basic solution for the restricted master problem (33) is also an optimal basic solution for the complete master problem (25) and therefore the corresponding

$$\mathbf{x} = \sum_{i\in\mathbf{I}^{t,0}} \lambda_i\mathbf{E}^i + \sum_{j\in\mathbf{J}^{t,0}} \mu_j\mathbf{C}^j + \sum_{k\in K^{t,0}} \sigma_k\mathbf{S}^k$$

is optimal for the original problem (24). If Eq. (36) has an optimal solution $\mathbf{E}^i$ for which $(\mathbf{c} - \boldsymbol{\pi}^t\mathbf{A})\,\mathbf{E}^i + \delta_t < 0$, the corresponding column $\binom{\alpha_i}{\mathbf{P}^i}$ where $\mathbf{P}^i = \mathbf{A}\mathbf{E}^i$, and $\alpha_i = \mathbf{c}\mathbf{E}^i$ is added to the restricted master (33) which is reoptimized to start the next iteration. If Eq. (36) has no finite minimum then a column corresponding to a $\mathbf{C}^j$ or a $\mathbf{L}^k$ is added to Eq. (33) in addition to the $\binom{\alpha_i}{\mathbf{P}^i}$ corresponding to the $\mathbf{E}^i$ which minimizes ${}_{i\in\mathbf{I}}(\mathbf{c} - \boldsymbol{\pi}^t\mathbf{A})\,\mathbf{E}^i + \delta_t$. Actually in the latter case the column $\binom{\alpha_i}{\mathbf{P}^i}$ need not be added every time; it need be added only periodically.[1]

In Van Slyke[40] it is shown that there is a subsequence of the $\mathbf{x}$'s and $(\boldsymbol{\pi}, \delta)$'s generated so that the $\mathbf{x}$'s converge to an optimal solution for

[1] In practice except where K is polyhedral it is usually the case that K is compact; then $\mathscr{C}$ and $\mathscr{L}$ are empty and Eq. (36) always has a finite minimum. For polyhedral K one is definitely interested in unbounded K, in particular for the dual decomposition method of Sec. 3-2-8.

Eq. (24) and the limit of the (π, δ)'s satisfies the optimality condition $\min (\mathbf{c} - \pi\mathbf{A})\,\mathbf{x} + \delta \geqslant 0$. Of course the z values are monotone non-increasing so that any convergent subsequence converges to an optimal solution. A very useful application of the general decomposition theorem is in convex programming.

3-2-7 Convex Programming

Let $\mathbf{f}_0(\mathbf{x}), \mathbf{f}_i(\mathbf{x}),\ i = 1,\ldots, m$, be continuous convex functions of the n-vector $\mathbf{x}$. Consider the problem

$$\begin{aligned} &\min_{\mathbf{z}} \mathbf{f}_0(\mathbf{x}) \\ &\text{subject to } \mathbf{f}_i(\mathbf{x}) \leqslant 0 \qquad i = 1,\ldots, m \end{aligned} \tag{37}$$

which is equivalent to

$$\begin{aligned} &\max_{\mathbf{y}} x_0 \\ &\text{subject to } \mathbf{U}_0 x_0 + \mathbf{P} = \mathbf{0} \\ &\qquad\qquad \mathbf{P} \in \mathbf{K} \end{aligned} \tag{38}$$

where $\mathbf{K} = \{(y_0, y_i, \ldots, y_m) \mid y_i \geqslant f_i(\mathbf{x}),\ i = 0,\ldots, m\}$ is also closed and convex. Thus Eq. (37) can be represented as a problem of the form (23) and the generalized L.P. method can be applied if the subproblem is easily solvable. Let $\hat{\pi} = (1, \pi_1, \ldots, \pi_m)$ be the dual variables associated with the restricted master program associated with Eq. (38) at some iteration. Then the *subproblem* is

$$\begin{aligned} &\min_{\mathbf{y}} y_0 + \sum \pi_i y_i \\ &\text{subject to } \mathbf{y} \in \mathbf{K} \end{aligned} \tag{39}$$

It can be shown that $\pi_i \geqslant 0$ for $i = 1,\ldots, m$. Thus Eq. (39) is equivalent to

$$\min f_0(\mathbf{x}) + \sum \pi_i f_i(\mathbf{x}) \tag{40}$$

which is an *unconstrained* minimization of a convex function. Thus at each iteration one finds the optimum of a linear program and the minimum of an unconstrained convex function.

It is assumed when the method is applied that there is a separate technique available for finding an unconstrained minimum of a convex function of the form (40) for any given values of $\pi_i \geqslant 0$. For more details of the convex programming algorithm and an application to chemical equilibrium problems see Dantzig,[11] chap. 24.

A special case of convex programming is the convex quadratic program. It is

$$\min_{\mathbf{x}} \mathbf{cx} + \tfrac{1}{2}\mathbf{x}^t\mathbf{Qx}$$
$$\text{subject to } \mathbf{Ax} \leqslant \mathbf{b} \tag{41}$$
$$\mathbf{x} \geqslant \mathbf{0}$$

where $\mathbf{Q}$ is a positive semidefinite matrix. Because the gradient of the objective is linear the first-order optimality conditions are essentially linear and Eq. (41) can be solved by using algorithms which terminate in a finite number of steps rather than by using ones in which only convergence is guaranteed such as the general convex programming algorithm we described here. We will say no more about quadratic programming algorithms; descriptions can be found in Kunzi and Krelle.[30] In Whinston[43] a decomposition method for quadratic programming is spelled out.

3-2-8 Dual Decomposition Methods

Finally we consider dual decomposition algorithms. These appear under different names in Cheney and Goldstein,[6] Kelley,[26] Benders,[3] and Van Slyke and Wets.[41] We follow here the treatment in Van Slyke and Wets.[41] Consider the linear program

$$\min_{\mathbf{x},\mathbf{y}} \mathbf{cx} + \mathbf{dy}$$
$$\text{subject to } \mathbf{Ex} \qquad = \mathbf{b} \tag{42}$$
$$\mathbf{Fx} + \mathbf{Gy} = \mathbf{h}$$
$$\mathbf{x} \geqslant \mathbf{0} \qquad \mathbf{y} \geqslant \mathbf{0}$$

where $\mathbf{E}$ is an $m_1 \times n_1$ matrix, $\mathbf{F}$ is an $m_2 \times n_1$ matrix, $\mathbf{G}$ is an $m_2 \times n_2$ matrix, and $\mathbf{c}$, $\mathbf{d}$, $\mathbf{b}$, and $\mathbf{h}$ are vectors of appropriate dimension. We will first give the algorithm for this problem and then show its relation to the decomposition algorithm. To motivate the algorithm let us suppose for the time being that $\mathbf{d} = 0$, and $\mathbf{G} = [\mathbf{I} \mid -\mathbf{I}]$ where $\mathbf{I}$ is the $m_2 \times m_2$ identity matrix. Then to solve Eq. (42) it would only be necessary to solve

$$\min_{\mathbf{x}} \mathbf{cx}$$
$$\text{subject to } \mathbf{Ex} = \mathbf{b} \tag{43}$$
$$\mathbf{x} \geqslant \mathbf{0}$$

The reason for this is that no matter what $\mathbf{x}$ is chosen there is a $\mathbf{y} \geqslant 0$ such that $[\mathbf{I} \mid -\mathbf{I}]\, \mathbf{y} = \mathbf{h} - \mathbf{Fx}$. Moreover since $\mathbf{d} = \mathbf{0}$, the cost does not depend on $\mathbf{y}$, so that the last m_2 equations involving $\mathbf{y}$ are essentially irrelevant. The algorithm we propose is for situations where the last m_2 equations are "almost" irrelevant. In particular in Sec. 3-4 we will use this algorithm to solve optimal control problems with state constraints. The last m_2 equations will correspond to the state constraints and we assume that for most times the state will not be constrained in the optimal solution. In this case our algorithm is useful.

Since, by hypothesis, we are not particularly interested in $\mathbf{y}$ we can rewrite Eq. (42) in the equivalent form:

$$\min_{\mathbf{x}} \mathbf{cx} + Q(\mathbf{x})$$

$$\text{subject to } \mathbf{Ex} = \mathbf{b} \tag{44}$$

$$\mathbf{x} \geqslant \mathbf{0}$$

$$\mathbf{x} \in K$$

where $\mathbf{K} = \{\mathbf{x} \mid \text{there exists } \mathbf{y} \geqslant \mathbf{0} \text{ such that } \mathbf{Fx} + \mathbf{Gy} = \mathbf{h}\}$ and $Q(\mathbf{x}) = \min \{\mathbf{dy} \mid \mathbf{y} \geqslant \mathbf{0}, \mathbf{Gy} = \mathbf{h} - \mathbf{F}(\mathbf{x})\}$. Then the algorithm can be stated as follows:

Step 0. Solve Eq. (43) to obtain $\mathbf{x}^0 \geqslant \mathbf{0}$. Set $\theta_0 = -\infty$. If Eq. (43) is infeasible then the entire problem (42) is infeasible. In general it is possible that $\mathbf{cx}$ in Eq. (43) is unbounded below. We assume for simplicity of exposition that this does not happen. What to do if this does happen is treated in detail in Van Slyke and Wets.[41] Assuming that Eq. (43) has a finite optimum set the iteration index k to zero and go to step 1.

Step 1. Check to see if $\mathbf{x}^k \in \mathbf{K}$; that is, we try to find $\mathbf{y} \geqslant 0$ such that

$$\mathbf{Gy} = \mathbf{h} - \mathbf{Fx}^k \tag{45}$$

This is accomplished using the phase 1 procedure of the simplex method. The phase 1 problem can be written

$$\min_{\mathbf{y},\mathbf{v}^+,\mathbf{v}^-} w = \mathbf{ev}^+ + \mathbf{ev}^-$$

$$\text{subject to } \mathbf{Gy} + \mathbf{Iv}^+ - \mathbf{Iv}^- = \mathbf{h} - \mathbf{Fx}^k \tag{46}$$

$$\mathbf{y} \geqslant \mathbf{0} \qquad \mathbf{v}^+ \geqslant \mathbf{0} \qquad \mathbf{v}^- \geqslant \mathbf{0}$$

where $e = (1, \ldots, 1)$ and $\mathbf{I}$ is the $m_2 \times m_2$ identity matrix. The problem (46) differs from the phase 1 problem (10) described earlier because in this case we do not wish to require that the right-hand side $\mathbf{h} - \mathbf{F}\mathbf{x}^k$ be nonnegative. If min $w = 0$, that is, if Eq. (45) is solvable and $\mathbf{x}^k \in K$, then go to step 2. If min $w > 0$ there exists an optimal dual vector (see Sec. 3-2-2) $\boldsymbol{\sigma} = [\sigma_1, \ldots, \sigma_{m_2}]$ for the phase 1 problem such that

$$\boldsymbol{\sigma}\mathbf{G} \leqslant 0 \qquad \text{and} \qquad \boldsymbol{\sigma}(\mathbf{h} - \mathbf{F}\mathbf{x}^k) = \min w > 0 \tag{47}$$

The relations (47) determine a "cut" on $\mathbf{x}$. Notice that if $\mathbf{G}\mathbf{y} = \mathbf{h} - \mathbf{F}\mathbf{x}$, $\mathbf{y} \geqslant \mathbf{0}$, then $\boldsymbol{\sigma}\mathbf{G}\mathbf{y} = \boldsymbol{\sigma}(\mathbf{h} - \mathbf{F}\mathbf{x}) \leqslant \mathbf{0}$; thus in order for $\mathbf{x} \in \mathbf{K}$, $\mathbf{x}$ must satisfy

$$(\boldsymbol{\sigma}\mathbf{F})\mathbf{x} \geqslant \boldsymbol{\sigma}\mathbf{h} \tag{48}$$

Notice that $\mathbf{x}^k$ does not satisfy Eq. (48). Then inequality (47) is now added to the master problem (53) in step 3.

Step 2. Calculate $Q(\mathbf{x}^k)$ by solving

$$\begin{aligned} &\min_{\mathbf{y}} \mathbf{d}\mathbf{y} \\ &\text{subject to } \mathbf{G}\mathbf{y} = \mathbf{h} - \mathbf{F}\mathbf{x}^k \\ &\mathbf{y} \geqslant 0 \end{aligned} \tag{49}$$

We already know that Eq. (49) has feasible solutions because we arrived here from step 1. If in Eq. (49) $\mathbf{d}\mathbf{y}$ is unbounded below for $\mathbf{y}$ satisfying $\mathbf{G}\mathbf{y} = \mathbf{h} - \mathbf{F}\mathbf{x}^k$, $\mathbf{y} \geqslant 0$, then the original problem (42) has an objective function which is unbounded below and we can stop. The final case occurs when Eq. (49) has a finite optimum $\mathbf{y}^k$. At the optimum by the duality theorem there is a dual vector $\boldsymbol{\pi} = [\pi_1, \ldots, \pi_{m_2}]$ satisfying

$$\boldsymbol{\pi}\mathbf{G} \leqslant \mathbf{d} \tag{50}$$

$$\boldsymbol{\pi}(\mathbf{h} - \mathbf{F}\mathbf{x}^k) = \min \mathbf{d}\mathbf{y} = Q(\mathbf{x}^k) \tag{51}$$

If $Q(\mathbf{x}^k) = \theta^k$ we terminate with the optimal solution $\mathbf{x}^k$, $\mathbf{y}^k$. Otherwise notice that $\boldsymbol{\pi}(\mathbf{h} - \mathbf{F}\mathbf{x}) \leqslant Q(\mathbf{x})$ for any $\mathbf{x}$ since $\boldsymbol{\pi}$ is a feasible solution to the dual of Eq. (49) and $Q(\mathbf{x})$ is equal to the maximum with respect to $\boldsymbol{\pi}$ of $\boldsymbol{\pi}(\mathbf{h} - \mathbf{F}\mathbf{x})$ for all $\boldsymbol{\pi}$ satisfying Eq. (50). If we let θ represent the value of $Q(\mathbf{x})$ then $\theta \geqslant \boldsymbol{\pi}(\mathbf{h} - \mathbf{F}\mathbf{x})$ and we add the constraint

$$(\boldsymbol{\pi}\mathbf{F})\mathbf{x} + \theta \geqslant \boldsymbol{\pi}\mathbf{h} \tag{52}$$

to the master (53) in step 3.

Step 3. We look for θ^{k+1}, $\mathbf{x}^{k+1}$ by solving

$$\begin{aligned} &\min_{\mathbf{x}} \mathbf{cx} + \theta \\ &\text{subject to } \mathbf{Ex} = \mathbf{b} \\ &\qquad (\boldsymbol{\sigma}^l\mathbf{F})\mathbf{x} \geqslant \boldsymbol{\sigma}^l\mathbf{h} \qquad l = 1,\dots,L \\ &\qquad (\boldsymbol{\mu}^l\mathbf{F})\mathbf{x} + \theta \geqslant \boldsymbol{\pi}^l\mathbf{h} \qquad l = 1,\dots,L_0 \\ &\qquad \mathbf{x} \geqslant \mathbf{0} \end{aligned} \tag{53}$$

where $L + L_0 = k$. Increase k by 1 and go to step 1.

In order to relate this algorithm to the previous ones let us examine the dual of Eq. (42) which is

$$\begin{aligned} &\max_{\boldsymbol{\xi},\boldsymbol{\eta}} \boldsymbol{\xi}\mathbf{b} + \boldsymbol{\eta}\mathbf{h} \\ &\text{subject to } \boldsymbol{\xi}\mathbf{E} + \boldsymbol{\eta}\mathbf{F} \leqslant \mathbf{c} \\ &\qquad \boldsymbol{\eta}\mathbf{G} \leqslant \mathbf{d} \end{aligned} \tag{54}$$

Now let $\mathbf{Y} = \{\boldsymbol{\eta} \mid \boldsymbol{\eta}\mathbf{G} \leqslant \mathbf{d}\}$. This is a closed convex polyhedral set, so that there is a finite set of vectors $\{\boldsymbol{\pi}^1,\dots,\boldsymbol{\pi}^{\mathbf{I}}\}$ and a finite set $\{\boldsymbol{\sigma}^1,\dots,\boldsymbol{\sigma}^{\mathbf{J}}\}$ such that $\boldsymbol{\eta} \in \mathbf{Y}$ if and only if

$$\boldsymbol{\eta} = \sum_{i=1}^{\mathbf{I}} \lambda_i \boldsymbol{\pi}^i + \sum_{j=1}^{\mathbf{J}} \mu_j \boldsymbol{\sigma}^j \qquad \sum_{i=1}^{\mathbf{I}} \lambda_i = 1$$

$\lambda_i \geqslant 0$, $i = 1,\dots,\mathbf{I}$, $\mu_j \geqslant 0$, $j = 1,\dots,\mathbf{J}$. If there is a translate of a linear space in $\mathbf{Y}$ we include it in

$$\mathscr{C} = \left\{\mathbf{y} \mid \mathbf{y} = \sum_{j=1}^{\mathbf{J}} \mu_j \boldsymbol{\sigma}^j,\ \mu_j \geqslant 0, j = 1,\dots,\mathbf{J}\right\}$$

Then Eq. (54) is equivalent to

$$\begin{aligned} &\max_{\boldsymbol{\xi},\boldsymbol{\lambda},\boldsymbol{\mu}} \boldsymbol{\xi}\mathbf{b} + \sum_{i=1}^{\mathbf{I}} \lambda_i(\boldsymbol{\pi}^i\mathbf{h}) + \sum_{j=1}^{\mathbf{J}} \mu_j(\boldsymbol{\sigma}^j\mathbf{h}) \\ &\text{subject to } \boldsymbol{\xi}\mathbf{E} + \sum_{i=1}^{\mathbf{I}} \lambda_i(\boldsymbol{\pi}^i\mathbf{F}) + \sum \mu_j(\boldsymbol{\sigma}^j\mathbf{F}) \leqslant \mathbf{c} \\ &\qquad \sum_{i=1}^{\mathbf{I}} \lambda_i = 1 \\ &\qquad \lambda_i \geqslant 0 \quad i = 1,\dots,\mathbf{I} \qquad \mu_j \geqslant 0 \quad j = 1,\dots,\mathbf{J} \end{aligned} \tag{55}$$

Finally we take the dual of Eq. (55) using the nonnegative n_1-vector $\mathbf{x}$ as the dual variables associated with all but the last equation $\mathbf{x}_0$ which we assign the unrestricted variable θ. The dual is

$$\min_{\theta,\mathbf{x}} \mathbf{cx} + \theta$$

$$\begin{aligned} \text{subject to } (\boldsymbol{\pi}^i\mathbf{F})\mathbf{x} + \theta &\geqslant \boldsymbol{\pi}^i\mathbf{h} \qquad i = 1,\dots, \mathbf{I} \\ (\boldsymbol{\sigma}^j\mathbf{F})\mathbf{x} \qquad &\geqslant \boldsymbol{\sigma}^j\mathbf{h} \qquad j = 1,\dots, \mathbf{J} \\ \mathbf{x} &\geqslant 0 \end{aligned}$$

which is of the form (53). So in effect the dual decomposition algorithms are equivalent to applying decomposition to the *dual* of the original problem. An important special case of Eq. (42) occurs when $\mathbf{G} = \mathbf{I}$, in other words, when $\mathbf{Fx} + \mathbf{Gy} = \mathbf{h}$ is equivalent to $\mathbf{Fx} \leqslant \mathbf{h}$. It is this form which we will use in Sec. 3-4. In this case step 1 is trivial because Eq. (46) becomes

$$\min_{\mathbf{y},\mathbf{v}^+,\mathbf{v}^-} \mathbf{w} = \mathbf{ev}^+ + \mathbf{ev}^-$$

$$\text{subject to } \mathbf{Iy} + \mathbf{Iv}^+ - \mathbf{Iv}^- = \mathbf{h} - \mathbf{Fx}^k$$

which has the obvious optimal solution

$$y_i = \max\{(\mathbf{h} - \mathbf{Fx}^k)_i, 0\} \qquad v_i^- = \max\{(\mathbf{Fx}^k - \mathbf{h})_i, 0\}$$

where y_i, $(\mathbf{h} - \mathbf{Fx}^k)_i$, y_i^- are the ith components of the respective vectors. An optimal dual solution $\boldsymbol{\sigma}$ is defined by

$$\sigma_i = \begin{cases} -1 & \text{if } (\mathbf{h} - \mathbf{Fx}^k)_i < 0 \\ \;\;0 & \text{if } (\mathbf{Fx}^k - \mathbf{h})_i \geqslant 0 \end{cases} \tag{57}$$

which can be verified by direct substitution.

3-3 THE DIRECT APPROACH TO CONTINUOUS-TIME LINEAR OPTIMAL CONTROL PROBLEMS

We consider Eq. (1) which was

$$\min_{\mathbf{u}} x_0(T)$$

$$\begin{aligned} \text{subject to } \frac{d}{dt}\mathbf{x}(t) &= \mathbf{A}(t)\mathbf{x}(t) + \mathbf{B}(t)\mathbf{u}(t) \\ \mathbf{u}(t) &\in \mathbf{U}(t) \qquad\qquad t \in [0, T] \\ \mathbf{x}(0) &= \mathbf{x}^0 \\ x_i(T) &= x_i^T \qquad\qquad i = 1,\dots, n \end{aligned} \tag{1}$$

where $\mathbf{x}(t) = (x_0(t),..., x_n(t))$ is an $n + 1$ component state vector function of time, $\mathbf{u}(t) = (u_1(t),..., u_r(t))$ is an r-component control vector function, and $\mathbf{A}(t)$ and $\mathbf{B}(t)$ are respectively $(n + 1) \times (n + 1)$ and $(n + 1) \times r$ matrix functions. Further $\mathbf{U}(t)$ will be a closed convex subset of $\mathbf{E}_r$ for each $t \in [0, T]$. We first treat the problem formally to illustrate the application of generalized linear programming in the simplest context. We then illustrate how the more general problem (3) can be treated also in a formal manner. Finally we give additional hypotheses on the problem sufficient to justify the formal operations performed.

We want to find a control vector $\mathbf{u}(t) = \mathbf{u}^*(t)$ which achieves a minimum for the component x_0 of the state vector $\mathbf{x}$. For a given control $\mathbf{u}$ the differential equations can be integrated yielding

$$\mathbf{x}(T) = \mathbf{\Phi}(T)\mathbf{x}^0 + \int_0^T \mathbf{\Phi}(T)\mathbf{\Phi}^{-1}(t)\mathbf{B}(t)\mathbf{u}(t)\,dt \tag{58}$$

where $\mathbf{\Phi}(t)$ is an $(n + 1) \times (n + 1)$ matrix function satisfying Eq. (2)

$$\begin{aligned} \frac{d}{dt}\mathbf{\Phi}(t) &= \mathbf{A}(t)\mathbf{\Phi}(t) \\ \mathbf{\Phi}(0) &= \mathbf{I} \end{aligned} \tag{59}$$

where $\mathbf{I}$ is the $(n + 1) \times (n + 1)$ identity matrix. For the special case where $\mathbf{A}(t) = \mathbf{A}$ is a constant matrix,

$$\mathbf{\Phi}(t) = e^{\mathbf{A}t} = \mathbf{I} + \mathbf{A}t + \left(\frac{1}{2!}\right)\mathbf{A}^2t^2 + \cdots$$

(see Zadeh and Desoer,[44] chaps. 5 and 6, for details). Using Eq. (1), Eq. (58) can be reformulated as the generalized program

$$\begin{aligned} &\max_{\mathbf{P}} x_0 \\ &\text{subject to } x_0\mathbf{U}_0 + \mathbf{P}^{\mathbf{u}} = \mathbf{b} \\ &\qquad\qquad \mathbf{P}^{\mathbf{u}} \in \mathscr{P} \end{aligned} \tag{60}$$

where $\mathbf{U}_0 = (1, 0,..., 0)'$,

$$\mathbf{P}^{\mathbf{u}} = \int_0^T \mathbf{\Phi}(T)\mathbf{\Phi}^{-1}(t)\mathbf{B}(t)\mathbf{u}(t)\,dt$$

$$\mathbf{b} = (0, x_1{}^T,..., x_n{}^T)' - \mathbf{\Phi}(T)\mathbf{x}^0$$

and finally $\mathscr{P} = \{\mathbf{P}^{\mathbf{u}} \mid \mathbf{u}(t) \in \mathbf{U}(t), 0 \leqslant t \leqslant T\}$.

The first observation we make is that $\mathscr{P}$ is a convex set, for if $\mathbf{P}^{\mathbf{u}_1}$ and $\mathbf{P}^{\mathbf{u}_2}$ are in $\mathscr{P}$ and $\lambda \in (0, 1)$, then

$$\begin{aligned}\mathbf{P}^{\lambda \mathbf{u}_1+(1-\lambda)\mathbf{u}_2} &= \int_0^T \mathbf{\Phi}(T)\mathbf{\Phi}^{-1}(t)\mathbf{B}(t)[\lambda \mathbf{u}_1(t) + (1-\lambda)\mathbf{u}_2(t)]\, dt \\ &= \lambda \int_0^T \mathbf{\Phi}(T)\mathbf{\Phi}^{-1}(t)\mathbf{B}(t)\mathbf{u}_1(t)\, dt + (1-\lambda) \int_0^T \mathbf{\Phi}(T)\mathbf{\Phi}^{-1}(t)\mathbf{B}(t)\mathbf{u}_2(t)\, dt \\ &= \lambda \mathbf{P}^{\mathbf{u}_1} + (1-\lambda)\mathbf{P}^{\mathbf{u}_2}\end{aligned}$$

At the kth step of the generalized linear programming algorithm (as applied to this problem) we have a set of dual multipliers $\boldsymbol{\pi}^k = (\pi_0{}^k, \ldots, \pi_{n+1}^k)$, μ_k and the subproblem is

$$\begin{aligned}&\min_{\mathbf{u}}\ (\boldsymbol{\pi}^k, \mu^k)\begin{pmatrix}\mathbf{P}^{\mathbf{u}}\\ 1\end{pmatrix} = \boldsymbol{\pi}^k \int_0^T \mathbf{\Phi}(T)\mathbf{\Phi}^{-1}(t)\mathbf{B}(t)\mathbf{u}(t)\, dt + \mu_k \\ &\text{subject to } \mathbf{u}(t) \in \mathbf{U}(t) \qquad t \in [0, T]\end{aligned} \tag{61}$$

Let $\boldsymbol{\pi}^k(t) = -\boldsymbol{\pi}^k\mathbf{\Phi}(T)\,\mathbf{\Phi}^{-1}(t)$. Then Eq. (61) is equivalent to

$$\begin{aligned}&\max_{\mathbf{u}} \int_0^T \boldsymbol{\pi}^k(t)\mathbf{B}(t)\mathbf{u}(t)\, dt \\ &\text{subject to } \mathbf{u}(t) \in \mathbf{U}(t) \qquad t \in [0, T]\end{aligned} \tag{62}$$

Clearly if there exists a $\mathbf{u}(t)$ satisfying the constraints which maximizes $\boldsymbol{\pi}^k(t)\,\mathbf{B}(t)\,\mathbf{u}(t)$ *pointwise* for almost every t in $[0, T]$, then the function $\mathbf{u}(t)$ solves Eq. (62) if the integration can be performed. If $\mathbf{U}(t)$ is polyhedral for each t the pointwise problem becomes a linear program. Moreover if $\mathbf{U}(t)$ does not depend on time the linear program in variables $\mathbf{u}$, $\min \boldsymbol{\pi}^k(t)\,\mathbf{B}(t)\mathbf{u}$ subject to $\mathbf{u} \in \mathbf{U}$ has fixed constraints independent of time and only the objective function changes with time (in an absolutely continuous manner). Finally if $\mathbf{A}(t) \equiv \mathbf{A}$ and $\mathbf{B}(t) \equiv \mathbf{B}$ are constant, $\boldsymbol{\pi}(t)\mathbf{B}$ has components which are simply polynomials in t times exponential functions of time. Jizmagian[24] has developed a parametric technique for solving this problem and has successfully implemented it on the computer. However, returning to the general problem, we have as the restricted master problem

$$\begin{aligned}&\max_{\lambda}\ x_0 \\ &\text{subject to } x_0\mathbf{U}_0 + \sum_i \lambda_i \begin{pmatrix}\mathbf{P}^{\mathbf{u}_i}\\ 1\end{pmatrix} = \begin{pmatrix}\mathbf{b}\\ 1\end{pmatrix} \\ &\lambda_i \geqslant 0\end{aligned} \tag{63}$$

assuming that $\mathscr{P}$ is bounded. At the optimum if one is attained in a finite number of steps or else on the limit on a subsequence, we obtain $\boldsymbol{\pi}^*$, μ^* as dual variables with the properties

$$\boldsymbol{\pi}^*\mathbf{P}^{\mathbf{u}} + \mu^* = \int_0^T \boldsymbol{\pi}^*\boldsymbol{\Phi}(T)\boldsymbol{\Phi}^{-1}(t)\mathbf{B}(t)\mathbf{u}(t)\,dt + \mu^* \geqslant 0 \tag{64}$$

for all $\mathbf{u} \in \{\mathbf{u}(t) \in \mathbf{U}(t),\ t \in [0, T]\}$, and

$$\boldsymbol{\pi}^*\mathbf{P}^{\mathbf{u}^*} + \mu^* = \int_0^T \boldsymbol{\pi}^*\boldsymbol{\Phi}(T)\boldsymbol{\Phi}^{-1}(t)\mathbf{B}(t)\mathbf{u}^*(t)\,dt + \mu^* = 0 \tag{65}$$

for an optimal control $\mathbf{u}^*$; also, $\boldsymbol{\pi}^*\mathbf{U}_0 = \pi_0{}^* = 1$. If we then take the difference of Eqs. (64) and (65) we obtain

$$\int_0^T \boldsymbol{\pi}^*\boldsymbol{\Phi}(T)\boldsymbol{\Phi}^{-1}(t)\mathbf{B}(t)[\mathbf{u}(t) - \mathbf{u}^*(t)]\,dt \leqslant 0 \tag{66}$$

for all admissible $\mathbf{u}$. Equation (66) is the integral form of the Pontryagin maximum principle.

If the set of admissible $\mathbf{u}$'s is sufficiently rich, the inequality (66) must hold pointwise; thus

$$\boldsymbol{\pi}^*\boldsymbol{\Phi}(T)\boldsymbol{\Phi}^{-1}(t)\mathbf{B}(t)\mathbf{u}^*(t) = \min_{\mathbf{u}(t)\in\mathbf{U}(t)} \boldsymbol{\pi}^*\boldsymbol{\Phi}(T)\boldsymbol{\Phi}^{-1}(t)\mathbf{B}(t)\mathbf{u}(t)$$

or equivalently if we set $\boldsymbol{\pi}^*(t) = -\boldsymbol{\pi}^*\boldsymbol{\Phi}(T)\,\boldsymbol{\Phi}^{-1}(t)$

$$\boldsymbol{\pi}^*(t)\mathbf{B}(t)\mathbf{u}^*(t) = \max_{\mathbf{u}\in\mathbf{U}(t)} \boldsymbol{\pi}^*(t)\mathbf{B}(t)\mathbf{u}(t) \tag{67}$$

for almost all $t \in [0, T]$, which is precisely the Pontryagin maximum principle.[34] Note that $\boldsymbol{\pi}^*(t) = -\boldsymbol{\pi}^*\Phi(T)\,\Phi^{-1}(t)$ satisfies the adjoint equation of (1), that is,

$$\frac{d}{dt}\,\boldsymbol{\pi}^*(t) = -\frac{d}{dt}\,\boldsymbol{\pi}^*\boldsymbol{\Phi}(T)\boldsymbol{\Phi}^{-1}(t)\mathbf{A}(t) = -\boldsymbol{\pi}^*(t)\mathbf{A}(t)$$

since $(d/dt)\,\boldsymbol{\Phi}^{-1}(t) = -\boldsymbol{\Phi}^{-1}\mathbf{A}(t)$, and note that $\boldsymbol{\pi}^*(T) = -\boldsymbol{\pi}^*$. Thus the identification connecting mathematical programming to the maximum principle approach is that the optimal dual variables for the generalized program are terminal boundary conditions for the adjoint system $(d/dt)\,\boldsymbol{\pi}(t)\,\mathbf{A}(t)$ involved in the maximum principle.

Let us give generalizations of the method before we proceed. Suppose instead of requiring that $\mathbf{x}(0) = \mathbf{x}^0$ and that $x_i(T) = x_i{}^T$ we allow the end points to vary within convex sets. Thus we allow $\mathbf{x}(0) \in \mathbf{X}^0$ a compact convex set and $\hat{\mathbf{x}} = (x_1, \ldots, x_n) \in \mathbf{X}^T$ a compact convex set (in one dimension less than $\mathbf{X}^0$). Then we get as an equivalent generalized linear program

$$\max_{\mathbf{P},\mathbf{S},\mathbf{T}} x_0$$

$$\text{subject to } x_0\mathbf{U}_0 + \mathbf{P}^u + \mathbf{S} + \mathbf{T} = 0 \tag{68}$$

$$\mathbf{P}^u \in \mathscr{P}$$

$$\mathbf{S} \in \mathscr{S}$$

$$\mathbf{T} \in \mathscr{T}$$

where $\mathbf{P}^u$, $\mathscr{P}$ are as previously defined, and

$$\mathbf{T} = (0, -x_1{}^{\mathrm{T}}, \ldots, -x_n{}^{\mathrm{T}})^{\mathrm{T}r} \in \{(0, -\hat{\mathbf{x}}) \mid \hat{\mathbf{x}} \in \mathbf{X}^{\mathrm{T}}\} = \mathscr{T}$$

and

$$\mathbf{S} = \boldsymbol{\Phi}(\mathbf{T})\mathbf{x}^0 \in \{\mathbf{S} \mid \mathbf{S} = \boldsymbol{\Phi}(\mathbf{T})\mathbf{x}^0, \mathbf{x}^0 \in \mathbf{X}^0\} = \mathscr{S}$$

This generalized program has three variable columns **P**, **S**, and **T**. Let the optimal dual variables be $\boldsymbol{\pi}^*$, $\boldsymbol{\mu}^*$ where $\boldsymbol{\mu}^* = (\mu_1{}^*, \mu_2{}^*, \mu_3{}^*)$. Then $\boldsymbol{\pi}^*$ and $\mu_1{}^*$ satisfy Eqs. (64) and (65) and $\boldsymbol{\pi}^*$, $\mu_2{}^*$, and $\mu_3{}^*$ satisfy

$$\boldsymbol{\pi}^*\mathbf{S} + \mu_2{}^* \geqslant 0 \qquad \text{for all} \quad \mathbf{S} \in \mathscr{S}$$

$$\boldsymbol{\pi}^*\mathbf{T} + \mu_3{}^* \geqslant 0 \qquad \text{for all} \quad \mathbf{T} \in \mathscr{T}$$

and

$$\boldsymbol{\pi}^*\mathbf{S}^* + \mu_2{}^* = 0 \qquad \text{for an optimal } \mathbf{S}^* \text{ in } \mathscr{S}$$

$$\boldsymbol{\pi}^*\mathbf{T}^* + \mu_3{}^* = 0 \qquad \text{for an optimal } \mathbf{T}^* \text{ in } \mathscr{T}$$

Thus

$$[\boldsymbol{\pi}^*\Phi(\mathbf{T})]\mathbf{x}^0 + \mu_2{}^* \geqslant 0 \qquad \text{for all} \quad \mathbf{x}^0 \in \mathbf{X}^0$$

and

$$[\boldsymbol{\pi}^*\Phi(\mathbf{T})]\mathbf{x}^{0*} + \mu_2{}^* = 0$$

for an optimum starting point $\mathbf{x}^{0*} \in \mathbf{X}^0$. Thus $[\boldsymbol{\pi}^*\boldsymbol{\Phi}(\mathbf{T})]$ and $\boldsymbol{\mu}^*$ determine a support for $\mathbf{X}^0$ at $\mathbf{x}^{0*}$. Similarly

$$-\pi^*(0, \hat{\mathbf{x}}) + \mu_3{}^* \geqslant 0 \qquad \text{for all} \quad \hat{\mathbf{x}} \in \mathbf{X}^T$$

and

$$-\boldsymbol{\pi}^*(0, \hat{\mathbf{x}}^*) + \mu_3{}^* = 0$$

for some optimal $\hat{\mathbf{x}}^* \in \mathbf{X}^{\mathrm{T}}$. These are the transversality conditions for this problem.[34]

The main advantage of this approach to linear optimal control problems is that one avoids the necessity of solving two-point boundary-value problems, which is a difficulty with many other methods of solving the same problem. Moreover we do not get involved in questions of normality.[1] To get the initial restricted master we solve the auxiliary problem (4), which is of interest in its own right.

The generalized program that we consider first is

$$\begin{aligned} \max_{\mathbf{P},\mathbf{s}^+,\mathbf{s}^-} \quad & x_0 \\ \text{subject to } \mathbf{P}^{\mathbf{u}} + \mathbf{I}\mathbf{s}^+ - \mathbf{I}\mathbf{s}^- &= \mathbf{b} \\ x_0 + \mathbf{e}\mathbf{s}^+ + \mathbf{e}\mathbf{s}^- &= 0 \\ \mathbf{P}^{\mathbf{u}} &\in \mathscr{P} \end{aligned} \tag{69}$$

where $\mathbf{P}^{\mathbf{u}} = \int_0^T \mathbf{\Phi}(T)\mathbf{\Phi}^{-1}(t)\mathbf{B}(t)\mathbf{u}(t)\,dt$, $\mathbf{b} = \mathbf{x}^T\mathbf{\Phi}(T)\,\mathbf{x}^0$, $\mathbf{s}^+$ and $\mathbf{s}^-$ are m-vectors, $\mathbf{e} = (1,\ldots, 1)$, and thus $-x^0 = \sum \mathbf{s}_i^+ + \sum \mathbf{s}_i^-$. Then $-x^0$ may be interpreted as the l_1 norm of the distance of the system state at time T from $\mathbf{x}^T$.

If $\mathbf{x}^T$ is an *interior* point of the set of reachable states

$$\mathbf{S}_T = \left\{\mathbf{x} \mid \mathbf{x} = \mathbf{\Phi}(T)\mathbf{x}^0 + \int_0^T \mathbf{\Phi}(T)\mathbf{\Phi}^{-1}(t)\mathbf{B}(t)\mathbf{u}(t)\,dt\right\}$$

then the generalized linear program reaches an optimal solution of zero in a finite number of steps. This gives us a starting basis for Eq. (63). Details and computational experience with such an algorithm is given in Jizmagian.[24] Note that solving the subproblems (63) and (69) does not require satisfying of the terminal conditions of lying on the line $x_i(T) = x_i(T)$ for $i = 1,\ldots, n$, x_0 arbitrary. This is guaranteed in the solution of the *master* problem.

3-3-1 Problems with Convex Objectives

As another generalization we can consider the case of a convex objective function

$$\begin{aligned} \min_{\mathbf{u}} \int_0^T & c(\mathbf{x}, \mathbf{u}, t)\,dt \\ \text{subject to } \frac{d}{dt}\mathbf{x}(t) &= \mathbf{A}(t)\mathbf{x}(t) + \mathbf{B}(t)\mathbf{u}(t) \\ \mathbf{u}(t) &\in \mathbf{U}(t) \qquad t \in [0, T] \\ \mathbf{x}(0) &= \mathbf{x}^0 \\ \mathbf{x}(T) &= \mathbf{x}^T \end{aligned} \tag{70}$$

[1] Normality is a condition which essentially guarantees that optimal solutions are unique. For a further discussion of this point see Lee and Markus,[32] especially p. 76.

where $\mathbf{x}$ is now an n-vector $(x_1, \ldots, x_n)$ and $c(\mathbf{x}, \mathbf{u}, t)$ is convex in $\mathbf{x}$ and $\mathbf{u}$ for all t. As before we formulate our problem as a generalized linear program. We define in $n + 1$ space the convex set

$$\mathscr{P} = \left\{ (y_0, \mathbf{y}) \mid y_0 \geqslant \int_0^T c(\mathbf{x}, \mathbf{u}, t)\, dt \qquad \mathbf{y} = \mathbf{x}(T) - \mathbf{\Phi}(T)\mathbf{x}^0 \right.$$

$$\left. \mathbf{x}(T) = \mathbf{\Phi}(T)\mathbf{x}^0 + \int_0^T \mathbf{\Phi}(T)\mathbf{\Phi}^{-1}(s)\mathbf{B}(s)\mathbf{u}(s)\, ds \qquad \mathbf{u}(t) \in \mathbf{U}(t) \right\}$$

The corresponding generalized linear program is

$$\begin{aligned} &\max_{\mathbf{P}} x_0 \\ &\text{subject to } \mathbf{U}_0 x_0 + \mathbf{P} = \mathbf{b} \\ &\qquad \mathbf{P} \in \mathscr{P} \quad \text{where} \quad \mathbf{b} = \begin{pmatrix} 0 \\ \mathbf{x}^T - \mathbf{\Phi}(T)\mathbf{x}^0 \end{pmatrix} \end{aligned} \tag{71}$$

The master program is

$$\begin{aligned} &\max x_0 \\ &\text{subject to } \mathbf{U}_0 x_0 + \sum_i \lambda_i \mathbf{P}^i = \mathbf{b} \\ &\qquad \sum_i \lambda_i = 1 \\ &\qquad \lambda_i \geqslant 0 \end{aligned}$$

assuming that $\mathscr{P}$ is bounded. If the dual variables are $\boldsymbol{\pi} = (1, \hat{\boldsymbol{\pi}})$ and δ at some iteration then the subproblem is

$$\begin{aligned} \min_{\mathbf{u}} &\left\{ \int_0^T \mathbf{c} \left[\mathbf{\Phi}(t)\mathbf{x}^0 + \int_0^t \mathbf{\Phi}(t)\mathbf{\Phi}^{-1}(s)\mathbf{B}(s)\mathbf{u}(s)\, ds, \mathbf{u}(t), t \right] dt \right. \\ &\left. + \boldsymbol{\pi} \int_0^T \mathbf{\Phi}(T)\mathbf{\Phi}^{-1}(s)\mathbf{B}(s)\mathbf{u}(s)\, ds + \delta \right\} \\ &\text{subject to } \mathbf{u}(t) \in \mathbf{U}(t) \qquad t \in [0, T] \end{aligned} \tag{72}$$

This in itself is a rather difficult problem for which a special algorithm is required, and so we will not go into more detail here. We simply observe that in the important special case where $\mathbf{c}(\mathbf{x}, \mathbf{u}, t)$ does not depend on $\mathbf{x}$ Eq. (72) can at least conceptually be optimized pointwise in t. That is, for each time $t \in [0, T]$ we have the convex programming problem

$$\begin{aligned} &\min_{\mathbf{u}} \mathbf{c}(\mathbf{u}(t), t) + \boldsymbol{\pi}\mathbf{\Phi}(T)\mathbf{\Phi}^{-1}(t)\mathbf{B}(t)\mathbf{u}(t) + \delta \\ &\text{subject to } \mathbf{u}(t) \in \mathbf{U}(t) \end{aligned} \tag{73}$$

Simplifications accumulate if respectively $\mathbf{U}(t)$ is polyhedral, $\mathbf{A}(t)$, $\mathbf{B}(t)$ are constant, and $\mathbf{c}(\mathbf{u}(t), t) = \mathbf{u}(t)\, \underset{\sim}{Q}\mathbf{u}(t)$ is positive semidefinite. In the simplest case Eq. (14) is a quadratic program for each t with an objective function with fixed quadratic part and an exponentially varying linear part. The analysis in this case has been carried out by Jizmagian.[24]

So at least formally we can attack the problem (3).

3-3-2 Mathematical Justification of Formal Steps

Now let us review the formal analysis and give sufficient conditions so that the steps can be carried out. The first step we wish to justify is the integral form of the differential equation (58). Moreover we will require that for any bounded time interval the solution $\mathbf{x}(t)$ will be uniquely defined and bounded for any fixed $\mathbf{u}$ we allow. Sufficient conditions[32] for this are

ASSUMPTION 1:

$\mathbf{A}(t)$ and $\mathbf{B}(t)$ are measurable on $[0, T]$ and their norms are integrable.

ASSUMPTION 2:

$\mathbf{u}(t)$ is bounded and measurable on $[0, T]$. We of course require that the differential equations appearing need only be satisfied on a subset of $[0, T]$ of full measure. Similarly unless otherwise stated all inequalities and equations need only hold almost everywhere.

Next in order for the theory of generalized linear programming to apply $\mathscr{P}$ in Eq. (60) must be closed, convex, and compact enough. For simplicity here we abandon the generality of the compact enough assumption and give conditions sufficient to imply that $\mathscr{P}$ is a compact convex set.

ASSUMPTION 3:

$\mathbf{U}(t)$ is convex and compact for each t and continuous in t with respect to the Hausdorf metric.[1]

Assumption 3 guarantees that $\mathscr{P}$ is closed and compact. A proof can be based on a result in Lee and Markus,[32] p. 69, which is stated for $\mathbf{U}(t)$ constant. The modification required for time varying $\mathbf{U}(t)$ is to generalize Lemma 1A of Lee and Markus,[32] p. 157. Since $\mathbf{U}(t)$ is continuous $\mathbf{U}(t)$ is uniformly contained in a compact convex set $\bar{\mathbf{U}}$. Since $\{\mathbf{u} \mid \mathbf{u}(t) \in \mathbf{U}(t)\}$ is weakly closed Lemma 1A is valid for $\mathbf{U}(t)$, a continuous function of time (Dunford and Schwartz,[15] p. 422).

[1] The distance between two sets $\mathbf{A}$ and $\mathbf{B}$ in the Hausdorf metric is given by $d(\mathbf{A}, \mathbf{B}) = \sup\,[d(\mathbf{x}, \mathbf{B}), d(\mathbf{y}, \mathbf{A}) \mid \mathbf{x} \in \mathbf{A}, \mathbf{y} \in \mathbf{B}]$ where $d(\mathbf{x}, B)$ is the ordinary Euclidean distance of the point $\mathbf{x}$ from the set $\mathbf{B}$.

The next operation we wish to justify is the pointwise solution of Eq. (66) to obtain $\mathbf{u}^*$. For this we need no additional assumptions. The relevant proofs are formed in Lee and Markus[32] in the appendix to chap. 1, especially in Lemmas 2A and 3A. Again simple modifications must be made for $\mathbf{U}(t)$ time varying.

To be able to carry out the generalizations it suffices to require that

ASSUMPTION 4:

$\mathbf{X}^0$ and $\mathbf{X}^T$ are compact and convex.

ASSUMPTION 5:

$c(\mathbf{x}, \mathbf{u}, t)$ is jointly convex and continuous in $\mathbf{u}$ and $\mathbf{x}$ for each t and integrable in t.

In order to use the direct approach to solving optimal control problems one must be able to calculate the fundamental matrix solution $\mathbf{\Phi}(\mathbf{t})$ given by Eq. (59). Also one must in some way carry out the pointwise optimization of Eqs. (66) and (72). If $\mathbf{A}$ and $\mathbf{B}$ are constant and $\mathbf{U}(t)$ is constant and polyhedral this can be carried out analytically.[24] In more general cases one usually resorts to the digital computer. In most cases this involves discretization. In the next section we study approaches resulting from direct discretization of the optimal control problem.

3-4 SOLUTION OF CONTINUOUS-TIME LINEAR OPTIMAL PROBLEMS BY DISCRETIZATION

We will build up the generality of the optimal control problems we can solve by mathematical programming in steps. First, however, we illustrate the *formal* steps one might use in discretizing a system of the form (1) to get a difference equation problem of the form (2). We divide the interval $[0, T]$ into K parts each of length $\Delta = T/K$. We first make the approximation[1]

$$\frac{d\mathbf{x}(k\Delta)}{dt} - \frac{\mathbf{x}^{k+1} - \mathbf{x}^k}{\Delta} \tag{74}$$

[1] Two comments should be made at this point. First, it is known that for certain linear optimal control problems the discretization is invalid in the sense that as $\Delta \to 0$ the solution of the discrete problems can converge to something which is not a solution of the continuous problem.[8] The question of the validity of discretization is not completely settled; however, a résumé of the known results is given in Sec. 3-5. We will assume that the discretization is valid. Second, the approximation (74) does not lead to the best numerical integration scheme; however, the implementation of more sophisticated integration procedures[37] requires only minor modifications to the theory.

where we denote $\mathbf{x}(k\Delta)$ by $\mathbf{x}^k$ and denote by $\mathbf{U}^k$ the set $\mathbf{U}(k\Delta)$. Then $\mathbf{x}^{k+1} \cong \mathbf{A}^k\mathbf{x}^k + \mathbf{B}^k\mathbf{u}^k$, $k = 0, 1, \ldots, K-1$, where $\mathbf{A}^k = [\mathbf{I} + \Delta\mathbf{A}(k\Delta)]$, $\mathbf{B}^k = \Delta\mathbf{B}(k\Delta)$, and Eq. (1) is approximated by a system of the form (2).

We now consider a very simple problem:

$$\begin{aligned} &\min x_0(T) \\ &\underset{\mathbf{u}}{\text{subject to}}\ \frac{d\mathbf{x}}{dt} = \mathbf{A}(t)\mathbf{x}(t) + \mathbf{B}(t)\mathbf{u}(t) \\ &\mathbf{x}(0) = \mathbf{x}^0 \\ &\mathbf{x}(T) = [x_0(T), \mathbf{x}^T] \qquad \mathbf{x}^T \text{ given} \\ &\mathbf{u}(t) \geqslant \mathbf{0} \end{aligned} \tag{75}$$

Instead of treating Eq. (75) we work on its discrete analog

$$\begin{aligned} &\min_{\mathbf{u}} x_0{}^T \\ &\text{subject to } \mathbf{x}^{k+1} = (\mathbf{I} + \Delta\mathbf{A}^k)\mathbf{x}^k + \Delta\mathbf{B}^k\mathbf{u}^k \qquad k = 0, 1, \ldots, K-1 \\ &\mathbf{x}^0 = \mathbf{x}^0 \\ &\mathbf{x}^K = (x_0{}^T, \mathbf{x}^T) \\ &\mathbf{u}^k \geqslant \mathbf{0} \qquad k = 0, 1, \ldots, K-1 \end{aligned} \tag{76}$$

For a given function $\mathbf{u}(t)$ the solution of Eq. (75) is

$$\mathbf{x}(t) = \mathbf{\Phi}(t; 0)\mathbf{x}(0) + \int_0^t \mathbf{\Phi}(t; \tau)\mathbf{B}(\tau)\mathbf{u}(\tau)\, d\tau \tag{77}$$

where the *state transition matrix* $\mathbf{\Phi}(t; \tau)$ is the matrix solution of

$$\frac{d}{dt}\mathbf{\Phi}(t; \tau) = \mathbf{A}(t)\mathbf{\Phi}(t; \tau)$$

with the initial condition

$$\mathbf{\Phi}(\tau; \tau) = \mathbf{I}$$

Similarly, the solution of the difference equation in Eq. (76) is

$$\mathbf{x}^k = \mathbf{\Phi}_{k,0}\mathbf{x}^0 + \sum_{\tau=1}^{k} \mathbf{\Phi}_{k,\tau}B^{\tau-1}\mathbf{u}^{\tau-1} \tag{78}$$

where

$$\begin{aligned} \mathbf{\Phi}_{k,\tau} &= (\mathbf{I} + \Delta\mathbf{A}^{k-1})(\mathbf{I} + \Delta\mathbf{A}^{k-2}) \cdots (\mathbf{I} + \Delta\mathbf{A}^{\tau}) \\ \mathbf{\Phi}_{\tau,\tau} &= \mathbf{I} \end{aligned}$$

satisfies the backward difference equation

$$\mathbf{\Phi}_{k+1,\tau} = (\mathbf{I} + \Delta\mathbf{A}^k)\Phi_{k,\tau}$$
$$\mathbf{\Phi}_{\tau,\tau} = 1$$

Now if we make the substitutions

$$\mathbf{b} = \begin{pmatrix} 0 \\ \mathbf{x}^T \end{pmatrix} - \mathbf{\Phi}_{K,0}\mathbf{x}^0$$
$$\tilde{\mathbf{A}} = [\mathbf{\Phi}_{K,1}\mathbf{B}^0, \mathbf{\Phi}_{K,2}\mathbf{B}^1, \ldots, \mathbf{\Phi}_{K,K}\mathbf{B}^{K-1}]$$
$$\mathbf{u} = (\mathbf{u}^0, \mathbf{u}^1, \ldots, \mathbf{u}^K)$$
$$Z = -\mathbf{x}_0{}^K$$

then Eq. (76) is equivalent to

$$\begin{aligned} &\max Z \\ &\text{subject to } \mathbf{U}_0 Z + \tilde{\mathbf{A}}\mathbf{u} = \mathbf{b} \\ &\mathbf{u} \geqslant 0 \end{aligned} \tag{79}$$

which is a linear program.

Notice that the number of equations in Eq. (79) depends only on the dimension of the state vector and does not depend on the number of time divisions K. Since the computation required in the simplex method depends much more strongly on the number of equations than on the number of columns we can use a relatively large number of time divisions at relatively small computational cost. We will in all cases try to preserve this advantage.

At the optimal solution we have a vector $\boldsymbol{\pi}$ of dual variables satisfying

$$\begin{aligned} \boldsymbol{\pi}\mathbf{U}_0 &= \boldsymbol{\pi}_0 = 1 \qquad & \boldsymbol{\pi}\tilde{\mathbf{A}}\mathbf{u} &= 0 \\ \boldsymbol{\pi}\tilde{\mathbf{A}} &\leqslant 0 & \boldsymbol{\pi}\mathbf{b} &= \max Z \end{aligned} \tag{80}$$

Interpreting Eq. (80) in terms of the original system, we have

$$\boldsymbol{\pi}\tilde{\mathbf{A}} \leqslant 0 \qquad \text{implies} \quad \boldsymbol{\pi}\mathbf{\Phi}_{K,\tau}\mathbf{B}^{\tau-1} \leqslant 0$$

and equality holds for each element j corresponding to $\mathbf{u}_j^{\tau-1} > 0$. Notice that $\mathbf{\Phi}_{K,\tau}$ is a matrix solution of the backward adjoint difference equation

$$\mathbf{\Phi}_{K,\tau} = \mathbf{\Phi}_{K,\tau+1}(\mathbf{I} + \Delta\mathbf{A}^\tau) \qquad \mathbf{\Phi}_{K,K} = \mathbf{I}$$

and $\boldsymbol{\pi}\mathbf{\Phi}_{K,\tau}$ is a vector solution of the same equation. Let $\phi(\tau) = \boldsymbol{\pi}\mathbf{\Phi}_{K,\tau}$,

$\mathbf{v}^\tau = \mathbf{B}^\tau \mathbf{u}^\tau$, and $\mathbf{V}^\tau = \{\mathbf{v} \mid \mathbf{v} = \mathbf{B}^\tau \mathbf{u}, \mathbf{u} \geqslant \mathbf{0}\}$. If $\mathbf{u}^*$ is an optimal solution to Eq. (6) there exists $\boldsymbol{\pi}^*$ such that $\pi_0{}^* = 1$ and

$$\boldsymbol{\phi}^*(\tau)\mathbf{v}^{*\tau} \geqslant \boldsymbol{\phi}^*(\tau)\mathbf{v}^\tau \qquad \text{for all} \quad \mathbf{v}^\tau \in \mathbf{V}^\tau \tag{81}$$

where $\boldsymbol{\phi}^*(\tau) = \boldsymbol{\pi}^*\boldsymbol{\Phi}_{K,\tau}$ satisfies the adjoint system

$$\boldsymbol{\phi}(\tau) = \boldsymbol{\phi}(\tau + 1)(\mathbf{I} + \Delta \mathbf{A}^\tau)$$
$$\boldsymbol{\phi}(K) = \boldsymbol{\pi}^*$$

This is the maximum principle[25] for Eq. (76).

3-4-1 The Bang-Bang Control Problem

We next consider a slightly more complicated problem. Suppose we replace the constraint $\mathbf{u}^k \geqslant 0$ by $\underline{\mathbf{u}}^k \leqslant \mathbf{u}^k \leqslant \bar{\mathbf{u}}^k$, $k = 0,..., K - 1$. $\underline{\mathbf{u}}^k$ and $\bar{\mathbf{u}}^k$ are respectively vectors of lower and upper bounds on the elements of $\mathbf{u}^k$. If we make the same substitutions as before we arrive at Eq. (79) with $\underline{\mathbf{u}} \leqslant \mathbf{u} \leqslant \bar{\mathbf{u}}$ replacing $\mathbf{u} \geqslant 0$, $\underline{\mathbf{u}} = (\underline{\mathbf{u}}^0,..., \underline{\mathbf{u}}^{K-1})$, $\bar{\mathbf{u}} = (\bar{\mathbf{u}}^0,..., \bar{\mathbf{u}}^{K-1})$. If we then make the substitutions

$$\tilde{u}_i{}^k = \frac{u_i{}^k - \underline{u}_i{}^k}{\bar{u}_i{}^k - \underline{u}_i{}^k}$$

then $0 \leqslant \tilde{\mathbf{u}}_i{}^k \leqslant 1$ and we have a linear program with upper-bounded variables. We can then apply the simplex method modified to handle upper-bounded variables. The fact that an optimal basic solution exists yields a switching point theorem. Namely, there exists an optimal solution with at most n variables $u_i{}^k$ whose values are not at the extremes plus one or minus one. For this reason the problem is often called the *bang-bang problem*. The remarkable fact is that this is independent of the fineness of the grid, i.e., independent of K in $\Delta = T/K$.

Applying the duality theorem in this case, again we obtain a maximum principle.[25] From Eq. (20) it follows that at the optimum we obtain a vector $\boldsymbol{\pi} = \pi_0, ..., \pi_m$ of dual variables satisfying

$$\pi_0 = 1$$

$$[\boldsymbol{\pi}\boldsymbol{\Phi}_{K,k}\mathbf{B}^{k-1}]_i = 0 \qquad \text{for} \quad \underline{u}_i^{k-1} < u_i^{k-1} < \bar{u}_i^{k-}$$

$$[\boldsymbol{\pi}\boldsymbol{\Phi}_{K,k}\mathbf{B}^{k-1}]_i \leqslant 0 \qquad \text{for} \quad u_i^{k-1} = \underline{u}_i^{k-1}$$

$$[\boldsymbol{\pi}\boldsymbol{\Phi}_{K,k}\mathbf{B}^{k-1}]_i \geqslant 0 \qquad \text{for} \quad u_i^{k-1} = \bar{u}_i^{k-1}$$

Thus if we let $\boldsymbol{\phi}(k) = \boldsymbol{\pi}\boldsymbol{\Phi}_{K,k}$ as before we obtain $\boldsymbol{\phi}^*(k)\, \mathbf{v}^{*k} \geqslant \boldsymbol{\phi}^*(k)\, \mathbf{v}^k$

for all $\mathbf{v}^k \in \mathbf{V}^k = \{\mathbf{v} \mid \mathbf{v} = \mathbf{B}^k\mathbf{u}^k,\ \underline{\mathbf{u}}^k \leqslant \mathbf{u}^k \leqslant \bar{\mathbf{u}}^k\}$, which is the maximum principle for this problem.

As our next generalization we consider the case where $\mathbf{u}^k \in \mathbf{U}^k$, where $\mathbf{U}^k$ is a convex polyhedral set defined by $\mathbf{F}^k\mathbf{u}^k \geqslant \mathbf{f}^k$. Then the appropriate linear program becomes

$$\max Z$$

$$\text{subject to } \mathbf{U}_0 Z + \sum \mathbf{\Phi}_{K,k}\mathbf{B}^{k-1}\mathbf{u}^{k-1} = \begin{pmatrix} 0 \\ \mathbf{x}^T \end{pmatrix} - \mathbf{\Phi}_{K,0}\mathbf{x}^0 \tag{82}$$

$$\mathbf{F}^k\mathbf{u}^k \geqslant \mathbf{f}^k \qquad k = 0, 1, \ldots, k-1$$

The structure of the matrix of coefficients, which is represented schematically in Fig. 3-4, is such that the decomposition algorithm can

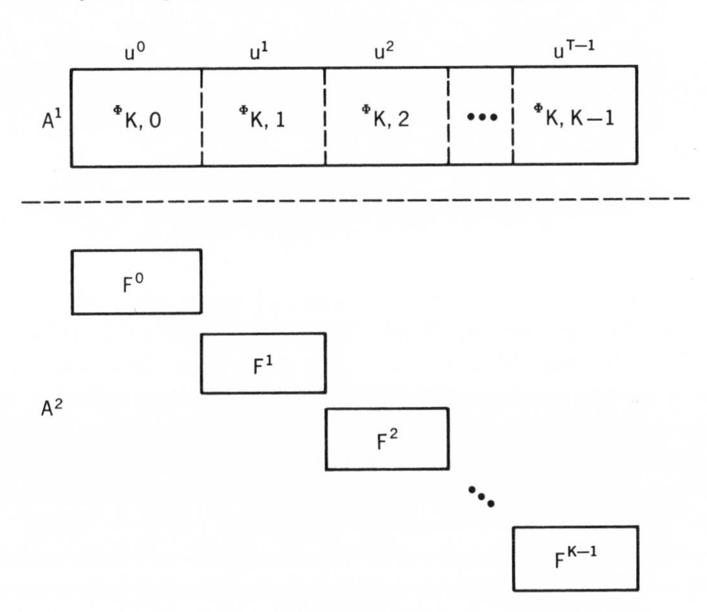

FIG. 3-4.

be applied. The subproblem corresponds to satisfying control vector constraints. It breaks into K separate subproblems which are solved separately. Note that the master program will consist of $n + 2$ rows independently of the value of K. Thus if our original problem arises as the result of a discretization of a continuous problem, the size of the grid, that is, K, will not affect the number of rows in the master problem. This property is very desirable since the effectiveness of the decomposition procedure appears to depend strongly on the number of rows in the master problem. As we saw in Sec. 3-2-5, the decomposition algorithm is finite. If $\mathbf{U}^k$ is merely a closed convex set rather than a convex polyhedral set, the same type of algorithm may be used but the algorithm while converging under rather general assumptions may require an infinite number of steps.

3-4-2 Problems with State Space Constraints

Next we consider the addition of state space restrictions. We restrict ourselves to the situation for which $\mathbf{x}^k$ is required to lie in a convex polyhedral set; i.e.,

$$\mathbf{G}^k\mathbf{x}^k \leqslant \mathbf{g}^k \qquad k = 1,\dots, T-1 \tag{83}$$

To simplify the description of the appropriate algorithm we assume that the control vectors $\mathbf{u}^k$ are only required to be nonnegative; this is no essential limitation. The restrictions (83) are implicit in the sense that they are expressed in terms of the state variables $\mathbf{x}^k$ rather than in terms of the independent variables $\mathbf{u}^k$. To express the bounds (83) in terms of the $\mathbf{u}^k$, Eq. (78) must be utilized. Thus $\mathbf{G}^k\mathbf{x}^k \leqslant \mathbf{g}^k$ can be expressed in terms of the control $\mathbf{u}$ as

$$\mathbf{G}^k\left(\mathbf{\Phi}_{k,0}\mathbf{x}^0 + \sum_{\tau=1}^{k} \mathbf{\Phi}_{k,\tau}\mathbf{B}^{\tau-1}\mathbf{u}^{\tau-1}\right) \leqslant \mathbf{g}^k$$

or

$$\sum_{\tau=1}^{k} (\mathbf{G}^k\Phi_{k,\tau}\mathbf{B}^{\tau-1})\mathbf{u}^{\tau-1} \leqslant (\mathbf{g}^k - \mathbf{G}^k\mathbf{\Phi}_{t,0}\mathbf{x}^0) \tag{84}$$

The schematic representation of the constraint matrix after this is carried out is depicted in Fig. 3-5.

Here decomposition methods are not immediately applicable and while basis partition methods[14] might offer some help, we turn to another approach.[41] If we assume on physical grounds that the state space constraints will rarely if ever be violated then a dual decomposition method is appropriate. In this case, the matrix $\mathbf{G}$ in Eq. (42) corresponds to the identity matrix of slack variables for the inequality constraints (84). Since the slack variables have no cost, d in Eq. (42) is zero. As we noted in Sec. 3-2-8, this simplifies the algorithm considerably. The dual decomposition algorithm applied to the state constrained optimal problem becomes: first solve the problem ignoring the state space constraints; if the resulting solution happens to satisfy the state constraints the solution is optimal for the problem with state space constraints; if not we add the *sum* of the *violated* constraints of Eq. (84) and solve the optimal control with the additional constraint. If this satisfies the state space constraints, we are finished and so on. This is a finite process leading to the optimal solution. Notice that the added

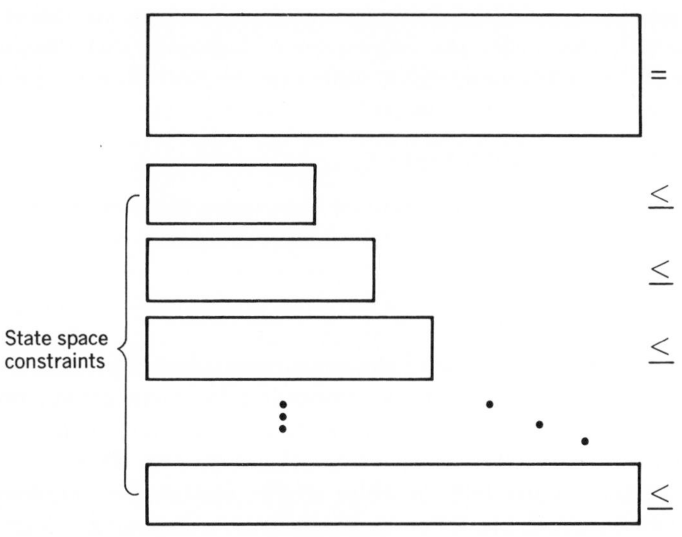

FIG. 3-5.

constraints have the essential feature that the current solution does not satisfy the added constraint but any control feasible for the original problem does.

All the problems we have considered in Sec. 3-3 have involved linear objectives. If convex nonlinear objective functions are considered the convex programming algorithm described in Sec. 3-2-7 can be used as well as many others.[1,6,19,26,46] A particular case for which special methods are available is for problems with quadratic objectives. For these problems, quadratic programming algorithms can be used.[30]

3-5 FINAL COMMENTS

Computational results using mathematical programming techniques to solve optimal control problems have been reported by Jizmagian,[24] Canon and Cullum,[5] and Rosen.[35,36]

Solving a continuous optimal control problem on a digital computer requires at some stage a discretization process. For problems with linear dynamics one can discretize the differential equation as in Sec. 3-4 or one can discretize the solution formula (58). For nonlinear problems only the first option is available (another approach often used is to linearize about the current trajectory to find an improved one; in this case the linearized problem can be discretized in both ways). Unfortunately, the validity of either approach is not clear in general. That is, it is not necessarily true that the solutions of the discrete problems converge to a solution of the continuous problem as the discretization mesh gets smaller. For an example see Cullum.[8] The discretization problem is closely connected to the question of whether an optimal control problem is "well posed," i.e., whether the solution is stable with respect to small perturbations in the data. Krasovskii[29] and then Neustadt[33] proved theorems establishing convergence for linear time optimal systems where the solution equation (58) is discretized. This is equivalent to solving the continuous problem restricted to using piecewise constant controls. Tyndall[38] established convergence of the solutions of the discretized difference equation approximations to an optimal solution for a very special class of linear optimal control problems, continuous linear programs; related results appear in Tyndall[39] and Grinold.[18] However, the strongest results seem to be those due to Cullum.[7–10] They stem from a study of how well posed an optimal control problem is (Kirillova;[27,28] see also Lee and Markus,[32] p. 416). The results of most interest to us here are that for problems with completely controllable dynamics linear in the state and control and with convex objective either approximation works. For problems linear in the control but not in the state if the values of the cost function for the difference equation approximations approach the cost of continuous problems, there is a convergent subsequence of controls and trajectories converging to the optimal ones (Cullum[8]).

Finally, we consider one last significant difference between continuous and discrete optimal control problems. Suppose the system dynamics for a continuous problem are $\dot{\mathbf{x}} = f(\mathbf{x}, \mathbf{u}, t)$, where the allowable controls are all measurable functions $\mathbf{u}(t)$ such that $\mathbf{u}(t) \in \mathbf{U}$ for almost all t where $\mathbf{U}$ is convex and compact. Let $\mathbf{R}(\mathbf{x}, t) = \{f(\mathbf{x}, \mathbf{u}, t) \mid \mathbf{u} \in \mathbf{U}\}$. Then under certain regularity conditions on f, Varaiya[42] has shown that the set of points reachable from $\mathbf{x}^0$ at time to using admissible controls is closed if and only if $R(\mathbf{x}, t)$ is convex for all t and every reachable $\mathbf{x}$. Suppose now that $\mathbf{R}(\mathbf{x}, t)$ is compact but not necessarily convex; for example, $\mathbf{U}$ may be a finite number of isolated points. Let $\hat{\mathbf{R}}(\mathbf{x}, t)$ be the closed convex hull of $\mathbf{R}(\mathbf{x}, t)$. Then the points reachable using velocities in $\mathbf{R}(\mathbf{x}, t)$ are dense in the points reachable using $\hat{\mathbf{R}}(\mathbf{x}, t)$; in fact, for linear problems the two sets are the same (see Lee and Markus,[32] p. 164). The reason for this can easily be seen in at least a heuristic sense. In order to achieve the same results as a velocity trajectory in $\hat{\mathbf{R}}(\mathbf{x}, t)$ by using velocities in $R(\mathbf{x}, t)$, one can oscillate rapidly between various points in $\mathbf{R}(\mathbf{x}, t)$ so that the velocity "averages" out to the desired trajectory in $\hat{\mathbf{R}}(\mathbf{x}, t)$. However, for discrete problems this averaging cannot be carried out since we have only as many "oscillations" available as there are time periods in the discretization. All the methods discussed here depend in some way or other on the convexity of the problems.

For problems which are inherently discrete, we may not have the required convexity; however, for discrete problems arising from approximations of continuous control problems, we can assume that the velocity space is convex due to the averaging characteristic of the approximated continuous problem. For a more detailed discussion of this see Halkin.[20,21]

REFERENCES:

1. Abadie, J. (ed.): "Nonlinear Programming," Wiley, 1967.
2. Balinskii, M. L.: An Algorithm for Finding All Verticles of Convex Polyhedral Sets, *SIAM J.*, vol. 9, no. 1, pp. 72–88, 1961.
3. Benders, J. F.: Partitioning Procedures for Solving Mixed-variables Programming Problems, *Numer. Math.*, vol. 4, pp. 238–252, 1962.
4. Broise, P., P. Huard, and J. Sentenac: "Décomposition des programmes mathématiques," Dunod, 1968.
5. Canon, M., and J. Cullum: The Numerical Solution of Linear Continuous Control Problems by Discrete Approximations, to appear.
6. Cheney, E. W., and A. A. Goldstein: Newton's Method for Convex Programming and Tchebycheff Approximation, *Numer. Math.*, vol. 1, pp. 253–268, 1959.
7. Cullum, J.: Perturbations of Optimal Control Problems, *SIAM J. Control*, vol. 4, no. 3, 1966.
8. Cullum, J.: Discrete Approximations to Continuous Optimal Control Problems, *IBM Watson Res. Cen. Rept.* RC 1858, Yorktown Heights, N. Y., 1967. *SIAM J. Control*, vol. 17, no. 1, 1969.
9. Cullum, J.: Perturbation and Approximation of Continuous Optimal Control Problems, in A. V. Balakrishnan and L. W. Neustadt (eds.), "Mathematical Theory of Control," Academic, 1967.
10. Cullum, J.: An explicit procedure for discretizing continuous optimal control problems, *IBM Watson Res. Cen. Rept.* RC 2705, Yorktown Heights, N. Y., 1969.
11. Dantzig, G. B.: "Linear Programs and Extensions," Princeton University Press, 1963.
12. Dantzig, G. B.: Linear Control Processes and Mathematical Programming, *SIAM J. Control*, vol. 4, 1966.
13. Dantzig, G. B., W. P. Blattner, and M. R. Rao: Finding a Cycle in a Graph with Minimum Cost to Time Ratio with Application to a Ship Routing Problem, in "Théorie de graphes," Dunod, 1967.
14. Dantzig, G. B.: Large Scale Linear Programming, in "Mathematics in the Decision Sciences," vol. 11 of "Lectures in Applied Mathematics," American Mathematical Society, 1968.
15. Dunford, N., and J. T. Schwartz: "Linear Operators," pt. I, Interscience, 1957.
16. Ford, L. R., and D. R. Fulkerson: "Flows in Networks," Princeton University Press, 1962.
17. Gomory, R. E.: Large and Nonconvex Problems in Linear Programming, *Proc. Symp. Interact. between Math. Res. High Speed Computing*, American Mathematical Society, vol. 15, 1963.
18. Grinold, R.: "Continuous Programming," doctoral dissertation, ORC 68-14, Operations Research Center, University of California, Berkeley, 1968.
19. Hadley, G.: "Nonlinear and Dynamic Programming," Addison-Wesley, 1964.
20. Halkin, H.: Optimal Control for Systems Described by Difference Equations, in C. T. Leondes (ed.), "Advances in Control Systems," pp. 173–196, Academic, 1964.

21. Halkin, H.: A Maximum Principle of the Pontryagin Type for Systems Described by Nonlinear Difference Equations, *SIAM J. Control*, vol. 4, no. 1, 1966.
22. Hellerman, E.: "Large Scale Linear Programs: Theory and Computation," C.E.I.R., Bethesda, Md.
23. Hirsch, W. M., and A. J. Hoffman: Extreme Varieties, Concave Functions, and the Fixed Charge Problem, *Commun. Pure Appl. Math.*, vol. 14, pp. 355–369, 1961.
24. Jizmagian, S.: "Generalized Programming Solution of Continuous Time Linear System Optimal Control Problems," doctoral dissertation, Operations Research Department, Stanford University, Stanford, Calif., 1968.
25. Jordan, B. W., and E. Polak: Optimal Control Aperiodic Discrete-time Systems, *SIAM J. Control*, vol. 2, no. 3, pp. 332–346, 1964.
26. Kelley, J. E.: The Cutting-plane Method for Solving Convex Programs, *SIAM J.*, vol. 8, pp. 703–712, 1960.
27. Kirillova, F. M.: On the Correct Statement of One Problem on Optimal Regulation, *Izv. Vysshikh a Zavedenii Mat.*, vol. 4, no. 5, pp. 113–126, 1958.
28. Kirillova, F. M.: On the Correctness of the Formulation of an Optimal Control Problem, *SIAM J. Control*, vol. 1, pp. 224–239, 1963.
29. Krasovskii, N. N.: On an Optimal Control Problem, *Prikl. Matic. Mekh.*, vol. 21, no. 5, pp. 670–677, 1957.
30. Kunzi, H. P., and W. Krelle: "Nonlinear Programming," Ginn and Blaisdell, 1966.
31. Kunzi, H. P., H. G. Tzschach, and C. A. Zehnder: "Numerical Methods of Mathematical Optimization," Academic, 1968.
32. Lee, E. B., and L. Markus: "Foundations of Optimal Control Theory," Wiley, 1968.
33. Neustadt, L.: Discrete Time Optimal Control Systems, in J. P. LaSalle and S. Lefshetz (eds.), "Nonlinear Differential Equations and Nonlinear Mechanics," Academic, 1963.
34. Pontryagin, L. S., V. G. Boltyanskii, R. V. Gamkrelidze, and E. F. Mishenko: "The Mathematical Theory of Optimal Processes," Interscience, 1962.
35. Rosen, J. B.: Iterative Solution of Nonlinear Optimal Control Problems, *SIAM J. Control*, vol. 4, no. 1, 1966.
36. Rosen, J. B.: Optimal Control and Convex Programming, in J. Abadie (ed.), "Nonlinear Programming," Wiley, 1967.
37. Todd, John (ed.): "Survey of Numerical Analysis," McGraw-Hill, 1962.
38. Tyndall, W. F.: A Duality Theorem for a Class of Continuous Linear Programming Problems, *SIAM J. Appl. Math.*, vol. 13, pp. 644–666, 1965.
39. Tyndall, W. F.: An Extended Duality Theorem for Continuous Linear Programming Problems, *SIAM J. Appl. Math.*, vol. 15, pp. 1294–1298, 1967.
40. Van Slyke, R. M.: Mathematical Programming and Optimal Control, *Univ. Calif. Operations Res. Cen. Rept.* ORC 69-21, Berkeley, 1968.
41. Van Slyke, R. M., and R. Wets: *L*-Shaped Linear Programs with Applications to Optimal Control and Stochastic Programming, *SIAM J. Control*, vol. 17, no. 4, pp. 638–663, 1969.
42. Varaiya, P.: On the Trajectories of a Differential System, in A. V. Balakrishnan and L. W. Neustadt (eds.), "Mathematical Theory of Control," Academic, 1967.
43. Whinston, A.: A Decomposition Algorithm for Quadratic Programming, *Yale Univ., Cowles Found., Discussion Paper* 172.
44. Zadeh, L. A., and C. A. Desoer: "Linear System Theory," McGraw-Hill, 1963.
45. Zadeh, L., and B. H. Whalen: On Optimal Control and Linear Programming, *IRE Trans. Autom. Control*, vol. 7, pp. 45–46, 1962.
46. Zoutendijk, G.: "Methods of Feasible Directions," Elsevier, 1960.

4

DYNAMIC DECOMPOSITION TECHNIQUES

J. D. PEARSON

4-1 INTRODUCTION AND PROBLEM STATEMENT

This chapter is concerned with the dynamic optimization of systems composed of subsystems coupled together. When the performance functional and constraints of the overall problem can be separated into components applying to each subsystem, we shall show that the solution can be found in two stages. The first stage involves setting up and solving subproblems posed for the subsystems. The second stage involves coordinating the solutions to form the solution to the overall problem.

In general, it is not a trivial matter to identify the widest class of systems that can be handled by decomposition. Rather we shall present a problem posed for an ideal system structure and indicate on the way the properties that the structure should possess. The technique for handling a more general problem would then be to transform it into the case considered here.

4-1-1 The Subsystems

Suppose the overall *integrated dynamic optimization problem*, or *integrated problem*, consists of N subsystems each defined in terms of an n_i-dimensional state vector $\mathbf{x}_i$, an r_i-dimensional control vector $\mathbf{m}_i$, and an s_i-dimensional input vector $\mathbf{z}_i$, and represented by a system of n_i ordinary differential equations in vector form.

$$\dot{\mathbf{x}}_i = \mathbf{F}_i(\mathbf{x}_i, \mathbf{m}_i, \mathbf{z}_i, t) \qquad \mathbf{x}_i(t_0) = \mathbf{x}_{i0} \qquad i = 1, 2, \ldots, N \tag{1}$$

Each vector $\mathbf{x}_i$, $\mathbf{m}_i$, $\mathbf{z}_i$ is a function of time t over some interval of interest. At each time t the variables $\mathbf{x}_i$, $\mathbf{m}_i$, $\mathbf{z}_i$ are constrained by a set of l_i inequalities.

$$\mathbf{R}_i(\mathbf{x}_i, \mathbf{m}_i, \mathbf{z}_i, t) \geqslant 0 \qquad i = 1, 2, \ldots, N \tag{2}$$

The coupling between subsystems is as follows. The input variables $\mathbf{z}_i$ are defined by interconnections with other subsystems, using simple permutations.

$$\mathbf{z}_i = \sum_{j=1}^{N} \mathbf{L}_{ij}\mathbf{y}_j \tag{3}$$

Here $\mathbf{z}_i$ can depend on $\mathbf{y}_i$, for this is the simplest coupling "feedback." The composite matrix $\mathbf{L}$ will be a permutation matrix, as discussed later in this section. The vector $\mathbf{y}_j$ is the observable output from subsystem j, an s_j-dimensional vector, and is given by the relation

$$\mathbf{y}_j = \mathbf{G}_j(\mathbf{x}_{j\cdot}, \mathbf{m}_j, t) \tag{4}$$

Equations (1) to (4) define a very general nonlinear system in which each subsystem can be a function of the variables in every other subsystem.

The problem also may have terminal constraints such that the end state $\mathbf{x}_i(t_1)$ of each system must lie on a surface $\mathscr{T}_i$ defined by

$$\begin{gathered}[\mathbf{x}_i(t_1), t_1] \in [\mathbf{x}_{i1}, t_1 ; \mathbf{h}_i(\mathbf{x}_{i1}, t_1) = 0] \\ i = 1, 2, \ldots, N\end{gathered} \tag{5}$$

where $\mathbf{h}_i$ is a p_i-dimensional vector function. The terminal conditions are much less general because it is supposed that they are *independent* for each subsystem. In addition to problem constraints (1) to (5) there is an objective function.

$$J_I(\mathbf{m}, \mathbf{z}) = \sum_{i=1}^{N} \left[g_i(\mathbf{x}_{i1}, t_1) + \int_{t_0}^{t_1} f_i(\mathbf{x}_i, \mathbf{m}_i, \mathbf{z}_i, t)\, dt \right] \tag{6}$$

The problem will be to minimize $J_I(\mathbf{m}, \mathbf{z})$. Let $\mathscr{D}_i$ be a bounded region of the $(n_i + 1)$-dimensional $(\mathbf{x}_i, t)$ space, and let $\mathscr{M}_i$ be a bounded region of r_i-dimensional $\mathbf{m}_i$ space, for each i. Then $\mathbf{F}_i$, $\mathbf{R}_i$, $\mathbf{G}_i$ will be real vector-valued functions of class C'' over $\mathscr{D}_i \times \mathscr{M}_i$. Let $\mathscr{Y}_i$ be the image of $\mathscr{D}_i \times \mathscr{M}_i$ under the map $\mathbf{G}_i$ and let $\mathscr{Z}_i$ be the union of the images of $\mathscr{Y}_j$. $j = 1, 2, \ldots, N$ each under the mapping $\mathbf{L}_{ij}$.

For the objective function, g_i, f_i will be real-valued scalar functions of class C'' over $\mathscr{D}_i$ and $\mathscr{D}_i \times \mathscr{M}_i \times \mathscr{Z}_i$ respectively for each i.

Let $\mathbf{h}_i$ be the p_i-dimensional real vector-valued function of class C'' over $\mathscr{D}_i$ and let $\mathscr{T}_i$ be the $(n - p_i + 1)$-dimensional terminal surface defined by $\mathbf{h}_i(\mathbf{x}, t) = 0$.

Consider the class of pairs of functions $\mathbf{m}_i = \mathbf{m}_i(t)$, $\mathbf{z}_i = \mathbf{z}_i(t)$ which are piecewise C'' for $(\cdot, t) \in \mathscr{D}_i$ and whose range is contained in $\mathscr{M}_i$ and $\mathscr{Z}_i$. For each such pair Eq. (1) has a continuous solution which defines a curve $\mathcal{K}_i$ with possible corners in $\mathcal{D}_i$. Let $\mathcal{A}_i$ be the subclass of functions which are *admissible* in the following sense:

1. $\mathcal{K}_i$ is interior to $\mathcal{D}_i$ on $t_0 \leqslant t \leqslant t_1$ where t_1 is the first point of contact of $\mathcal{K}_i$ with $\mathcal{T}_i$.
2. Along $\mathcal{K}_i$ $\mathbf{x}_i$, $\mathbf{m}_i$, $\mathbf{z}_i$ satisfy $\mathbf{R}_i(\mathbf{x}_i, \mathbf{m}_i, \mathbf{z}_i, t) \geqslant 0$ for each t.

4-1-2 The Integrated System

Let $\mathbf{x}$, $\mathbf{m}$, $\mathbf{z}$, and $\mathbf{y}$ be vectors formed by adjoining their components $\mathbf{x}_i$, $\mathbf{m}_i$, $\mathbf{z}_i$, and $\mathbf{y}_i$ in order. Let n, r, and s be the sums of n_i, r_i, and s_i, respectively.

Let $\mathcal{K}$ denote the composite trajectory of $\mathbf{x}(t)$ when Eqs. (1) to (5) are satisfied. Let $\mathcal{D}$, $\mathcal{M}$, $\mathcal{Y}$, $\mathcal{Z}$, and $\mathcal{T}$ be the cartesian products of $\mathcal{D}_i$, $\mathcal{M}_i$, $\mathcal{Y}_i$, $\mathcal{Z}_i$, and $\mathcal{T}_i$, respectively. We shall say that $(\mathbf{m}, \mathbf{z})$ is admissible if each component is admissible. As defined, if *each* $\mathbf{m}_i$ is admissible and $\mathbf{z}_i$ is defined by Eqs. (3) and (4) then $\mathbf{z}_i$ will also be admissible.

Let us now adjoin each subscripted subsystem to form a system.

$$\dot{\mathbf{x}} = \mathbf{F}(\mathbf{x}, \mathbf{m}, \mathbf{z}, t) \tag{7}$$

$$\mathbf{R}(\mathbf{x}, \mathbf{m}, \mathbf{z}, t) \geqslant 0 \tag{8}$$

$$\mathbf{z} = \mathbf{L}\mathbf{y} \tag{9}$$

$$\mathbf{y} = \mathbf{G}(\mathbf{x}, \mathbf{m}, t) \tag{10}$$

$$\mathbf{h}(\mathbf{x}_1, t_1) = 0 \tag{11}$$

where $\mathbf{F}' = (\mathbf{F}_1', \mathbf{F}_2', \ldots, \mathbf{F}_N')$, $\mathbf{R}' = (\mathbf{R}_1', \mathbf{R}_2', \ldots, \mathbf{R}_N')$, $\mathbf{G}' = (\mathbf{G}_1', \mathbf{G}_2', \ldots, \mathbf{G}_N')$, $\mathbf{h}' = (\mathbf{h}_1', \ldots, \mathbf{h}_N')$, and $\mathbf{L}$ is the $s \times s$ matrix whose ijth partitioned block is $\mathbf{L}_{ij}$. The objective function is similarly

$$J_I(\mathbf{m}, \mathbf{z}) = g(\mathbf{x}_1, t_1) + \int_{t_0}^{t_1} f(\mathbf{x}, \mathbf{m}, \mathbf{z}, t)\, dt \tag{12}$$

where

$$g = g_1 + g_2 + \cdots + g_N \qquad \text{and} \qquad f = f_1 + f_2 + \cdots + f_N$$

4-1-3 Conditions

The vector functions $\mathbf{R}$ and $\mathbf{G}$ must satisfy a *constraint condition.* At any time t, let $\hat{\mathbf{R}}$ designate the constraints (2) which are binding; that is, $\mathbf{R}_{ij} = 0$ on $\mathcal{K}$ nontrivially (where double subscripts indicate components). Along the composite trajectory $\mathcal{K}$ of Eqs. (1) to (4) we require that (1) if $l_i > r_i$, $i = 1, 2, \ldots, N$, then no more than r_i components of $\mathbf{R}_i$ may vanish at any t. (2) At each point of $\mathscr{K}$, the trajectory of Eqs. (1) to (4), the matrix

$$\left[\begin{array}{c|c} \left[\dfrac{\partial \hat{\mathbf{R}}_i}{\partial \mathbf{m}_j}\right] & \left[\dfrac{\partial \hat{\mathbf{R}}_i}{\partial \mathbf{z}_i}\right] \\ \hline \left[\mathbf{L}_{ij}\dfrac{\partial \mathbf{G}_j}{\partial \mathbf{m}_j}\right] & -\mathbf{I}_s \end{array}\right]$$

must have maximal rank. $\mathbf{I}_s$ is the $s \times s$ identity matrix.

By pivoting on the lower-right block the upper-right block becomes null and the resulting upper left block is

$$\left[\frac{\partial \hat{\mathbf{R}}_i}{\partial \mathbf{m}_j}\delta_{ij} + \frac{\partial \hat{\mathbf{R}}_i}{\partial \mathbf{z}_i}\mathbf{L}_{ij}\frac{\partial \mathbf{G}_j}{\partial \mathbf{m}_j}\right]$$

This gives the Jacobian of $\hat{\mathbf{R}}$ when the interconnections $\mathbf{z}$ are eliminated. An equivalent condition is that this matrix must have maximal rank. When the $\mathbf{m}_i$, $\mathbf{z}_i$ variables are found independently, Eqs. (3) and (4) are relaxed and we require merely that (2′) at each point of $\mathscr{K}_i$, the trajectory of Eq. (1), the matrix $[\partial \hat{\mathbf{R}}_i/\partial \mathbf{m}_j, \partial \hat{\mathbf{R}}_i/\delta \mathbf{z}_i]$ must have maximal rank. If no constraints are binding, then from (2) it follows that the matrix $[\mathbf{LG}_\mathbf{m}, -\mathbf{I}_s]$ must have rank s in order to satisfy the coupled system constraint condition, which it obviously has.

The interconnection matrix $\mathbf{L}$ is of dimension $s \times s$ and in this application will consist only of 1s and 0s and be a permutation matrix, i.e., a rearrangement of the rows of $\mathbf{I}_s$. Consequently, $\mathbf{L}^{-1}$ exists and $\mathbf{L}^{-1} \equiv \mathbf{L}'$.

In order to avoid questions of variational normality to come, we shall assume that the integrated system and each subsystem is *completely locally controllable* about any admissible trajectory $\bar{\mathbf{x}}$ defined by inputs $\bar{\mathbf{m}}, \bar{\mathbf{z}} \in \mathscr{A}$. More precisely linearize about $\bar{\mathbf{x}}, \bar{\mathbf{m}}, \bar{\mathbf{z}}$, to obtain $\mathbf{x} = \Delta\mathbf{x} + \bar{\mathbf{x}}$, $\mathbf{m} = \Delta\mathbf{m} + \bar{\mathbf{m}}$, $\mathbf{z} = \Delta\mathbf{z} + \bar{\mathbf{z}}$, such that for the subsystems we have

$$\Delta\dot{\mathbf{x}} = \bar{\mathbf{F}}_\mathbf{x}\Delta\mathbf{x} + \bar{\mathbf{F}}_\mathbf{m}\Delta\mathbf{m} + \bar{\mathbf{F}}_\mathbf{z}\Delta\mathbf{z} \qquad \Delta\mathbf{x}(t_0) = 0$$

Let $\bar{\mathbf{\Phi}}(t, t_0)$ satisfy $\dot{\bar{\mathbf{\Phi}}} = \bar{\mathbf{F}}_\mathbf{x}\,\bar{\mathbf{\Phi}}$, $\bar{\mathbf{\Phi}}(t_0, t_0) = \mathbf{I}_n$. Then define $\bar{\mathbf{w}}(t_1, t_0)$ as

$$\bar{\mathbf{w}}(t_1, t_0) = \int_{t_0}^{t_1} \bar{\mathbf{\Phi}}(t, t_0)[\bar{\mathbf{F}}_\mathbf{m}, \bar{\mathbf{F}}_\mathbf{z}][\bar{\mathbf{F}}_\mathbf{m}, \bar{\mathbf{F}}_\mathbf{z}]'\bar{\mathbf{\Phi}}(t_0, t_0)'\,dt$$

Then the system (7) *is said to be completely locally controllable if and only if*

$\bar{\mathbf{w}}(t_1, t_0)$ *has rank n* (Kalman, Ho, and Narendra[8]). Similarly for the integrated system, define

$$\bar{\mathbf{w}}_I(t_1, t_0) = \int_{t_0}^{t_1} \mathbf{\Phi}_\mathrm{I}(t, t_0)[\bar{\mathbf{F}}_\mathrm{m} + \mathbf{L}\bar{\mathbf{G}}_\mathrm{m}][\bar{\mathbf{F}}_\mathrm{m} + \mathbf{L}\bar{\mathbf{G}}_\mathrm{m}]' \, \mathbf{\Phi}_\mathrm{I}(t, t_0) \, dt$$

Then $\bar{\mathbf{w}}_\mathrm{I}(t_1, t_0)$ *must have rank n*, where $\bar{\mathbf{\Phi}}_\mathrm{I}(t_1, t_0)$ is a solution of $\dot{\bar{\mathbf{\Phi}}}_\mathrm{I} = (\mathbf{F}_\mathrm{x} - \mathbf{L}\mathbf{G}_\mathrm{x})\bar{\mathbf{\Phi}}_\mathrm{I}$, $\bar{\mathbf{\Phi}}_\mathrm{I}(t_0, t_0) = \mathbf{I}_n$. There appear to be no results available concerning the relation between these two conditions, $\mathbf{L}$ and the rank of the Jacobian of $[\mathbf{LG} - \mathbf{z}]$.

This completes the definition of the problem. In a later section we will require additional convexity conditions in order to use sufficiency arguments.

4-1-4 Notation

The remaining notation conventions will be as follows. When possible, functional arguments will be omitted or indicated by superscripts; that is, $f(\mathbf{x}^*) \equiv f^*$. Partial derivatives will be denoted by $f_\mathbf{x} \equiv \partial f / \partial \mathbf{x}$, a *row* vector, while $(\mathbf{G}_k)_\mathbf{x}$ is the matrix $[\partial \mathbf{G}_{ki} / \partial \mathbf{x}_{kj}]$ evaluated at $\mathbf{x}_k(t)$, $\mathbf{m}_k(t)$, t and $\mathbf{G}_k$ has s_i components $\mathbf{G}_k^{\,1}, \mathbf{G}_k^{\,2}, \ldots, \mathbf{G}_k^{s_i}$. For vectors $\mathbf{x}$ and $\mathbf{y}$, $\mathbf{x}'\mathbf{y}$ denotes an inner product and $\int_{t_0}^{t_1} \mathbf{x}(t)'\mathbf{y}(t)\, dt$ will be $\langle \mathbf{x}, \mathbf{y} \rangle$. The quadratic form $\mathbf{x}'\mathbf{Q}\mathbf{x}$ will often be written as $\| \mathbf{x} \|^2 \mathbf{Q}$ using pseudonorm notation for compactness. The symbol $\| \mathbf{x} \|^2$ is equivalent to $\mathbf{x}'\mathbf{x}$.

4-2 VARIATIONAL APPROACH TO DECOMPOSITION

The simplest approach to decomposition is to write down the variational conditions for a solution to the integrated problem. These consist of constraints and conditions on subproblems, together with constraints and conditions on interconnections.

By appropriate relaxation of the interconnection equations or conditions it is possible to separate the subproblems for independent solution. To pursue this approach we first set up the necessary conditions for the integrated problem solution.

Let $\mathbf{m}^*$, $\mathbf{z}^*$ be admissible optimal control and input variables and let $\mathscr{X}^*$ be the resulting trajectory of $\mathbf{x}^*$ defined by Eqs. (1) to (5) or (7) to (11). The classical approach is to consider a Lagrange function $\mathscr{L}(\mathbf{x}, \mathbf{m}, \mathbf{z}, \lambda_0, \boldsymbol{\lambda}, \boldsymbol{\mu}, \boldsymbol{\pi}, t)$ defined as follows

$$\begin{aligned} \mathscr{L} &= \lambda_0 f + \boldsymbol{\lambda}'(\mathbf{F} - \dot{\mathbf{x}}) - \boldsymbol{\mu}'\mathbf{R} + \boldsymbol{\pi}'(\mathbf{LG} - \mathbf{z}) \\ &= \sum_{i=1}^{N} \left[\lambda_0 f_i + \boldsymbol{\lambda}_i'(\mathbf{F}_i - \dot{\mathbf{x}}_i) - \boldsymbol{\mu}_i'\mathbf{R}_i + \boldsymbol{\pi}_i' \left(\sum_{i=1}^{N} \mathbf{L}_{ij}\mathbf{G}_j - \mathbf{z}_i \right) \right] \end{aligned}$$

where λ_0, $\boldsymbol{\lambda}$, $\boldsymbol{\mu}$, and $\boldsymbol{\pi}$ are multipliers. The following theorem by Berkovitz[1] clarifies the role of the multipliers.

Theorem 4-1: There exists a constant $\lambda_0^* \geqslant 0$, a continuous n-dimensional vector $\boldsymbol{\lambda}^*(t)$, an m-dimensional vector $\boldsymbol{\mu}^*(t) \geqslant 0$, and an s-dimensional vector $\boldsymbol{\pi}^*(t)$ all defined on an interval $[t_0, t_1]$, such that $\boldsymbol{\mu}^*$ and $\boldsymbol{\pi}^*$ are continuous except perhaps at corners of $\mathscr{K}^*$ where they have unique left- and right-hand limits and such that $(\lambda_0^*, \boldsymbol{\lambda}_0^*(t))$ is not zero. Along $\mathscr{K}^*$ a function $\mathscr{H}$ of $\mathbf{x}$, $\mathbf{m}$, $\mathbf{z}$, λ_0, $\boldsymbol{\lambda}$, $\boldsymbol{\mu}$, $\boldsymbol{\pi}$ satisfies the following conditions 1, 2, and 3.

Let $\mathscr{H}$ and $\boldsymbol{\kappa}$ be defined by

$$\begin{aligned} \mathscr{H} &= \lambda_0 f + \boldsymbol{\lambda}'\mathbf{F} - \boldsymbol{\mu}'\mathbf{R} \\ \boldsymbol{\kappa} &= \mathbf{L}'\boldsymbol{\pi} \end{aligned} \tag{13}$$

CONDITION 1:

Along $\mathscr{K}^*$ $\mathbf{x}^*$, $\mathbf{m}^*$, $\mathbf{z}^*$, λ_0^*, $\boldsymbol{\lambda}^*$, $\boldsymbol{\mu}^*$, $\boldsymbol{\pi}^*$ are given by the following equations:

$$\dot{\mathbf{x}} = \mathscr{H}_{\boldsymbol{\lambda}} \tag{14}$$

$$\dot{\boldsymbol{\lambda}} = -\mathscr{H}_{\mathbf{x}} - \mathbf{G}_{\mathbf{x}}'\,\boldsymbol{\kappa} \tag{15}$$

$$\mathscr{H}_{\mathbf{m}} + \mathbf{G}_{\mathbf{m}}'\,\boldsymbol{\kappa} = 0 \tag{16}$$

$$\mathscr{H}_{\mathbf{z}} = \boldsymbol{\pi} \tag{17}$$

$$\boldsymbol{\mu}'\mathbf{R} = 0 \qquad \boldsymbol{\mu}(t) \geqslant 0 \tag{18}$$

$$\mathbf{LG} - \mathbf{z} = 0 \tag{19}$$

At the end point $(\mathbf{x}_1^*, t_1)$, where t_1 is the first point of contact of $\mathscr{K}^*$ with $\mathscr{T}$, a transversality condition holds.

$$\lambda_0(g_{\mathbf{x}}d\mathbf{x} + g_t dt) + (\mathscr{H} + \boldsymbol{\kappa}'\mathbf{G} - \boldsymbol{\pi}'\mathbf{z})\,dt_1 - \boldsymbol{\lambda}'\,d\mathbf{x}_1 = 0 \tag{20}$$

for all values of $d\mathbf{x}_1$, dt_1 on $\mathscr{T}$ satisfying

$$0 = \mathbf{h}_{\mathbf{x}}d\mathbf{x}_1 + \mathbf{h}_t dt_1$$

Along $\mathscr{K}^*$, the function $\mathscr{H}(\mathbf{x}, \mathbf{m}, \mathbf{z}, \lambda_0, \boldsymbol{\lambda}, \boldsymbol{\mu}, t)$ is continuous.

CONDITION 2:

For every element $(\mathbf{x}^*, \mathbf{m}^*, \mathbf{z}^*, \lambda_0^*, \boldsymbol{\lambda}^*, \boldsymbol{\mu}^*, \boldsymbol{\pi}^*)$ of $\mathscr{K}^*$ and every admissible $\mathbf{m}$, $\mathbf{z}$ satisfying Eqs. (9) and (10)

$$\mathscr{H}(\mathbf{x}^*, \mathbf{m}, \mathbf{z}, \lambda_0^*, \boldsymbol{\lambda}^*, \boldsymbol{\mu}^*, \mathbf{t}) \geqslant \mathscr{H}^* \tag{21}$$

CONDITION 3:

At each point of $\mathscr{K}^*$ let $\hat{\mathbf{R}}$ denote the vector formed from the binding components of $\hat{\mathbf{R}}$, and let $\mathbf{E}$ and $\nabla^2\mathscr{H}$ be defined as

$$\mathbf{E} = \begin{bmatrix} \hat{\mathbf{R}}_{\mathbf{m}} & \hat{\mathbf{R}}_{\mathbf{z}} \\ \mathbf{LG}_{\mathbf{m}} & -\mathbf{I}_s \end{bmatrix} \qquad \nabla^2\mathscr{H} = \begin{bmatrix} (\mathscr{H} + \boldsymbol{\kappa}'\mathbf{G})_{\mathbf{mm}} & \mathscr{H}_{\mathbf{mz}} \\ \mathscr{H}_{\mathbf{zm}} & \mathscr{H}_{\mathbf{zz}} \end{bmatrix} \tag{22}$$

Then, for all nontrivial $\mathbf{e}$ vectors satisfying $\mathbf{Ee}^* = 0$, $\| \mathbf{e} \|^2 \nabla^2\mathscr{H} \geqslant 0$.

Proof: This theorem is established from the standard theorems[1] by noting that the interconnections (3) and (4) can be handled as extra inequalities, $\mathbf{R}^+ = \mathbf{z} - \mathbf{LG} \geqslant 0$ and $\mathbf{R}^- = \mathbf{LG} - \mathbf{z} \geqslant 0$. The corresponding multipliers $\boldsymbol{\pi}^+$ and $\boldsymbol{\pi}^-$, both nonnegative, occur together and $\boldsymbol{\pi}^*$ is defined by $\boldsymbol{\pi}^* = \boldsymbol{\pi}^+ - \boldsymbol{\pi}^-$. Consequently, $\boldsymbol{\pi}^*$ can have arbitrary sign but inherits the continuity properties of $\boldsymbol{\pi}^+$ and $\boldsymbol{\pi}^-$.

In condition 2 the fact that $\mathbf{m}$, $\mathbf{z}$ must be admissible with respect to the additional inequalities $\mathbf{R}^+$ and $\mathbf{R}^-$ implies that Eqs. (9) and (10) must be satisfied. This completes the proof.

We say that the solution is normal if $\lambda_0 \neq 0$ and consequently can be normalized to 1. Mitter[14] has shown that the assumptions of complete local controllability imply that a problem is normal. Another interpretation is that a nonnormal system either has a set of solutions to its system equations (1) to (5) which do not intersect the terminal surface $\mathscr{T}$ or has a unique admissible solution. Consequently, it is not controllable along any admissible linearized trajectory.

4-2-1 Subproblems

In order to construct subproblems we use the fact that, due to the problem structure, the Lagrange function can be rearranged to form a sum of expressions connected with only one subsystem.

$$\mathscr{L} = \sum_{i=1}^{N} \left[\lambda_0 f_i + \boldsymbol{\lambda}_i'(\mathbf{F}_i - \dot{\mathbf{x}}_i) - \boldsymbol{\mu}_i'\mathbf{R}_i + \sum_{j=1}^{N} \boldsymbol{\pi}_j'\mathbf{L}_{ji}\mathbf{G}_i - \boldsymbol{\pi}_i'\mathbf{z}_i\right]$$

$$= \sum_{i=1}^{N} [(\lambda_0 f_i + \boldsymbol{\kappa}_i'\mathbf{G}_i - \boldsymbol{\pi}_i'\mathbf{z}_i) + \boldsymbol{\lambda}_i'(\mathbf{F}_i - \dot{\mathbf{x}}_i) - \boldsymbol{\mu}_i'\mathbf{R}_i] \tag{23}$$

where $\boldsymbol{\kappa}_i = \sum_{j=1}^{N} \mathbf{L}_{ji}\boldsymbol{\pi}_j$ is a commonly occurring quantity, the multiplier associated with the output $\mathbf{y}_i$. If $\boldsymbol{\kappa}_i$ and $\boldsymbol{\pi}_i$ are scaled by λ_0 or if we assume normality, each argument of the sum can be considered as the Lagrangian of a parametric subproblem.

Thus taking advantage of structure define N *parametric subproblems* associated with each subsystem (1), constraint system (2), and terminal condition (5) of minimizing an objective function $\mathbf{J}_i(\mathbf{m}_i, \mathbf{z}_i; \boldsymbol{\pi})$.

$$J_i(\mathbf{m}_i, \mathbf{z}_i; \boldsymbol{\pi}) = g_i + \int_{t_0}^{t_1} \left(f_i + \sum_{j=1}^{N} \boldsymbol{\pi}_j' \mathbf{L}_{ji} \mathbf{G}_i - \boldsymbol{\pi}_i' \mathbf{z}_i \right) dt \tag{24}$$

Here $\boldsymbol{\pi} = (\boldsymbol{\pi}_1', \boldsymbol{\pi}_2', \ldots, \boldsymbol{\pi}_N')$ is a given piecewise continuous vector function (by virtue of Theorem 4-1). Let $\boldsymbol{\pi} \in \Omega$ be the set of piecewise continuous functions such that J_i has a minimum subject to Eqs. (1), (2), and (5).

The following theorem due to Macko[13] indicates that the optimal $\boldsymbol{\pi}^*$ for the integrated problems will cause the subproblem solutions to solve the integrated problem.

Theorem 4-2: If a normal minimizing solution exists to the integrated problems and to each of the subproblems, then a $\boldsymbol{\pi}^*$ exists such that solutions to the necessary conditions of the subproblems also satisfy the necessary conditions of the integrated problem.

Proof: Suppose that each $\boldsymbol{\pi}_i \in \Omega_i$ and denote the solutions to the subproblems by $\mathbf{x}_i^*(\boldsymbol{\pi})$, $\mathbf{m}_i^*(\boldsymbol{\pi})$, $\mathbf{z}_i^*(\boldsymbol{\pi})$ with $\mathbf{x}_i^*(\boldsymbol{\pi})$ denoted by the curve $\mathscr{K}_i^*(\boldsymbol{\pi})$. Define $\mathscr{H}_i$ and $\boldsymbol{\kappa}_i$, $\boldsymbol{\kappa} = (\boldsymbol{\kappa}_1, \boldsymbol{\kappa}_2, \ldots, \boldsymbol{\kappa}_N)$

$$\begin{aligned} \mathscr{H}_i &= f_i + \boldsymbol{\lambda}_i' \mathbf{F}_i - \boldsymbol{\mu}_i' \mathbf{R}_i \\ \boldsymbol{\kappa}_i &= \sum_{j=1}^{N} \mathbf{L}_{ji}' \boldsymbol{\pi}_j \end{aligned} \tag{25}$$

Then when the problems are all normal Eqs. (13) and (25) require

$$\mathscr{H} = \sum_{i=1}^{N} \mathscr{H}_i \tag{26}$$

The necessary conditions for subproblem i are now ás follows.

CONDITION 1′:

Along $\mathscr{K}_i^*(\boldsymbol{\pi})$, $\mathbf{x}_i^*$, $\mathbf{m}_i^*$, $\mathbf{z}_i^*$, $\boldsymbol{\lambda}_i^*$, and $\boldsymbol{\mu}_i^*$ are solutions of the following equations for each $\boldsymbol{\pi}$:

$$\mathbf{x}_i = (\mathscr{H}_i)'_{\boldsymbol{\lambda}_i} \tag{27}$$

$$\dot{\boldsymbol{\lambda}}_i = -(\mathscr{H}_i)'_{\mathbf{x}_i} - (\mathbf{G}_i)'_{\mathbf{x}_i} \boldsymbol{\kappa}_i \tag{28}$$

$$(\mathscr{H}_i)'_{\mathbf{m}_i} + (\mathbf{G}_i)'_{\mathbf{m}_i} \boldsymbol{\kappa}_i = 0 \tag{29}$$

$$(\mathscr{H}_i)'_{\mathbf{z}_i} = \boldsymbol{\pi}_i \tag{30}$$

$$\boldsymbol{\mu}_i'\mathbf{R}_i = 0 \qquad \boldsymbol{\mu}_i(t) \geqslant 0 \tag{31}$$

At the end point $[\mathbf{x}_{1i}(\boldsymbol{\pi}), t_{1i}(\boldsymbol{\pi})]$, where $t_{1i}(\boldsymbol{\pi})$ is the first point of contact of $\mathscr{K}_i^*(\boldsymbol{\pi})$ with $\mathscr{T}_i$, a transversality condition holds.

$$[(g_i)_{\mathbf{x}_i}\, d\mathbf{x}_{1i} + (g_i)_t\, dt_{1i}] + (\mathscr{H}_i + \boldsymbol{\kappa}_i'\mathbf{G}_i - \boldsymbol{\pi}_1'\mathbf{z}_i)\, dt_{1i} - \boldsymbol{\lambda}_i'\, d\mathbf{x}_{1i} = 0 \tag{32}$$

for all $d\mathbf{x}_{1i}$, dt_{1i} on $\mathscr{T}_i$ satisfying

$$(\mathbf{h}_i)'_{\mathbf{x}_i}\, d\mathbf{x}_{1i} + (\mathbf{h}_i)_t\, dt_{1i} = 0$$

Along $\mathscr{K}_i^*(\boldsymbol{\pi})$ the function $(\mathscr{H}_i + \boldsymbol{\kappa}_i'\mathbf{G}_i - \boldsymbol{\pi}_i'\mathbf{z}_i)$ is continuous.

CONDITION 2′:

For every element of $\mathscr{K}_i^*(\boldsymbol{\pi})$ and every pair $(\mathbf{m}_i, \mathbf{z}_i) \in \mathscr{A}_i$

$$\begin{aligned} &\mathscr{H}_i(\mathbf{x}_i^*(\boldsymbol{\pi}), \mathbf{m}_i, \mathbf{z}_i, \boldsymbol{\lambda}_i^*(\boldsymbol{\pi}), t) + \boldsymbol{\kappa}_i'\mathbf{G}_i(\mathbf{x}_i^*(\boldsymbol{\pi}), \mathbf{m}_i, t) - \mathbf{z}_i'\boldsymbol{\pi}_i \\ &\qquad \geqslant \mathscr{H}_i^*(\boldsymbol{\pi}) + \boldsymbol{\kappa}_i'\mathbf{G}_i^*(\boldsymbol{\pi}) - \mathbf{z}_i^*(\boldsymbol{\pi})'\boldsymbol{\pi} \end{aligned} \tag{33}$$

CONDITION 3′:

At each point of $\mathscr{K}_i^*(\boldsymbol{\pi})$ let $\hat{\mathbf{R}}_i$ denote the binding components of $\mathbf{R}_i$ and define $\mathbf{E}_i(\boldsymbol{\pi})$ as

$$\mathbf{E}_i(\boldsymbol{\pi}) = [(\hat{\mathbf{R}}_i)_{\mathbf{m}}, (\hat{\mathbf{R}}_i)_{\mathbf{z}}], \ \nabla^2\mathscr{H}_i = \begin{bmatrix} (\mathscr{H}_i + \boldsymbol{\kappa}_i'\mathbf{G}_i)_{\mathbf{m}_i\mathbf{m}_i} & (\mathscr{H}_i)_{\mathbf{m}_i\mathbf{z}_i} \\ (\mathscr{H}_i)_{\mathbf{z}_i\mathbf{m}_i} & (\mathscr{H}_i)_{\mathbf{z}_i\mathbf{z}_i} \end{bmatrix} \tag{34}$$

Then for all nontrivial solutions to $\mathbf{E}_i(\boldsymbol{\pi})\mathbf{e}_i = 0$, $\| \mathbf{e}_i \|^2\ \nabla^2\mathscr{H}_i \geqslant 0$.

This completes the statement of the necessary conditions for the subproblems. We need to show that if $\boldsymbol{\pi} = \boldsymbol{\pi}^*$ then conditions 1′, 2′, and 3′ imply conditions 1, 2, and 3.

Because of the structure of $\mathbf{G}$ and $\mathscr{H}$, Eqs. (14) to (18) have the same form as Eqs. (27) to (31). Similarly because of the structure of $\mathbf{g}$, $\mathbf{G}$, and $\mathscr{H}$, Eqs. (20) and (32) are the same. Thus, if $\boldsymbol{\pi} = \boldsymbol{\pi}^*$ and $\boldsymbol{\kappa} = \mathbf{L}\boldsymbol{\pi}^*$, the set of solutions to both will coincide. However, since Eqs. (14) to (21) have a solution $\mathbf{x}^*$, $\mathbf{m}^*$, $\mathbf{z}^*$, $\lambda_0^* = 1$, $\boldsymbol{\lambda}^*$, $\boldsymbol{\mu}^*$, $\boldsymbol{\pi}^*$, when Eq. (19) is enforced solutions to Eqs. (27) to (31) for $\boldsymbol{\pi} = \boldsymbol{\pi}^*$, $\boldsymbol{\kappa} = \mathbf{L}\boldsymbol{\pi}^*$, must exist which also satisfy the interconnection constraint (19) repeated below.

$$\sum_{j=1}^{N} \mathbf{L}_{ij}\mathbf{G}_j(\mathbf{x}_j^*(\boldsymbol{\pi}^*), \mathbf{m}_j^*(\boldsymbol{\pi}^*) = \mathbf{z}_i^*(\boldsymbol{\pi}^*) \tag{35}$$

Then continuity of $\mathscr{H}_i^* + \boldsymbol{\kappa}_i'\mathbf{G}_i^* - \boldsymbol{\pi}_i'\mathbf{z}_i^*$ implies continuity of $\mathscr{H}$ for $\boldsymbol{\pi} = \boldsymbol{\pi}^*$. Thus, conditions 1 and 1′ are identical for $\boldsymbol{\pi} = \boldsymbol{\pi}^*$.

Now suppose Eq. (33) is satisfied for $\boldsymbol{\pi} = \boldsymbol{\pi}^*$ in addition to condition 1′. Then since $\mathscr{A}_i$ contains an $\mathbf{m}_i$ and $\mathbf{z}_i$ such that for $\mathbf{i} = 1, 2, \ldots, N$,

$$\mathbf{z}_i = \sum_{j=1}^{N} \mathbf{L}_{ij}\mathbf{G}_j(\mathbf{x}_j^*(\boldsymbol{\pi}^*), \mathbf{m}_j, t) \tag{36}$$

and Eq. (33) is satisfied, we may add inequalities (33) for $i = 1, 2, \ldots, N$ and thereby cancel the spurious terms with Eqs. (35) and (36) to derive condition 2. Thus, if condition 2′ is satisfied for each i, then this is sufficient to enforce condition 2.

In condition 3′, since condition 1′ is satisfied for $\boldsymbol{\pi} = \boldsymbol{\pi}^*$, the trajectories and thus the sets of binding inequalities in $\hat{\mathbf{R}}$ and $\hat{\mathbf{R}}_i$, $i = 1, 2, \ldots, N$, will be the same. This means that the upper half of E in Eq. (22) can be rearranged to consist of a block diagonal matrix whose ith block is E_i in Eq. (34), and if $\mathbf{e} = (\mathbf{e}_i', \mathbf{e}_2', \ldots, \mathbf{e}_N')$ satisfies $\mathbf{E}^*\mathbf{e} = 0$, it will also satisfy $\mathbf{E}_i^*\mathbf{e}_i = 0$, $i = 1, 2, \ldots, N$. In addition, however, $\nabla^2\mathscr{H}$ of Eq. (22) is composed of $\nabla^2\mathscr{H}_i$, $i = 1, 2, \ldots, N$, similarly rearranged in the order of $\mathbf{m}_i$ and $\mathbf{z}_i$, that is,

$$\| \mathbf{e} \|^2 \, \nabla^2\mathscr{H} \equiv \sum_{i=1}^{N} \| \mathbf{e}_i \|^2 \, \nabla^2\mathscr{H}_i \tag{37}$$

Consequently, when $\boldsymbol{\pi} = \boldsymbol{\pi}^*$, condition 3′ implies condition 3.

This completes the proof. In summary it can be seen that the main problem of coordinating subproblem solutions so that they solve the integrated problem is in the determination of $\boldsymbol{\pi}^*$. Since adjusting $\boldsymbol{\pi}$ adjusts the subproblem performance indices or goals, this is known as *goal coordination.*

A second method of coordination, *model coordination*, operates by adjusting the interconnection variables $\mathbf{y}$, $\mathbf{z}$ directly. This second method has the advantage that when it can be used, all the system constraints are satisfied continuously, whereas for the goal-coordination scheme the interconnection constraint $\mathbf{z}^*(\boldsymbol{\pi}) = \mathbf{LG}(\mathbf{x}^*(\boldsymbol{\pi}), \mathbf{m}^*(\boldsymbol{\pi}))$ is satisfied when $\boldsymbol{\pi} = \boldsymbol{\pi}^*$ and not necessarily for any other values of $\boldsymbol{\pi}$.

Before the variational subproblems can be stated we need some properties of the dual problem because fixing $\mathbf{z} = \mathbf{Ly}$ is dual to fixing $\boldsymbol{\pi}$, the multiplier associated with the interconnection constraint. The dual is discussed in Sec. 4-3 and the application to model coordination is in Sec. 4-7.

Theorems 4-1 and 4-2 are almost sufficient to show that $\mathscr{K}^*$ is a minimizing arc when the variational problem is well behaved. Specifically, when the inequalities of conditions 2 and 3 are strict throughout a neighborhood of $\mathscr{K}^*$ and there is no point s_1 conjugate to t_0 in $[t_0, t_1]$, then $\mathscr{K}^*$ is a locally minimizing arc. A conjugate point t_1 is one at which the auxiliary or second variational problem, about $\mathscr{K}^*$, fails to have a minimum over the interval $[t_0, t_1 \geqslant s_1]$. (This is precisely analogous to minimizing the distance on a globe between two points which are steadily separated until they become antipodes, for at this point the antipode is conjugate to the starting point.) This is summarized next.

Theorem 4-3: If necessary and sufficient conditions for a local minimum are satisfied along $\mathscr{K}_i^*(\pi^*)$ for each subproblem, and if neither the subproblems nor the integrated problems have conjugate points, then this is sufficient to establish that $\mathscr{K}^*$ is a local minimizing arc for the integrated problem.

Proof: If the strict forms of conditions 2′ and 3′ are satisfied throughout some neighborhood of $\mathscr{K}_i^*(\pi^*)$, then from the proof of Theorem 4-2 this implies the same strictness of conditions 2 and 3 along $\mathscr{K}^*$. These conditions are sufficient only in the absence of conjugate points, completing the proof.

4-2-2 Goal Coordination of the Subproblems

We must consider the location of $\boldsymbol{\pi}^* \in \Omega$ so that the subproblem solutions are manipulated into solving the integrated problem. If $\boldsymbol{\pi} \neq \boldsymbol{\pi}^*$, then the interconnection constraints are not satisfied, so that it is rational to examine the effect of $\boldsymbol{\pi}$ on the interconnection error, which will be defined as

$$\boldsymbol{\xi}_i(\boldsymbol{\pi}, t) = \sum_{j=1}^{N} \mathbf{L}_{ij}\mathbf{G}_j(\mathbf{x}_j^*(\boldsymbol{\pi}), \mathbf{m}_j^*(\boldsymbol{\pi}), t) - \mathbf{z}_i^*(\boldsymbol{\pi}) \tag{38}$$

The usefulness of the error can be seen in the following lemma.

Lemma 4-1: If the solution $\mathbf{x}_i^*(\boldsymbol{\pi})$, $\mathbf{m}_i^*(\boldsymbol{\pi})$, $\mathbf{z}_i^*(\boldsymbol{\pi})$ and each subproblem is unique for $\boldsymbol{\pi}_i \in \Omega_i$, then the error functions $\boldsymbol{\xi}_i(\boldsymbol{\pi})$, $i = 1, 2, \ldots, N$, are zero if and only if $\boldsymbol{\pi} = \boldsymbol{\pi}^*$.

Proof: Uniqueness serves to exclude the possibility that for $\boldsymbol{\pi} = \boldsymbol{\pi}^*$ there are solutions which do not satisfy the interconnection constraint. If $\boldsymbol{\pi} = \boldsymbol{\pi}^*$ then from condition 1′ of Theorem 4-2 the interconnection constraint (35) is satisfied and $\boldsymbol{\xi}_i(\boldsymbol{\pi}^*) = 0$, $i = 1, 2, \ldots, N$. On the other hand, if $\boldsymbol{\xi}_i(\boldsymbol{\pi}) = 0$ for some $\boldsymbol{\pi}$, then all the conditions of Theorem 4-1 are satisfied, hence $\boldsymbol{\pi} = \boldsymbol{\pi}^*$ by definition.

A simple algorithm can now be proposed for the adjustment of $\boldsymbol{\pi}$ to obtain $\boldsymbol{\pi}^*$. One method of doing this is to choose $\boldsymbol{\pi}$ so as to minimize $\sum_{i=1}^{N} \langle \boldsymbol{\xi}_i(\boldsymbol{\pi}), \boldsymbol{\xi}_i(\boldsymbol{\pi}) \rangle$ a measure of the interconnection error. As we shall see, another method is to maximize the associated dual functional, and a development due to Lasdon and Schoeffler[11] turns out to do just that.

Lasdon and Schoeffler base their argument on an interesting economic interpretation. Imagine a competitive system of N subsystems which service the necessities of each other according to a price structure. Subsystem i has gross operating costs of

$$g_i + \int_{t_0}^{t_{1i}} f_i \, dt$$

and charges $\boldsymbol{\pi}_i'\mathbf{z}_i$ to subsystems $1, \ldots, N$ for accepting $\mathbf{z}_i$, while it pays $\boldsymbol{\kappa}_i'\mathbf{G}_i$ to have its output $\mathbf{y}_i = \mathbf{G}_i$ taken away. Thus, the total cost is $\mathbf{J}_i(\mathbf{m}_i, \mathbf{z}_i, \boldsymbol{\pi})$ and naturally the subsystem operates optimally to generate $\mathbf{x}_i^*(\boldsymbol{\pi})$, $\mathbf{m}_i^*(\boldsymbol{\pi})$, $\mathbf{z}_i^*(\boldsymbol{\pi})$, given $\boldsymbol{\pi}_i$ and $\boldsymbol{\kappa}_i = \sum_{j=1}^{N} \mathbf{L}_{ji}'\boldsymbol{\pi}_j$.

The coordination problem is solved by invoking the immutable law of supply and demand (overlooking the fact that each subsystem is a captive consumer, for this is a perfectly competitive system). If the optimal outputs from subsystems $1, 2, \ldots, N$, which are dependent on subsystem i's input, exceed $\mathbf{z}_i^*(\boldsymbol{\pi})$, then clearly this subsystem is not charging enough to accept that input. Increasing $\boldsymbol{\pi}_i$ increases $\mathbf{z}_i^*(\boldsymbol{\pi})$ and decreases $\sum_{j=1}^{N} \mathbf{L}_{ij}\mathbf{G}_j^*(\pi)$, in principle. Thus a suitable adjustment law is to make

$$\frac{d\boldsymbol{\pi}}{d\tau}(\tau, t) = k_i \left[\sum_{j=1}^{N} \mathbf{L}_{ij}\mathbf{G}_j(\mathbf{x}_j^*(\boldsymbol{\pi}), \mathbf{m}_j^*(\boldsymbol{\pi}), t) - \mathbf{z}_i^*(\boldsymbol{\pi}) \right]$$

$$= k_i \boldsymbol{\xi}_i^*(\boldsymbol{\pi}(\tau, t), t) \qquad k_i > 0 \tag{39}$$

Here $\boldsymbol{\pi}(\tau, t)$ represents the estimate of $\boldsymbol{\pi}(t)$ at adjustment time τ. The intention is that $\boldsymbol{\pi}(0, t)$ is the initial estimate of $\boldsymbol{\pi}^*(t)$ and that $\boldsymbol{\pi}(\tau, t) \to \boldsymbol{\pi}^*(t)$ for $t[t_0, t_1]$ as $\tau \to \infty$. In iterative calculation the dependence on τ will generally be implicitly assumed.

It is to be expected that the adjustment rule will converge if the functions in the problem definition are well behaved. A formal proof of convergence can be given under this blanket assumption for the general problem. As a computational technique the scheme has little value, but it has an interesting theoretical property that the adjustment rule is a steepest ascent rule for maximizing a functional $J(\boldsymbol{\pi})$ where

$$J(\boldsymbol{\pi}) = \sum_{i=1}^{N} J_i(\mathbf{m}_i^*(\boldsymbol{\pi}), \mathbf{z}_i^*(\boldsymbol{\pi}), \boldsymbol{\pi}) \tag{40}$$

This is the sum of the optimal subproblem performance indices for a given $\boldsymbol{\pi} \in \Omega$.

Lemma 4-2: If each subproblem has a unique minimum at $\mathbf{x}_i^*(\boldsymbol{\pi})$, $\mathbf{m}_i^*(\boldsymbol{\pi})$, $\mathbf{z}_i^*(\boldsymbol{\pi})$ for the given $\boldsymbol{\pi} \in \Omega$, then

$$J(\boldsymbol{\pi}) < J(\boldsymbol{\pi}^*)$$

if $\boldsymbol{\pi} \neq \boldsymbol{\pi}^*$.

Proof: Since the minimizing point is unique, we have for $\boldsymbol{\pi} \neq \boldsymbol{\pi}^*$

$$J_i(\mathbf{m}_i^*(\boldsymbol{\pi}), \mathbf{z}_i^*(\boldsymbol{\pi}), \boldsymbol{\pi}) < J_i(\mathbf{m}_i^*(\boldsymbol{\pi}^*), \mathbf{z}_i^*(\boldsymbol{\pi}^*), \boldsymbol{\pi})$$

Summing the inequalities and using the definition of $\mathbf{J}_i$ in Eq. (24)

$$J(\boldsymbol{\pi}) < \sum_{i=1}^{N} \left\{ g_i^* + \int_{t_0}^{t_1} \left[f_i^* + \sum_{j=1}^{N} \boldsymbol{\pi}_j' \mathbf{L}_{ji} \mathbf{G}_i^* - \boldsymbol{\pi}_i' \mathbf{z}_i^*(\boldsymbol{\pi}^*) \right] dt \right\}$$

$$= \sum_{i=1}^{N} \left(g_i^* + \int_{t_0}^{t_1} \left\{ f_i^* + \boldsymbol{\pi}_i' \left[\sum_{j=1}^{N} \mathbf{L}_{ij} \mathbf{G}_j^* - \mathbf{z}_i^*(\boldsymbol{\pi}^*) \right] \right\} dt \right) \tag{41}$$

However, according to Theorem 4-2, $\mathbf{x}_i^*(\boldsymbol{\pi}^*)$, $\mathbf{m}_i^*(\boldsymbol{\pi}^*)$, $\mathbf{z}_i^*(\boldsymbol{\pi}^*)$, $i = 1, 2, \ldots, N$, satisfies the interconnection constraint (35) and so the right-hand side of the preceding inequality is *independent* of $\boldsymbol{\pi}$ and we may substitute $\boldsymbol{\pi} = \boldsymbol{\pi}^*$, giving as desired

$$J(\pi) < J(\boldsymbol{\pi}^*)$$

If $\boldsymbol{\pi} = \boldsymbol{\pi}^*$, then equality is achieved and from Eq. (41) the value $J(\boldsymbol{\pi}^*)$ is equal to $J_I(\mathbf{m}^*, \mathbf{z}^*)$, the optimal integrated objective value.

The derivative $dJ(\boldsymbol{\pi})/d\boldsymbol{\tau}$ can be found at least formally by assuming that in some neighborhood of $\boldsymbol{\pi}^*$ the solutions to the subproblems are sufficiently well-behaved functions of $\boldsymbol{\pi}$. Along optimal curves $\mathscr{K}_i^*(\boldsymbol{\pi})$ we may write

$$J_i(\mathbf{m}_i^*(\boldsymbol{\pi}), \mathbf{z}_i^*(\boldsymbol{\pi}), \boldsymbol{\pi}) = \left[g_i + \int_{t_0}^{t_{1i}} (f_i + \boldsymbol{\kappa}_i' \mathbf{G}_i - \boldsymbol{\pi}_i' \mathbf{z}_i)\, dt \right]^{\mathscr{K}_i^*(\boldsymbol{\pi})}$$

$$= \left[g_i + \int_{t_0}^{t_{1i}} (\mathscr{H}_i + \boldsymbol{\kappa}_i' \mathbf{G}_i - \boldsymbol{\pi}_i' \mathbf{z}_i - \boldsymbol{\lambda}_i' \dot{\mathbf{x}}_i)\, dt \right]^{\mathscr{K}_i^*(\boldsymbol{\pi})}$$

Now assuming that $J(\boldsymbol{\pi})$ has a derivative with respect to τ

$$\frac{dJ}{d\tau}(\boldsymbol{\pi})$$

$$= -\sum_{i=1}^{N}\left\{\frac{dg_i}{d\tau}\bigg|^{t=t_{1i}} + (\mathscr{H}_i - \boldsymbol{\lambda}_i'\dot{\mathbf{x}}_i + \boldsymbol{\kappa}_i'\mathbf{G}_i - \boldsymbol{\pi}_i'\mathbf{z}_i)\bigg|^{t=t_{1i}}\frac{dt_{1i}}{d\tau}\right.$$

$$+ \int_{t_0}^{t_{1i}}\Big\{[(\mathscr{H}_i)_{\mathbf{x}_i} + \boldsymbol{\kappa}_i'(\mathbf{G}_i)_{\mathbf{x}_i}]\frac{d\mathbf{x}_i}{d\tau} - \boldsymbol{\lambda}_i'\frac{d\dot{\mathbf{x}}_i}{d\tau} + [(\mathscr{H}_i)_{\mathbf{m}_i} + \mathscr{H}_i'(\mathbf{G}_i)_{\mathbf{m}_i}]$$

$$\cdot\frac{d\mathbf{m}_i}{d\tau} + [(\mathscr{H}_i)_{\mathbf{z}_i} - \boldsymbol{\pi}_i']\frac{d\mathbf{z}_i}{d\tau}$$

$$\left. + [(\mathscr{H}_i)_{\boldsymbol{\lambda}_i} - \dot{\mathbf{x}}_i']\frac{d\boldsymbol{\lambda}_i}{d\tau} + [(\mathscr{H}_i)_{\boldsymbol{\mu}_i}]\frac{d\boldsymbol{\mu}_i}{d\tau} + \left(\mathbf{G}_i'\frac{d\boldsymbol{\kappa}_i}{d\tau} - \mathbf{z}_i\frac{d\boldsymbol{\pi}_i}{d\tau}\right)\Big\}\,dt\right\}^{\mathscr{K}_i^*(\boldsymbol{\pi})} \tag{42}$$

However at the end surface $\mathbf{h}_i = 0$ where

$$\frac{d}{d\tau}(\mathbf{h}_i)\bigg|^{\mathscr{T}_i} = 0$$

$$\frac{d\mathbf{x}_{1i}}{d\tau}(t_{1i})\bigg|^{\mathscr{T}_i} = \frac{d\mathbf{x}_{1i}}{d\tau}(t_{1i}) + \dot{\mathbf{x}}_i(t_{1i})\frac{dt_{1i}}{d\tau}$$

Similarly for the inequality constraint we have

$$\mathbf{R}_i'\frac{d\boldsymbol{\mu}_i}{d\tau} = [(\mathscr{H}_i)_{\boldsymbol{\mu}_i}]\frac{d\boldsymbol{\mu}_i}{d\tau} = 0$$

These conditions together with conditions 1′, which define $\mathscr{K}_i^*(\boldsymbol{\pi})$, show that most of Eq. (42) cancels, leaving

$$\frac{dJ(\boldsymbol{\pi})}{d\tau} = \sum_{i=1}^{N}\left[\int_{t_0}^{t_{1i}}\left(\mathbf{G}_i'\frac{d\boldsymbol{\kappa}_i}{d\tau} - \mathbf{z}_i'\frac{d\boldsymbol{\pi}_i}{d\tau}\right)dt\right]^{\mathscr{K}_i^*(\boldsymbol{\pi})}$$

with

$$\frac{d\boldsymbol{\kappa}_i}{d\tau} = \sum_{j=1}^{N}\mathbf{L}_{ji}'\frac{d\boldsymbol{\pi}_j}{d\tau}$$

Finally, the adjustment rule can be substituted, and after rearranging the summations,

$$\frac{dJ(\boldsymbol{\pi})}{d\tau} = \sum_{i=1}^{N}\left[\int_{t_0}^{t_{1i}}\boldsymbol{\xi}_i(\boldsymbol{\pi})'\,\boldsymbol{\xi}_i(\boldsymbol{\pi})\,dt\right]^{\mathscr{K}_i^*(\boldsymbol{\pi})} > 0 \tag{43}$$

It can be concluded then that the adjustment rule is a method of ascent in $J(\pi)$ and since $J(\pi)$ is bounded above by $J(\pi^*) = J_I(\mathbf{m}^*, \mathbf{z}^*)$ given by the unique solution to the integrated problem, then $\pi \to \pi^*$ as $\tau \to \infty$.

The preceding analysis shows that the adjustment rule, formally at least, should converge and generate the optimal coordinating π^*. However, Lemmas 4-1 and 4-2 are of greater interest for they show that the coordination problem is to maximize $J(\pi)$ and that the error functions $\xi_i(\pi)$ are of significance. Furthermore the primary property required in the process of decomposition is that each subproblem should have a well-defined minimum.

4-3 PRIMAL AND DUAL CONTROL PROBLEMS

Section 4-2 presented a very wide class of problems and was based essentially on the manipulation of the *necessary* conditions for the minimum. In this section the concept of the *dual problem* will be introduced. It will be shown that for a class of convex problems the solution of the original "primal" problem is equivalent to the solution of an associated "dual" problem. The interesting aspect of this is that the decomposition and coordination can be viewed as primal and dual operations. Since the presentation is based entirely on global sufficiency arguments, it is necessary to specialize the class of problems to those for which global sufficiency arguments are known.

In place of Eqs. (1) to (5) the following system will be used for each i.

$$\dot{\mathbf{x}}_i = \mathbf{A}_i(t)\,\mathbf{x}_i + \mathbf{B}_i(t)\,\mathbf{m}_i + \mathbf{C}_i(t)\,\mathbf{z}_i \qquad \mathbf{x}_i(t_0) = \mathbf{x}_{0i} \tag{44}$$

$$\mathbf{R}_i(\mathbf{x}_i, \mathbf{m}_i, \mathbf{z}_i, t) \geqslant 0 \tag{45}$$

$$\mathbf{z}_i = \sum_{j=1}^{N} \mathbf{L}_{ij}(t)\,\mathbf{y}_j \tag{46}$$

$$\mathbf{y}_i = \mathbf{M}_i(t)\,\mathbf{x}_i + \mathbf{N}_i(t)\,\mathbf{m}_i \tag{47}$$

$$t_1 \qquad \text{fixed} \tag{48}$$

Equations (44) to (48) define a standard linear system subject to inequality constraints.

Consider the composite system, analogous to Eqs. (7) to (10)

$$\dot{\mathbf{x}} = \mathbf{A}(t)\,\mathbf{x} + \mathbf{B}(t)\,\mathbf{m} + \mathbf{C}(t)\,\mathbf{z} \tag{49}$$

$$\mathbf{R}(\mathbf{x}, \mathbf{m}, \mathbf{z}, \mathbf{t}) \geqslant 0 \tag{50}$$

$$\mathbf{z} = \mathbf{L}(t)\,\mathbf{y} \tag{51}$$

$$\mathbf{y} = \mathbf{M}(t)\,\mathbf{x} + \mathbf{N}(t)\,\mathbf{m} \tag{52}$$

where **A, B, C, L, M,** and **N** are composed of the blocks $\mathbf{A}_i$, $\mathbf{B}_i$, $\mathbf{C}_i$, $\mathbf{L}_{ij}$, $\mathbf{M}_i$, $\mathbf{N}_i$ ordered according to the subsystem components $\mathbf{x}_i$, $\mathbf{m}_i$, $\mathbf{z}_i$ in **x, m, z**. We assume that the controllability and constraint conditions are satisfied by Eqs. (29) to (52).

Finally, consider as an objective function for fixed t_0, t_1

$$J_I(\mathbf{m}, \mathbf{z}) = g(\mathbf{x}(t_1)) + \int_{t_0}^{t_1} f(\mathbf{x}, \mathbf{m}, \mathbf{z}, t)\, dt$$

$$= \sum_{i=1}^{N} \left[g_i(\mathbf{x}_i(\mathbf{t}_1)) + \int_{t_0}^{t_1} f_i(\mathbf{x}_i, \mathbf{m}_i, \mathbf{z}_i, t)\, dt \right] \tag{53}$$

Certain convexity conditions must be imposed on g, f, and **R**. A differentiable function $\phi(\mathbf{x})$ will be said to be convex if for any two points $\mathbf{x}_1$, $\mathbf{x}_2$, $\phi(\mathbf{x}_1) \geqslant \phi(\mathbf{x}_2) + \phi_{\mathbf{x}}(\mathbf{x}_2)(\mathbf{x}_1 - \mathbf{x}_2)$ equivalently

$$\phi(\alpha\mathbf{x}_1 + \beta\mathbf{x}_2) \leqslant \alpha\phi(\mathbf{x}_1) + \beta\phi(\mathbf{x}_2) \qquad \alpha + \beta = 1 \qquad \alpha, \beta \geqslant 0$$

If the inequalities are strict when $\mathbf{x}_1 \neq \mathbf{x}_2$, the convexity is said to be strict.

The utility of this restrictive definition of the integrated problem is that the necessary conditions 1 of Theorem 4-1 are now also sufficient, as the next theorem will show.

Theorem 4-4: Suppose that g, f, and $-\mathbf{R}$ are convex C'' functions of **x, m,** and **z,** and that f is a strictly convex function of **m** and **z,** for each t. If a solution exists to the first-order necessary condition for Eqs. (49) to (53) it provides a unique globally minimizing arc $\mathscr{K}^*$.

Proof: Let $\mathscr{H} = f + \boldsymbol{\lambda}'(\mathbf{Ax} + \mathbf{Bm} + \mathbf{Cz}) - \boldsymbol{\mu}'\mathbf{R} + \boldsymbol{\pi}'\mathbf{L}(\mathbf{Mx} + \mathbf{Nm}) - \boldsymbol{\pi}'\mathbf{z}$, with normality assumed by virtue of the controllability assumption.

The necessary condition 1 of Theorem 4-1 requires that $\mathbf{x}^*$, $\mathbf{m}^*$, $\mathbf{z}^*$, $\boldsymbol{\lambda}^*$, $\boldsymbol{\mu}^*$, $\boldsymbol{\pi}^*$ should be a solution of the following

$$\begin{aligned} &\dot{\boldsymbol{\lambda}} + \mathscr{H}_{\mathbf{x}}' = 0 \\ &\mathscr{H}_{\mathbf{m}} = 0 \qquad \mathscr{H}_{\mathbf{z}} = 0 \\ &\boldsymbol{\mu}'\mathbf{R} = 0 \qquad \boldsymbol{\mu} \geqslant 0 \\ &\mathbf{x}(\mathbf{t}_0) = \mathbf{x}_0 \qquad \mathbf{p}(\mathbf{t}_1) = \mathbf{g}_{\mathbf{x}}(t_1) \end{aligned} \tag{54}$$

where **p** is a multiplier given below. Let **x, m, z** be any other solution of the system equations and constraints (49) to (52), and evaluate $J_I(\mathbf{m}, \mathbf{z})$ along this solution $\mathscr{K}$

$$J_I(\mathbf{m}, \mathbf{z}) = g(\mathbf{x}(t_1)) + \int_{t_0}^{t_1} f(\mathbf{x}, \mathbf{m}, \mathbf{z}, t)\, dt$$

Since $\boldsymbol{\mu}^*\mathbf{R} \geqslant 0$ along any $\mathscr{K}$

$$J_I(\mathbf{m}, \mathbf{z}) \geqslant g(\mathbf{x}(t_1) + \int_{t_0}^{t_1} [f + (\mathbf{p}^*)' (\mathbf{Ax} + \mathbf{Bm} + \mathbf{Cz} - \dot{\mathbf{x}}) - (\boldsymbol{\mu}^*)' \mathbf{R}$$

$$+ (\boldsymbol{\pi}^*)' \mathbf{L}(\mathbf{Mx} + \mathbf{Nm} - \mathbf{z})]\, dt$$

$$= g(\mathbf{x}(t_1)) + \int_{t_0}^{t_1} [\mathscr{H}(\mathbf{x}, \mathbf{m}, \mathbf{z}, \boldsymbol{\lambda}^*, \boldsymbol{\mu}^*, \boldsymbol{\pi}^*) - (\mathbf{p}^*)' \dot{\mathbf{x}}]\, dt$$

using the definition of $\mathscr{H}$.

Recalling the definition of $\mathscr{H}$ above, it is convex in $\mathbf{x}$, $\dot{\mathbf{x}}$ and strictly convex in $\mathbf{m}$, $\mathbf{z}$ for each t. Comparing the curves $\mathscr{K}$ and $\mathscr{K}^*$ using the convexity property gives

$$J_I(\mathbf{m}, \mathbf{z}) - J_I(\mathbf{m}^*, \mathbf{z}^*) \geqslant \mathbf{g}_{\mathbf{x}}^*[\mathbf{x}(t_1) - \mathbf{x}^*(t_1)]$$

$$+ \int_{t_0}^{t_1} [\mathscr{H}_{\mathbf{x}}^*(\mathbf{x} - \mathbf{x}^*) + \mathscr{H}_{\mathbf{m}}^*(\mathbf{m} - \mathbf{m}^*) + \mathscr{H}_{\mathbf{z}}^*(\mathbf{z} - \mathbf{z}^*) - (\mathbf{p}^*)' (\dot{\mathbf{x}} - \dot{\mathbf{x}}^*)]\, dt$$

Now substituting the value of $\mathscr{H}_{\mathbf{x}}$, $\mathscr{H}_{\mathbf{m}}$, $\mathscr{H}_{\mathbf{z}}$ along the optimal curve $\mathscr{K}^*$ and integrating the $\dot{\mathbf{x}}$ terms gives

$$J_I(\mathbf{m}, \mathbf{z}) - J_I(\mathbf{m}^*, \mathbf{z}^*) \geqslant 0 \tag{55}$$

Equality occurs if and only if $(\mathbf{m}, \mathbf{z}) \equiv (\mathbf{m}^*, \mathbf{z}^*)$. Thus any solution to the necessary conditions (49) to (52) and (54) gives a globally minimizing arc $\mathscr{K}^*$.

To show that Eqs. (49) to (52) and (54) admit a unique solution under the convexity condition, note that if $\mathbf{m}$ and $\mathbf{z}$ satisfy these equations, the system of equations (49), (51), and (52) has a unique solution because they are linear. Let $\mathbf{x}^*$, $\mathbf{m}^*$, $\mathbf{z}^*$ and $\mathbf{x}^0$, $\mathbf{m}^0$, $\mathbf{z}^0$ be two solutions to the system (49) to (52) and (54) which give equal minimal performance.

$$J_I(\mathbf{m}^*, \mathbf{z}^*) = J_I(\mathbf{m}^0, \mathbf{z}^0)$$

Now $J_I(\mathbf{m}, \mathbf{z})$ is a convex functional of $\mathbf{x}$, $\mathbf{m}$, and $\mathbf{z}$ and Eqs. (49) to (52) define a convex set of solutions; that is, $(\alpha\mathbf{x}^* + \beta\mathbf{x}^0$, $\alpha\mathbf{m}^* + \beta\mathbf{m}^0$,

$\alpha\mathbf{z}^* + \beta\mathbf{z}^0$) for $\alpha + \beta = 1$, $\alpha, \beta \geqslant 0$, is also a solution because of the linearity of Eqs. (49), (51), and (52) and the convexity of $-\mathbf{R}$ in Eq. (50). By definition

$$-\mathbf{R}(\alpha\mathbf{x}^* + \beta\mathbf{x}^0, \alpha\mathbf{m}^* + \beta\mathbf{m}^0, \alpha\mathbf{z}^* + \beta\mathbf{z}^0)$$
$$\leqslant -\alpha\mathbf{R}(\mathbf{x}^*, \mathbf{m}^*, \mathbf{z}^*) - \beta\mathbf{R}(\mathbf{x}^0, \mathbf{m}^0, \mathbf{z}^0) \leqslant 0$$

From this it follows that there is some combination of $\mathscr{K}^*$ and $\mathscr{K}^0$ which is better still, for using the strict convexity, we have

$$J_I(\alpha\mathbf{m}^* + \beta\mathbf{m}^0, \alpha\mathbf{z}^* + \beta\mathbf{z}^0) = g(\alpha\mathbf{x}^*(t_1) + \beta\mathbf{x}^0(t_1))$$
$$+ \int_{t_0}^{t_1} f(\alpha\mathbf{x}^* + \beta\mathbf{x}^0, \alpha\mathbf{m}^* + \beta\mathbf{m}^0, \alpha\mathbf{z}^* + \beta\mathbf{z}^0, t)\, dt$$
$$< \alpha g(\mathbf{x}^*(t_1)) + \beta g(\mathbf{x}^0(t_1))$$
$$+ \alpha \int_{t_0}^{t_1} f(\mathbf{x}^*, \mathbf{m}^*, \mathbf{z}^*, t)\, dt + \beta \int_{t_0}^{t_1} f(\mathbf{x}^0, \mathbf{m}^0, \mathbf{z}^0, t)\, dt$$
$$= \alpha J_I(\mathbf{m}^*, \mathbf{z}^*) + \beta J_I(\mathbf{m}^0, \mathbf{z}^0)$$
$$= J_I(\mathbf{m}^*, \mathbf{z}^*)$$

Thus, neither $\mathscr{K}^*$ nor $\mathscr{K}^0$ is as good as their combination, which is a contradiction of the optimality of $\mathscr{K}^*$ and $\mathscr{K}^0$ already established. Therefore $\mathscr{K}^*$ must be unique and the theorem is complete.

The *dual problem* concerns the objective function $J_I(\mathbf{m}, \mathbf{z})$ evaluated along the trajectory of the stationary conditions (54) alone. Due to the strict convexity of $f(\mathbf{x}, \mathbf{m}, \mathbf{z}, t)$ in $\mathbf{m}$ and $\mathbf{z}$, they can be eliminated in terms of $\mathbf{x}$, $\boldsymbol{\lambda}$, $\boldsymbol{\mu}$, $\boldsymbol{\pi}$ using $\mathscr{H}_{\mathbf{m}} = 0$, $\mathscr{H}_{\mathbf{z}} = 0$. Thus given $\boldsymbol{\pi}(t)$, $\mathbf{x}(t)$ and $\boldsymbol{\mu}(t) \geqslant 0$, for $t \in [t_0, t_1]$, $\dot{\boldsymbol{\lambda}} + \mathscr{H}_{\mathbf{x}} = 0$ defines $\boldsymbol{\lambda}$ and consequently $\mathbf{m}$ and $\mathbf{z}$.

Theorem 4-5: If f, $\mathbf{g}$, and $-\mathbf{R}$ are convex C'' functions of $\mathbf{x}$, $\mathbf{m}$, $\mathbf{z}$, and f is a strictly convex function of $\mathbf{m}$ and $\mathbf{z}$, then a dual functional $J^d(\mathbf{m}, \mathbf{z})$ ensures a global maximum at the unique solution to the necessary condition (49) to (52) and (54) of $\mathscr{K}^*$.

Define $J^d(\mathbf{m}, \mathbf{z})$ as follows:

$$J^d(\mathbf{m}, \mathbf{z}) = g + \int_{t_0}^{t_1} \{f + \boldsymbol{\lambda}'(\mathbf{Ax} + \mathbf{Bm} + \mathbf{Cz} - \dot{\mathbf{x}}) - \boldsymbol{\mu}'\mathbf{R}$$
$$+ \boldsymbol{\pi}'[\mathbf{L}(\mathbf{Mx} + \mathbf{Nm}) - \mathbf{z}]\}\, dt$$

Let $\mathscr{K}^*$ be defined by $\mathbf{x}^*$, $\mathbf{m}^*$, $\mathbf{z}^*$. Then along $\mathscr{K}^*$ $J_I(\mathbf{m}^*, \mathbf{z}^*) = J^d(\mathbf{m}, \mathbf{z}^*)$ and

$$J^d(\mathbf{m}^*, \mathbf{z}^*) = g(\mathbf{x}^*(t_1)) + \int_{t_0}^{t_1} [\mathscr{H}(\mathbf{x}^*, \mathbf{m}^*, \mathbf{z}^*, \boldsymbol{\lambda}^*, \boldsymbol{\mu}^*, t) - (\boldsymbol{\lambda}^*)' \, \dot{\mathbf{x}}^*] \, dt$$

$$\geqslant g(\mathbf{x}^*(t_1)) + \int_{t_0}^{t_1} [\mathscr{H}(\mathbf{x}^*, \mathbf{m}^*, \mathbf{z}^*, \boldsymbol{\lambda}, \boldsymbol{\mu}, t) - (\boldsymbol{\lambda})' \, \dot{\mathbf{x}}^*] \, dt$$

where $\boldsymbol{\mu} \geqslant 0$, $\mathbf{R}^* \geqslant 0$, and $(\boldsymbol{\lambda})'(\mathbf{Ax}^* + \mathbf{Bm} + \mathbf{Cz}^* - \dot{\mathbf{x}}^*) \equiv 0$. Again $\mathscr{H}$ is a convex function of $\mathbf{x}^*$, $\mathbf{m}^*$, $\mathbf{z}^*$, and $\dot{\mathbf{x}}^*$, so that by comparing the curve $\mathscr{K}^*$ with a curve $\mathscr{K}$ defined by any $(\mathbf{x}, \mathbf{m}, \mathbf{z}, \boldsymbol{\lambda}, \boldsymbol{\mu})$ which satisfies Eq. (54) above, we obtain

$$J(\mathbf{m}^*, \mathbf{z}^*) - J^d(\mathbf{m}, \mathbf{z}) \geqslant g_{\mathbf{x}}(\mathbf{x}^*(t_1) - \mathbf{x}(t_1))$$

$$+ \int_{t_0}^{t_1} [\mathscr{H}_{\mathbf{x}}(\mathbf{x}^* - \mathbf{x}) + \mathscr{H}_{\mathbf{m}}(\mathbf{m}^* - \mathbf{m}) + \mathscr{H}_{\mathbf{z}}(\mathbf{z}^* - \mathbf{z}) - (\boldsymbol{\lambda}^*)'(\dot{\mathbf{x}}^* - \dot{\mathbf{x}})] \, dt$$

Using Eq. (54) to substitute for $\mathscr{H}_{\mathbf{x}}$, $\mathscr{H}_{\mathbf{m}}$, $\mathscr{H}_{\mathbf{z}}$ and integrating the $\dot{\mathbf{x}}$ terms yields

$$J_I(\mathbf{m}^*, \mathbf{z}^*) = J^d(\mathbf{m}^*, \mathbf{z}^*) - J^d(\mathbf{m}, \mathbf{z}) \geqslant 0 \tag{56}$$

Due to the strict convexity of $\mathscr{H}$ with respect to $\mathbf{m}$ and $\mathbf{z}$, equality will occur if $(\mathbf{x}^*, \mathbf{m}^*, \mathbf{z}^*) \equiv (\mathbf{x}, \mathbf{m}, \mathbf{z})$. However, since f does not have strict convexity to $\mathbf{x}$ it is possible that there are other values $\mathbf{x}^0$, $\mathbf{m}^*$, $\mathbf{z}^*$ which satisfy Eq. (54), giving equality (56) but not satisfaction of Eqs. (49) to (52). This question of nonuniqueness is resolved in the next section by requiring that $\mathbf{x}$, $\mathbf{m}$, $\mathbf{z}$ satisfy the system equation (49).

Theorems 4-4 and 4-5 demonstrate an appealing symmetry between the two problems and show that Eqs. (49) to (52) and (54) constitute a reciprocal pair. Each set of equations provides necessary and sufficient conditions for a global optimum of an objective function when it is constrained by the other set. Another result that follows directly from the theorems is that

$$J^d(\mathbf{m}, \mathbf{z}) \leqslant J^d(\mathbf{m}^*, \mathbf{z}^*) = J_I(\mathbf{m}^*, \mathbf{z}^*) \leqslant J_I(\mathbf{m}, \mathbf{z}) \tag{57}$$

By this means the optimality of a given possible trajectory, which does not satisfy all the necessary conditions, can be estimated from the upper and lower bounds.

Example 4-1

The quadratic "regulator problem" provides a good example. Let $\mathbf{P}, \mathbf{Q}_1, \mathbf{Q}_2$, and $\mathbf{Q}_3$ be positive definite symmetric matrices and let $\mathbf{R}_1$ and $\mathbf{R}_2$ be $l \times n$ and $l \times r$ matrices such that $\mathbf{R}_1\mathbf{x} + \mathbf{R}_2\mathbf{m} \geqslant 0$ satisfies the constraint condition.

PRIMAL PROBLEM:

$$\min J_l(\mathbf{m}, \mathbf{z}) = \tfrac{1}{2} \| \mathbf{x}(t_1)\|^2 \mathbf{P} + \int_{t_0}^{t_1} \tfrac{1}{2}(\| \mathbf{x} \|^2 \mathbf{Q}_1 + \| \mathbf{m} \|^2 \mathbf{Q}_2 + \| \mathbf{z} \|^2 \mathbf{Q}_3)\, dt$$

$$\text{subject to } \dot{\mathbf{x}} = \mathbf{A}\mathbf{x} + \mathbf{B}\mathbf{m} + \mathbf{C}\mathbf{z} \qquad \mathbf{x}(t_0) = \mathbf{x}_0$$

$$\mathbf{y} = \mathbf{M}\mathbf{x} + \mathbf{N}\mathbf{m}$$

$$\mathbf{z} = \mathbf{L}\mathbf{y}$$

$$\mathbf{R}_1\mathbf{x} + \mathbf{R}_2\mathbf{m} \geqslant 0$$

The necessary conditions for this problem require that $\boldsymbol{\lambda}, \boldsymbol{\mu}, \boldsymbol{\pi}$ satisfy

$$\dot{\boldsymbol{\lambda}} + \mathbf{Q}_1\mathbf{x} + \mathbf{A}'\boldsymbol{\lambda} - \mathbf{R}_1'\boldsymbol{\mu} + \mathbf{M}'\mathbf{L}'\boldsymbol{\pi} = 0$$

$$\mathbf{Q}_2\mathbf{m} + \mathbf{B}'\boldsymbol{\lambda} - \mathbf{R}_2'\boldsymbol{\mu} + \mathbf{N}'\mathbf{L}'\boldsymbol{\pi} = 0$$

$$\mathbf{Q}_3\mathbf{z} + \mathbf{C}'\boldsymbol{\lambda} - \boldsymbol{\pi} = 0$$

$$\boldsymbol{\mu}'(\mathbf{R}_1\mathbf{x} + \mathbf{R}_2\mathbf{m}) = 0 \qquad \boldsymbol{\lambda}(t_1) = \mathbf{P}\mathbf{x}(t_1) \qquad \boldsymbol{\mu} \geqslant 0$$

Along the optimal trajectory $J^d(\mathbf{m}, \mathbf{z})$ is equivalent to

$$\tfrac{1}{2} \| \mathbf{x}(t_1)\|^2 \mathbf{P} + \int_{t_0}^{t_1} [\tfrac{1}{2}(\| \mathbf{x} \|^2 \mathbf{Q}_1 + \| \mathbf{m} \|^2 \mathbf{Q}_2 + \| \mathbf{z} \|^2 \mathbf{Q}_3) + \boldsymbol{\lambda}'(\mathbf{A}\mathbf{x} + \mathbf{B}\mathbf{m} + \mathbf{C}\mathbf{z} - \dot{\mathbf{x}}) - \boldsymbol{\mu}'(\mathbf{R}_1\mathbf{x} + \mathbf{R}_2\mathbf{m}) + \boldsymbol{\pi}'(\mathbf{L}\mathbf{M}\mathbf{x} + \mathbf{L}\mathbf{N}\mathbf{m} - \mathbf{z})]\, dt$$

Integrating by parts and substituting for $\dot{\boldsymbol{\lambda}}$ then leads to

$$\mathbf{x}_0'\boldsymbol{\lambda}(t_0) - \tfrac{1}{2} \| \mathbf{x}(t_1)\|^2 \mathbf{P} - \tfrac{1}{2} \int_{t_0}^{t_1} (\| \mathbf{x} \|^2 \mathbf{Q}_1 + \| \mathbf{m} \|^2 \mathbf{Q}_2 + \| \mathbf{z} \|^2 \mathbf{Q}_3)\, dt$$

DUAL PROBLEM:

$$\max J^d(\mathbf{x}, \mathbf{m}, \mathbf{z}, \boldsymbol{\mu}) = \mathbf{x}_0'\boldsymbol{\lambda}(t_0) - \tfrac{1}{2}\| \mathbf{x}(t_1)\|^2 \mathbf{P}$$

$$- \tfrac{1}{2}\int_{t_0}^{t_1} (\| \mathbf{x} \|^2 \mathbf{Q}_1 + \| \mathbf{m} \|^2 \mathbf{Q}_2 + \| \mathbf{z} \|^2 \mathbf{Q}_3)\, dt$$

subject to $\dot{\boldsymbol{\lambda}} + \mathbf{Q}_1\mathbf{x} + \mathbf{A}'\boldsymbol{\lambda} - \mathbf{R}_1'\boldsymbol{\mu} + \mathbf{M}'\boldsymbol{\kappa} = 0$

$$\mathbf{Q}_2\mathbf{m} + \mathbf{B}'\boldsymbol{\lambda} - \mathbf{R}_2'\boldsymbol{\mu} + \mathbf{N}'\boldsymbol{\kappa} = 0$$

$$\mathbf{Q}_3\mathbf{z} + \mathbf{C}'\boldsymbol{\lambda} - \boldsymbol{\pi} = 0$$

$$\boldsymbol{\lambda}(t_1) - \mathbf{P}\mathbf{x}(t_1) = 0$$

$$\boldsymbol{\kappa} = \mathbf{L}'\boldsymbol{\pi} \qquad \boldsymbol{\mu} \geqslant 0$$

The primal variables can be eliminated by using the first three equations to give more symmetry, as shown by Pearson.[15] The concavity of J^d is evident and it is straightforward to check that the necessary conditions for a maximum generate the original optimal constraint equations. This problem will be used in Sec. 4-7 in the discussion of model coordination.

4-4 APPLICATION OF DUALITY TO DECOMPOSITION

In Sec. 4-2 it was shown that the general structured optimal control problem could be decomposed into two parts, or levels of computation. Written in terms of the linear problem of Sec. 4-2, these are

LEVEL A:

Minimize $J_i(\mathbf{m}_i, \mathbf{z}_i, \boldsymbol{\pi}_i)$ for a given $\boldsymbol{\pi}$, where

$$J_i(\mathbf{m}_i, \mathbf{z}_i, \boldsymbol{\pi}_i)$$

$$= g_i(\mathbf{x}_{1i}) + \int_{t_0}^{t_1} [f_i(\mathbf{x}_i, \mathbf{m}_i, \mathbf{z}_i, t) + \boldsymbol{\kappa}_i(\mathbf{M}_i(t)\, \mathbf{x}_i + \mathbf{N}_i(t)\, \mathbf{m}_i) - \boldsymbol{\pi}_i\mathbf{z}_i]\, dt$$

subject to subsystem constraints

$$\dot{\mathbf{x}}_i = \mathbf{A}_1(t)\, \mathbf{x}_i + \mathbf{B}_i(t)\, \mathbf{m}_i + \mathbf{C}_1(\mathrm{t})\, \mathbf{z}_i \qquad \mathbf{x}_i(t_0) = \mathbf{x}_{i0} \tag{59}$$

$$\mathbf{R}_i(\mathbf{x}_i, \mathbf{m}_i, \mathbf{z}_i, t) \geqslant 0 \tag{60}$$

$$\boldsymbol{\kappa}_i = \sum_{j=1}^{N} \mathbf{L}_{ji}'\boldsymbol{\pi}_j \tag{61}$$

$$t_1 \quad \text{fixed}$$

Each of these subproblems for $i = 1, 2, \ldots, N$ is minimized. Assuming $\boldsymbol{\pi} \in \Omega$ and subject to the convexity conditions on g, f, and $-\mathbf{R}$, and constraint conditions on $\mathbf{R}$, a global minimum $\mathbf{x}_i^*(\boldsymbol{\pi})$, $\mathbf{m}_i^*(\boldsymbol{\pi})$, $\mathbf{z}_1^*(\boldsymbol{\pi})$ is obtained for $\boldsymbol{\pi} \in \Omega$, the class of piecewise continuous s-dimensional vector functions of time $t \in [t_0, t_1]$, for which J_i has a minimum.

After the subproblems are solved, a set of multipliers $\boldsymbol{\lambda}_i^*(\boldsymbol{\pi})$, $\boldsymbol{\mu}_i^*(\boldsymbol{\pi})$ are obtained which satisfy stationary conditions for the problem.

The dual to the integrated problem is given by $J^d(\mathbf{m}, \mathbf{z})$ in Theorem 4-5 and has the form

$$J^d(\mathbf{m}, \mathbf{z}) = g + \int_{t_0}^{t_1} \{f + \boldsymbol{\lambda}'(\mathbf{Ax} + \mathbf{Bm} + \mathbf{Cz} - \dot{\mathbf{x}}) - \boldsymbol{\mu}'\mathbf{R} + \boldsymbol{\pi}'[\mathbf{L}(\mathbf{Mx} + \mathbf{Nm}) - \mathbf{z}]\}\, dt \tag{62}$$

Along trajectories $\mathscr{K}_i^*(\boldsymbol{\pi})$, however, Eq. (59) is satisfied and similarly $\boldsymbol{\mu}_i^*(\boldsymbol{\pi})'\mathbf{R}_i(\mathbf{x}_i^*(\boldsymbol{\pi}), \mathbf{m}_i^*(\boldsymbol{\pi}), \mathbf{z}_i^*(\boldsymbol{\pi}), t) \equiv 0$ for $i = 1, 2, \ldots, N$. Consequently the dual problem is simplified

$$\begin{aligned} J^d(\mathbf{m}^*(\boldsymbol{\pi}), \mathbf{z}^*(\boldsymbol{\pi})) &= g(\mathbf{x}^*(\boldsymbol{\pi}, t_1)) + \int_{t_0}^{t_1} \{f(\mathbf{x}^*(\boldsymbol{\pi}), \mathbf{m}^*(\boldsymbol{\pi}), \mathbf{z}^*(\boldsymbol{\pi}), t) \\ &\quad + \boldsymbol{\pi}'[\mathbf{L}(\mathbf{Mx}^*(\boldsymbol{\pi}) + \mathbf{Nm}^*(\boldsymbol{\pi})) - \mathbf{z}^*(\boldsymbol{\pi})]\}\, dt \\ &= \sum_{i=1}^{N} J_i(\mathbf{m}_i^*(\boldsymbol{\pi}), \mathbf{z}_i^*(\boldsymbol{\pi}), \boldsymbol{\pi}) \end{aligned} \tag{63}$$

The last equation follows from the definition of J_i in Eq. (58). Thus the dual problem objective function is the sum of the subproblem optimum performance indices and corresponds to $J(\boldsymbol{\pi})$ defined by Eq. (40). It can now be seen why $J(\boldsymbol{\pi})$ should be maximized, for given an arbitrary $\boldsymbol{\pi}$ the necessary conditions 1′ for the subproblem coincide with the necessary conditions 1 for the integrated problem, except for the interconnection constraint. However, this constraint is a necessary condition for the dual problem which is therefore not optimum. Thus, choosing $\boldsymbol{\pi}$ to enforce $\mathbf{z}^*(\boldsymbol{\pi}) = \mathbf{L}[\mathbf{Mx}^*(\boldsymbol{\pi}) + \mathbf{Nm}^*(\boldsymbol{\pi})]$ completes the conditions for the maximization of the dual problem. This is summarized as follows:[17]

Theorem 4-6: If (1) g, f, and $-\mathbf{R}$ are convex functions of $\mathbf{x}$, $\mathbf{m}$, $\mathbf{z}$ for each t, (2) f is strictly convex in $\mathbf{m}$ and $\mathbf{z}$ for each t, (3) $\mathbf{R}$ satisfies the constraint condition, (4) the system (44), (46), and (47) is a bounded controllable linear system, and (5) solutions exist for the integrated problem and each subproblem for $\boldsymbol{\pi} \in \Omega$, then (*a*) each subproblem and the integrated problem have globally minimizing solutions $\mathbf{x}^*$, $\mathbf{m}^*$, $\mathbf{z}^*$

and $\mathbf{x}_i^*(\boldsymbol{\pi})$, $\mathbf{m}_i^*(\boldsymbol{\pi})$, and $\mathbf{z}_i^*(\boldsymbol{\pi})$, $i = 1, 2,\ldots, N$, which are unique, and (*b*) the coordination problem can be solved by maximizing $J(\boldsymbol{\pi}) = J^d(\mathbf{m}^*(\boldsymbol{\pi}), \mathbf{z}^*(\boldsymbol{\pi}))$ and its solution is unique.

Proof: The conditions 1 to 5 catalog the assumptions made during the course of the development and result in Theorem 4-4, which demonstrates part (*a*), when applied to the integrated problem and each subproblem.

Since the coordination problem of finding $\boldsymbol{\pi}$ is identified with the dual problem of Sec. 4-3, it can be seen that Theorem 4-5 applies only where $\mathbf{x}^*(\boldsymbol{\pi})$, $\mathbf{m}^*(\boldsymbol{\pi})$, $\mathbf{z}^*(\boldsymbol{\pi})$ are functions which satisfy the system constraint (59) for $i = 1, 2,\ldots, N$ and as a consequence $\mathbf{x}^*(\boldsymbol{\pi})$ is uniquely determined by $\mathbf{m}^*(\boldsymbol{\pi})$ and $\mathbf{z}^*(\boldsymbol{\pi})$. The dual problem then has a unique maximizing solution $\mathbf{x}^*$, $\mathbf{m}^*$, $\mathbf{z}^*$ which satisfies interconnection constraints as a necessary condition. This completes the proof.

It must be emphasized that the conditions are sufficient and it would be expected that in general these results would hold locally for a more general system. This is in essence the connection between the arguments in Sec. 4-2 and in Sec. 4-4.

It does not follow that if $J(\mathbf{m}, \mathbf{z})$ has a minimum then $J_i(\mathbf{m}_i, \mathbf{z}_i, \boldsymbol{\pi})$ will have a minimum, even under strict convexity assumptions, for consider $J_I(m, z)$ defined in the example by.

$$J_I(m, z) = \int_0^{1/2} [m^2 + z + x + \exp(-z - x) - 1]\, dt$$

with

$$\dot{x} = z + m \qquad x(0) = 0 \qquad z = x$$

Here the feedback in a first-order system is explicit. The integrand $m^2 + z + x + \exp(-z - x)$ is strictly convex in m, z, and x. The single subproblem obtained by severing the feedback is

$$J_1(m, z, \pi) = \int_0^{1/2} [m^2 + z + x + \exp(-z - x) - 1 + \pi(z - x)]\, dt$$

for $\dot{\mathbf{x}} = z + m$, $x(0) = 0$. Now the integrand is unbounded below in x or z for some values of π. Thus by choosing, for example,

$$z = a \qquad \pi = -2 \qquad m = 0 \qquad \text{for} \quad t \in [0, \tfrac{1}{2}]$$

$$J_1(0, a, -2) = -a \int^{1/2} \left(1 - 3t - \frac{e^{-a(1+t)}}{a}\right) dt$$

$$= \frac{e^{-a}(1 - e^{-a/2})}{a} - \frac{a}{8}$$

This is unbounded below as a function of a and can be made as small as desired by moving $a \to +\infty$. On the other hand it is easy to see that $x^* = m^* = z^* = \pi^* = 0$, and this is a global minimum. What has happened is that the integrand of $J_1(m, z, \pi)$ violates a common existence requirement that, for example,

$$f(x, m, z) > \alpha \| z \| + \beta \| m \|$$

for some $(\alpha, \beta) > 0$. Thus, as a parametric function of π, the integrand can be dominated by the linear part influenced by π and thus become unbounded.

4-4-1 Properties of $J(\pi)$

To complete this section, we will briefly investigate the properties of $J(\boldsymbol{\pi}) = \sum_{i=1}^{N} J_i(\mathbf{m}_i^*(\boldsymbol{\pi}), \mathbf{z}_i^*(\boldsymbol{\pi}), \boldsymbol{\pi})$ under more restrictive assumptions on the control sets. The results obtained are intended to be illustrative rather than completely general.

The following assumptions will be made for each subsystem.

ASSUMPTION 1:

The inputs $\mathbf{m}$, $\mathbf{z}$ to the noninteracting system will be drawn from convex compact subsets of a space of $\mathbf{L}^2$ functions defined on $[t_0, t_1]$ such that $\mathbf{x}$, $\mathbf{m}$, $\mathbf{z}$ defined by solutions to

$$\dot{\mathbf{x}} = \mathbf{A}(t)\,\mathbf{x} + \mathbf{B}(t)\,\mathbf{m} + \mathbf{C}\mathbf{z} \qquad \mathbf{x}(t_0) = \mathbf{x}_0$$

also satisfy

$$\mathbf{R}(\mathbf{x}, \mathbf{m}, \mathbf{z}, t) \geqslant 0$$

and $(\mathbf{x}, t)$, $\mathbf{m}$, $\mathbf{z}$ take values in $\mathscr{D}$, $\mathscr{M}$, and $\mathscr{Z}$ for each t.

ASSUMPTION 2:

The subsystems are uniformly completely controllable on $[t_0, t_1]$. See Kalman, Ho, and Narendra.[8]

ASSUMPTION 3:

The function of f, g, and $-\mathbf{R}$ will be convex C'' functions of $\mathbf{x}$, $\mathbf{m}$, $\mathbf{z}$ for each t with strict convexity of f in $\mathbf{m}$ and $\mathbf{z}$.

Assumptions 1 to 3 have the following immediate consequences.

Consider the linear system as a uniformly continuous mapping of the space of $\mathbf{L}^2[t_0, t_1]$ functions $(\mathbf{m}, \mathbf{z})$ to the space of $\mathbf{L}^2[t_0, t_1]$ functions $\mathbf{x}$. Since a continuous map of a compact set is compact, the range space $\mathbf{x}$ is compact. Similarly the product space formed from the compact spaces of $\mathbf{x}$, $\mathbf{m}$, $\mathbf{z}$ is compact. Furthermore, if $\mathbf{x}^1$, $\mathbf{m}^1$, $\mathbf{z}^1$ and $\mathbf{x}^2$, $\mathbf{m}^2$, $\mathbf{z}^2$ are points in the product space, it follows from the convexity of $-\mathbf{R}(\mathbf{x}, \mathbf{m}, \mathbf{z}, t)$ and the system linearity that any convex combination of these points is in the product space, which is consequently also compact.

The subproblem is to minimize the parametric problem $J_i(\mathbf{m}_i, \mathbf{z}_i, \boldsymbol{\pi})$ for $\boldsymbol{\pi} \in \Omega$, the set of piecewise C' functions defined on $[t_0, t_1]$ for which J_i has a minimum where

$$J_i(\mathbf{m}_i, \mathbf{z}_i, \boldsymbol{\pi}) = g_i(\mathbf{x}_i(t_1)) + \int_{t_0}^{t_1} [f_i(\mathbf{x}_i, \mathbf{m}_i, \mathbf{z}_i, t)$$

$$+ \boldsymbol{\kappa}_i'(\mathbf{M}_i\mathbf{x}_i + \mathbf{N}_i\mathbf{z}_i) - \boldsymbol{\pi}_i'\mathbf{z}_i]\, dt$$

and

$$\boldsymbol{\kappa}_i = \sum_{j=1}^{\mathbf{N}} \mathbf{L}'_{ji}\boldsymbol{\pi}_j$$

Under assumptions 1 to 3 an equivalent problem is to minimize a piecewise continuous functional of $\mathbf{x}_i$, $\mathbf{m}_i$, $\mathbf{z}_i$ on a compact space, for any given piecewise continuous $\boldsymbol{\pi} \in \Omega$. It is well known that a lower semicontinuous function achieves its lower bound on a compact metric space. Since J_i is of class C'' in $(\mathbf{x}_i, \mathbf{m}_i, \mathbf{z}_i)$, it is also lower semicontinuous and thus the subproblems have a solution for any $\boldsymbol{\pi}$, and Ω is the whole space of piecewise continuous functions. Thus compactness essentially assumes away the problem of existence of a solution to the subproblems.

For convenience let $\mathscr{X}$ be the set

$$\mathscr{X} = \{\mathbf{X} = (\mathbf{x}, \mathbf{m}, \mathbf{z}) : (\mathbf{m}, z) \in \mathbf{L}^2[t_0, t_1], \dot{\mathbf{x}} = \mathbf{A}(t)\,\mathbf{x} + \mathbf{B}(t)\,\mathbf{m}, +\mathbf{C}(t)\mathbf{z},$$

$$\mathbf{R}(\mathbf{x}, \mathbf{m}, \mathbf{z}, t) \geqslant 0, \mathbf{x}(t_0) = \mathbf{x}_0, t \in [t_0, t_1]\}$$

Let $\mathscr{X}_i = \{\mathbf{X}_i = (\mathbf{x}_i, \mathbf{m}_i, \mathbf{z}_i)\colon i = 1, 2, \ldots, N\}$, then $\mathscr{X}_i$ and $\mathscr{X}$ are compact convex subsets of a Banach space under assumptions 1 to 3.

The following results follow closely the work of Falk[5] for the similar problem posed in a finite dimensional space.

The next theorem replaces Lemma 4-2.

Theorem 4-7: The functional $J(\boldsymbol{\pi}) = \sum_{i=1}^{N} J_i(\mathbf{m}_i^*(\boldsymbol{\pi}), \mathbf{x}_i^*(\boldsymbol{\pi}), \boldsymbol{\pi})$ is concave over convex subsets of Ω.

Proof: Let $\boldsymbol{\pi}^1 \in \Omega$, $\boldsymbol{\pi}^2 \in \Omega$, choose $\alpha \geqslant 0$, $\beta \geqslant 0$, $\alpha + \beta = 1$, and define $\boldsymbol{\pi}^3 = \alpha\boldsymbol{\pi}^1 + \beta\boldsymbol{\pi}^2$ with the assumption $\boldsymbol{\pi}^3 \in \Omega$.

$$J(\boldsymbol{\pi}^3) = \min \left\{ \sum_{i=1}^{N} [J_i(\mathbf{m}_i, \mathbf{z}_i, \boldsymbol{\pi}^3), \mathbf{X}_i \in \mathscr{X}_i] \right\}$$

$$= \min \left\{ \sum_{i=1}^{N} [\alpha J_i(\mathbf{m}_i, \mathbf{z}_i, \boldsymbol{\pi}^1) + \beta J_i(\mathbf{m}_i, \mathbf{z}_i, \boldsymbol{\pi})^2; \mathbf{X}_i \in \mathscr{X}_i] \right\}$$

$$\geqslant \alpha \min \left\{ \sum_{i=1}^{N} [J_i(\mathbf{m}_i, \mathbf{z}_i, \boldsymbol{\pi}^1); \mathbf{X}_i \in \mathscr{X}_i] \right\}$$

$$+ \beta \min \left\{ \sum_{i=1}^{N} [J_i(\mathbf{m}_i, \mathbf{z}_i, \boldsymbol{\pi}^2); \mathbf{X}_i \in \mathscr{X}_i] \right\}$$

$$= \alpha J(\boldsymbol{\pi}^1) + \beta J(\boldsymbol{\pi}^2)$$

This is the definition of concavity of $\mathbf{J}(\boldsymbol{\pi})$.

The next important question is whether solutions $\mathbf{X}^*(\boldsymbol{\pi})$ are continuous functions of $\boldsymbol{\pi} \in \Omega$.

Theorem 4-8: Under assumptions 1 to 3 for each $\boldsymbol{\pi} \in \Omega$, the subproblem solutions $\mathbf{X}^*(\boldsymbol{\pi}) = (\mathbf{x}^*(\boldsymbol{\pi}), \mathbf{m}^*(\boldsymbol{\pi}), \mathbf{m}^*(\boldsymbol{\pi}), \mathbf{z}^*(\boldsymbol{\pi}))$ are continuous functions of $\boldsymbol{\pi}$.

Proof: As before let $\mathbf{X} = \mathbf{x}, \mathbf{m}, \mathbf{z}$ be an element of $\mathscr{X}$, the compact convex admissible set of solutions. Define $\| \mathbf{x} \|^2$ as

$$\left[\int_{t_0}^{t_1} [\mathbf{x}'(t)\,\mathbf{x}(t)]\, dt \right] \qquad \text{and} \qquad \| \mathbf{X} \| = \max[\| \mathbf{x} \|, \| \mathbf{m} \|, \| \mathbf{z} \|]$$

To prove continuity of $\mathbf{X}^*(\boldsymbol{\pi}^*)$ we have to show that for a fixed $\epsilon > 0$ and $\boldsymbol{\pi}^*$, there always exists a $\delta > 0$ such that if $\| \mathbf{X} - \mathbf{X}^*(\boldsymbol{\pi}^*) \| < \epsilon$ then $\| \boldsymbol{\pi} - \boldsymbol{\pi}^* \| < \delta$.

Let $\mathbf{X}^* \equiv \mathbf{X}^*(\boldsymbol{\pi})$ and define $\mathbf{k} > 0$ such that

$$\mathbf{k} > \max[\| (\mathbf{LMx} + \mathbf{LNm} - \mathbf{z}) - (\mathbf{LMx}^* + \mathbf{LNm}^* - \mathbf{z}^*) \| \cdot (t_1 - t_0)\, \mathbf{X} \in \mathscr{X} \text{ and } \| \mathbf{X} - \mathbf{X}^* \| \geqslant \epsilon]$$

Thus $\mathbf{k}$ is the biggest deviation of the interaction term throughout $\mathbf{X}$ outside of the ϵ region around $\mathbf{X}^*$.

Let $\delta > 0$ be any number such that

$$\left[\sum_{i=1}^{N} J_i(\mathbf{m}_i, \mathbf{z}_i, \boldsymbol{\pi}^*) - \sum_{i=1}^{N} J_i(\mathbf{m}_i^*(\boldsymbol{\pi}^*), \mathbf{z}_i^*(\boldsymbol{\pi}^*), \boldsymbol{\pi}^*) \right] > \mathbf{k} \cdot \delta$$

for $\mathbf{X} \in \mathscr{X}$ and $\| \mathbf{X} - \mathbf{X}^* \| \geqslant \epsilon$. Now choosing $\pi \in \Omega$ such that $\| \pi - \pi^* \| < \delta$ and using the Schwartz inequality

$$\mathbf{k} \cdot \delta > \|(\mathbf{LMx} + \mathbf{LNm} - \mathbf{z}) - (\mathbf{LMx}^* + \mathbf{LNm}^* - \mathbf{z}^*)\| \cdot \| \pi^* - \pi \| \cdot (t_1 - t_0)$$

$$\geqslant \int_{t_0}^{t_1} [(\mathbf{LMx} + \mathbf{LNm} - \mathbf{z}) - (\mathbf{LMx}^* + \mathbf{LNm}^* - \mathbf{z}^*)]' \, [\pi^* - \pi] \, dt$$

consequently substituting for J_i in the definition of δ, we have

$$\sum_{i=1}^{\mathbf{N}} \left\{ \mathbf{g}_i(\mathbf{x}_i(t_1)) + \int_{t_0}^{t_1} [f_i + (\pi^*)' \, (\mathbf{LMx} + \mathbf{LNm} - \mathbf{z})] \, dt \right\}$$

$$- \sum_{i=1}^{\mathbf{N}} \left\{ \mathbf{g}_i(\mathbf{x}_i^*(t_1)) + \int_{t_0}^{t_1} [f_i^* + (\pi^*)' \, (\mathbf{LMx}^* + \mathbf{LNm}^* - \mathbf{z}^*)] \, dt \right\} > \mathbf{k}\delta$$

$$> \int_{t_0}^{t_1} [(\mathbf{LMx} + \mathbf{LNm} - \mathbf{z}) - (\mathbf{LMx}^* + \mathbf{LNm}^* - \mathbf{z}^*)]' \, [\pi^* - \pi] \, dt$$

Rearranging and canceling terms in π^* leads to

$$\sum_{i=1}^{\mathbf{N}} J_i(\mathbf{m}_i, \mathbf{z}_i, \pi) > \sum_{i=1}^{\mathbf{N}} J_i(\mathbf{m}_i^*(\pi^*), \mathbf{z}_i^*(\pi^*), \pi)$$

for all $(\mathbf{x}, \mathbf{m}, \mathbf{z}) = \mathbf{X} \in \mathscr{X}$ such that $\| \mathbf{X} - \mathbf{X}^* \| \geqslant \epsilon$.

This shows that for π such that $\| \pi - \pi^* \| < \delta$, the solution $\mathbf{X}^* = \mathbf{X}^*(\pi^*)$ is better than any other admissible $\mathbf{X}$ outside of the ϵ region. However, the strict convexity of each $J_i(\mathbf{m}_i, \mathbf{z}_i, \pi)$ requires that

$$J_i(\mathbf{m}_i^*(\pi), \mathbf{z}_i^*(\pi), \pi) < J_i(\mathbf{m}_i^*(\pi^*), \mathbf{z}_i^*(\pi^*), \pi)$$

for $\pi \neq \pi^*$ but $\| \pi - \pi^* \| < \delta$. Thus $\mathbf{X}^*(\pi)$ must be inside the ϵ region $\| \mathbf{X}^*(\pi) - \mathbf{X}^*(\pi^*)\| < \epsilon$ for $\| \pi - \pi^* \| < \delta$, which establishes the continuity of $\mathbf{X}^*(\pi)$. It remains to prove that $J(\pi)$ has a directional derivative as computed in Sec. 4-3.

Theorem 4-9: Under assumptions 1 to 3 the directional derivative $dJ(\pi)/d\tau$ exists and is a continuous function of π throughout the interior of Ω.

Proof: Fix π in the interior of Ω and let η denote a piecewise continuous perturbation function of $t \in [t_0, t_1]$. For $\tau \neq 0$ we have

$$\frac{J(\pi + \tau \cdot \eta) - J(\pi)}{\tau}$$

$$\leqslant \sum_{i=1}^{N} \frac{J_i(\mathbf{m}_i^*(\pi), \mathbf{z}_i^*(\pi), \pi + \tau \cdot \eta) - J_i(\mathbf{m}_i^*(\pi), \mathbf{z}_i^*(\pi), \pi)}{\tau}$$

$$= \int_{t_0}^{t_1} \eta'[\mathbf{LMx}^*(\pi) + \mathbf{LMm}^*(\pi) - \mathbf{z}^*(\pi)]\, dt$$

since each J_i is linear in π.

On the other hand,

$$\frac{J(\pi + \tau \cdot \eta) - J(\pi)}{\tau} \geqslant \frac{1}{\tau} \sum_{i=1}^{N} [J_i(\mathbf{m}_i^*(\pi + \tau\eta), \mathbf{z}_i^*(\pi + \tau\eta), \pi + \tau\eta)$$

$$- J_i(\mathbf{m}_i^*(\pi + \tau\eta), \mathbf{z}_i^*(\pi + \tau\eta), \pi)]$$

$$= \int_{t_0}^{t_1} \eta'[\mathbf{LMx}^*(\pi + \tau\eta) + \mathbf{LNm}^*(\pi + \tau\eta)$$

$$- \mathbf{z}^*(\pi + \tau\eta)]\, dt$$

Since $\mathbf{x}^*(\pi)$, $\mathbf{m}^*(\pi)$, $\mathbf{z}(\pi)$ are continuous functions of π,

$$\frac{dJ(\pi)}{d\tau} = \lim_{\tau \to 0} \frac{J(\pi + \tau \cdot \eta) - J(\pi)}{\tau} = \int_{t_0}^{t_1} \eta'[\mathbf{LMx}^* + \mathbf{LNm}^* - \mathbf{z}^*)\, dt$$

and this is a continuous function of π, which completes the proof.

Since this $dJ(\pi)/d\tau$ is linear in η we will also define the Fréchet derivative $\delta J(\pi)/\delta\pi$ as

$$\frac{\delta J(\pi)}{\delta \pi} = \mathbf{LMx}^*(\pi) + \mathbf{LNm}^*(\pi) - \mathbf{z}^*(\pi)$$

This completes the theorem and this section.

4-5 COMPUTATIONAL CONSIDERATIONS

It can be seen that the decomposition technique has essentially recast the "min" problem as a "max-min" problem. The inside minimization depends on a parameter, the function π, and the outside maximization is over a class of functions defined implicitly in terms of π.

Ideally, methods are required which can manipulate parametric problems readily. For example, given the previous subproblem solution and sensitivity results, we find the new solutions with very little extra effort. As will be shown, only in the case of linear necessary conditions is this really possible.

At level 1, in the goal-coordination methods, the problem is to minimize $J_i(\mathbf{m}_i, \mathbf{z}_i, \boldsymbol{\pi})$ for a given $\boldsymbol{\pi}$, for each $i = 1, 2, \ldots, N$. This can be done by any suitable method for a particular subproblem. The result is that $\mathbf{x}_i^*(\boldsymbol{\pi})$, $\mathbf{m}_i^*(\boldsymbol{\pi})$, $\mathbf{z}_i^*(\boldsymbol{\pi})$, $\boldsymbol{\lambda}_i^*(\boldsymbol{\pi})$, $\boldsymbol{\mu}_i^*(\boldsymbol{\pi})$, and $\boldsymbol{\pi}$ are all available as functions of t.

At level 2 some method of adjusting $\boldsymbol{\pi}$ must be found which does not use subsystem constraint information directly. Ideally, the second level need be concerned only with violated constraints. We will now focus attention on the second-level problem.

The previous section has shown that given a direction or perturbation η, the total derivative of the effect of $\boldsymbol{\pi} + \tau\boldsymbol{\eta}$ is given by

$$\left.\frac{dJ(\boldsymbol{\pi} + \tau\boldsymbol{\eta})}{d\tau}\right|_{\tau=0} = \int_{t_0}^{t_1} [\mathbf{LMx}^*(\boldsymbol{\pi}) + \mathbf{LNm}^*(\boldsymbol{\pi}) - \mathbf{z}^*(\boldsymbol{\pi})]' \, \eta \, dt$$

while the variational or Fréchet derivative of $J(\boldsymbol{\pi})$ at $\boldsymbol{\pi}$ is given by

$$\frac{\delta J(\boldsymbol{\pi})}{\delta \boldsymbol{\pi}} = \mathbf{LMx}^*(\boldsymbol{\pi}) + \mathbf{LNm}^*(\boldsymbol{\pi}) - \mathbf{z}^*(\boldsymbol{\pi})$$

The interesting point about both derivatives is that they use only information from violated constraints. Since these derivatives are both "first order," we will examine first-order methods of maximizing $J(\boldsymbol{\pi})$.

The available first-order methods are as follows:

1. Methods of steepest ascent, or gradient methods
2. Conjugate gradient methods
3. Variable metric methods

Of the three methods, the method of steepest ascent is usually the slowest except on exceptionally messy problems. Conjugate gradient methods are much better and approximate the performance of second-order methods. Variable metric methods are usually good but are not always practical.

Let $\boldsymbol{\pi}_n$ denote the nth estimate of $\boldsymbol{\pi}(t)$ for $t \in [t_0, t_1]$ and let τ_n denote the step in Ω space.

4-5-1 Steepest Ascent

At iteration n choose a search direction $\delta J(\pi_n)/\delta\pi$, and step τ_n so that

$$\boldsymbol{\pi}_{n+1} = \boldsymbol{\pi}_n + \tau_n \frac{\partial J(\boldsymbol{\pi}_n)}{\partial \boldsymbol{\pi}}$$

and then repeat for $n + 1$ replacing n.

Generally speaking the control literature has suggested that $\boldsymbol{\tau}_n$ should be a small step such that an improvement is achieved. See Dreyfus,[4] Breakwell, Speyer, and Bryson,[2] and Kelley, Kopp, and Moyer.[9]

Experience seems to indicate that it is worth attempting to choose τ_n as the first maximum of $J(\boldsymbol{\pi}_{n+1})$

$$\max_{\tau_n \geqslant 0} J\left(\boldsymbol{\pi}_n + \tau_n \frac{\delta J(\boldsymbol{\pi}_n)}{\delta \boldsymbol{\pi}}\right)$$

evaluated over $\mathbf{x}^*(\pi)$, $\mathbf{m}^*(\pi)$, $\mathbf{z}^*(\pi)$.

This can be found by a Fibanacci method, Golden section, interpolation, etc. (Wilde[24]). However, there is a trade-off between this linear maximization accuracy and total computing time, which is important to remember. The more accurately that τ_n is located the longer it takes per step but the fewer the number of iterations. Less accurate values of τ_n lead to more iterations but often require a smaller total time. Experimentation is well justified.

4-5-2 Conjugate Gradient Methods

Recently these methods have been applied to control problems by Warren[23] and his ideas extend directly to the second-level problem here.

A set of direction functions $\mathbf{d}_n(t)$, $t \in [t_0, t_1]$, $n = 0, 1, 2, \ldots, \infty$, are said to be conjugate with respect to a metric $\mathbf{K}$ if

$$\int_{t_0}^{t_1} \int_{t_0}^{t_1} \mathbf{d}_i'(s)\, \mathbf{K}(s, t)\, \mathbf{d}_j(t)\, ds\, dt = 0 \qquad i \neq j$$

This remarkable kind of orthogonality has the property that the second variational operator, say $\mathbf{K}$, can be expanded in terms of the $\mathbf{d}_i$ and other coefficients which can be found independently. Furthermore, the solution can be expanded in the $\mathbf{d}_i$ and often requires only a very few terms of the series $\mathbf{d}_0, \mathbf{d}_1, \mathbf{d}_2, \ldots$ to achieve good accuracy.

The simplest method of generating the directions is to use the Fletcher–Reeves algorithm,[7] which can be derived by an interesting device due to Takahashi.[20] Let the search direction be

$$d_n = \frac{\delta J(\boldsymbol{\pi}_n)}{\delta \boldsymbol{\pi}} + \sigma_n d_{n-1}$$

and locate τ_n and σ_n to solve

$$\max_{\tau_n, \sigma_n} J\left[\boldsymbol{\pi}_n + \tau_n \left(\frac{\delta J(\boldsymbol{\pi}_n)}{\delta \boldsymbol{\pi}} + \sigma_n d_{n-1}\right)\right]$$

using the previous remarks. This is a two-dimensional maximization in which the *best step* in the *best direction* is found. Thus one automatically expects to do better than when $\sigma_n = 0$. On quadratic functions an equivalent algorithm is as follows. Start with $d_0 = \delta J(\boldsymbol{\pi}_0)/\delta\boldsymbol{\pi}$. Then find τ_n

$$\max_{\tau_n} J(\boldsymbol{\pi}_n + \tau_n \mathbf{d}_n)$$

and compute

$$\boldsymbol{\pi}_{n+1} = \boldsymbol{\pi}_n + \tau_n \mathbf{d}_n$$

Letting

$$\mathbf{g}_{n+1} = \frac{\delta J(\boldsymbol{\pi}_{n+1})}{\delta \boldsymbol{\pi}}$$

find σ_n

$$\sigma_n = \langle \mathbf{g}_{n+1}, \mathbf{g}_{n+1} \rangle / \langle \mathbf{g}_n, \mathbf{g}_n \rangle$$

and find d_{n+1}

$$\mathbf{d}_{n+1} = \sigma_n \mathbf{d}_n + \frac{\delta J(\boldsymbol{\pi}_{n+1})}{\delta \boldsymbol{\pi}}$$

The sequence $\mathbf{d}_0, \mathbf{d}_1, \ldots, \mathbf{d}_n$ is infinite in an infinite dimensional space (Artosciewiez[22]). However in practice τ_n, σ_n, etc., cannot be found exactly and the directions lose conjugacy as the iteration proceeds. It pays then to restart with $\mathbf{d}_m = \mathbf{g}_m$ for $m = n + 1$, say (n equals the problem state dimension). This kind of error (lack of conjugacy) can be checked by computing $\langle \mathbf{g}_{n+1}, \mathbf{d}_i \rangle$ for $i \leqslant n$. The answer should be zero, which rarely occurs, especially for nonlinear problems, since the metric $\mathbf{K}$ changes with the solution point.

4-5-3 Variable Metric Schemes

Such schemes can be based on the Davidon–Fletcher–Powell algorithm,[6] or more readily on the projection algorithm (Pearson[16]). These methods generate conjugate directions and also estimate the sensitivity of $\delta J(\boldsymbol{\pi}_n)/\delta\boldsymbol{\pi}$ to $\boldsymbol{\pi}$. Only the projection algorithm will be described here, but the generalization to other algorithms is obvious. Let

$$\mathbf{g}_n = \frac{\delta J(\boldsymbol{\pi}_n)}{\delta \boldsymbol{\pi}} \qquad \mathbf{y}_n = \mathbf{g}_{n+1} - \mathbf{g}_n$$

Let $\mathbf{P}_n(t, s)$ be a projection operator with the property that

$$\int_{t_0}^{t_1} \mathbf{P}_n(t, s)\, \mathbf{y}_i(s)\, ds = 0 \qquad = 0, 1, 2, \ldots, n - 1$$

Then the search direction $\mathbf{d}_n$ is given by

$$\mathbf{d}_n = \int_{t_0}^{t_1} \mathbf{P}_n(t, s) \frac{\delta J(\boldsymbol{\pi}_n)}{\delta \boldsymbol{\pi}}\, ds$$

and $\boldsymbol{\pi}_{n+1} = \boldsymbol{\pi}_n + \tau_n \mathbf{d}_n$ with τ_n found as before by the maximization of $J(\boldsymbol{\pi}_n + \tau_n \mathbf{d}_n) = J(\boldsymbol{\pi}_{n+1})$. A suitable set of projection operators can be generated as follows

$$\mathbf{P}_0(t, s) = \mathbf{I} \qquad (t, s) \in [t_0, t_1]$$

$$\mathbf{P}_{n+1}(t, s) = \mathbf{P}_n(t, s) - \left[\int_{t_0}^{t_1} \mathbf{P}_n(t, s)\, \mathbf{y}_n(s)\, ds\right] \frac{\int_t^{t_1} \mathbf{y}_n'(t)\, \mathbf{P}_n(t, s)\, dt}{\Delta}$$

where

$$\Delta = \int_{t_0}^{t_1} \int_{t_0}^{t_1} \mathbf{y}_n'(t)\, \mathbf{P}_n(t, s)\, \mathbf{y}_n(s)\, ds\, dt$$

Clearly these have the projection property referred to earlier, and the conjugacy follows from this (Pearson[16]).

The difficulty with computing with an operator is the horrendous storage problem which it presents. This can be avoided to some extent or even eliminated by expanding the unknown function $\boldsymbol{\pi}$ in a set of independent or orthogonal polynomials and finding a finite number of the coefficients using the finite dimensional version of the methods (Lynch[12]).

4-5-4 Linear Subproblems

For linear subproblems with quadratic objective functions and linear constraints, the sequence of minimizations to find τ_n is not required. This is because the necessary conditions are linear and consequently each variable $\mathbf{x}_i^*(\boldsymbol{\pi})$, etc., is a linear functional of $\boldsymbol{\pi}$. For example, the necessary conditions for the following problem

$$J(\mathbf{m}, \mathbf{z}, \boldsymbol{\pi}) = \| \mathbf{x}(t_1) \|^2 \mathbf{P} + \int_{t_0}^{t_1} [\| \mathbf{x} \|^2 \mathbf{Q}_1 + \| \mathbf{m}^2 \| \mathbf{Q}_2$$

$$+ \| \mathbf{z} \|^2 \mathbf{Q}_3 + \boldsymbol{\pi}'(\mathbf{L}\mathbf{x} - \mathbf{z})]\, dt$$

$$\dot{\mathbf{x}} = \mathbf{A}\mathbf{x} + \mathbf{B}\mathbf{m} + \mathbf{C}\mathbf{z} \qquad \mathbf{x}(t_0) = \mathbf{x}_0$$

$$\mathbf{R}_1\mathbf{x} + \mathbf{R}_2\mathbf{m} \geqslant 0$$

are in fact given by the additional equations

$$\dot{\boldsymbol{\lambda}} + \mathbf{A}'\boldsymbol{\lambda} + 2\mathbf{Q}_1\mathbf{x} - \mathbf{R}_1{}'\boldsymbol{\mu} = 0$$
$$2\mathbf{Q}_2\mathbf{m} + \mathbf{B}'\boldsymbol{\lambda} - \mathbf{R}_2{}'\boldsymbol{\mu} = 0$$
$$\mathbf{C}'\boldsymbol{\lambda} - \boldsymbol{\pi} + 2\mathbf{Q}_3\mathbf{z} = 0$$
$$\boldsymbol{\lambda}(t_1) = \mathbf{P}\mathbf{x}(t_1) \qquad \boldsymbol{\mu}'(\mathbf{R}_1\mathbf{x} + \mathbf{R}_2\mathbf{m}) = 0 \qquad \boldsymbol{\mu} \geqslant 0$$

This is a linear system, so that the solution for $\mathbf{x}(\boldsymbol{\pi})$, for example, has the property that

$$\mathbf{x}(\boldsymbol{\pi}_n + \tau_n\, \mathbf{d}_n) = \mathbf{x}(\boldsymbol{\pi}_n) + \tau_n \bar{\mathbf{x}}(\mathbf{d}_n)$$

where $\mathbf{x}(\boldsymbol{\pi}_n)$ is the solution for $\boldsymbol{\pi} = \boldsymbol{\pi}_n$ and $\bar{\mathbf{x}}(\mathbf{d}_n)$ is the homogeneous solution for $\boldsymbol{\pi} = \mathbf{d}_n$, $\mathbf{x}(t_0) = 0$. It follows that the Fréchet derivative is given by

$$\frac{\delta J(\boldsymbol{\pi}_n + \tau_n\, \mathbf{d}_n)}{\delta \boldsymbol{\pi}} = \frac{\delta J(\boldsymbol{\pi}_n)}{\delta \boldsymbol{\pi}} + \tau_n \frac{\overline{\delta J}(\mathbf{d}_n)}{\delta \boldsymbol{\pi}} = \frac{\delta J(\boldsymbol{\pi}_{n+1})}{\delta \boldsymbol{\pi}}$$

where $\overline{\delta J}(d_n)/\delta\boldsymbol{\pi}$ is the homogeneous form of the Fréchet derivative with $\boldsymbol{\pi} = \mathbf{d}_n$ and all nonvariable terms identically zero (initial conditions, disturbance inputs, reference signals, etc.). The choice of τ_n is now explicit since max $J(\boldsymbol{\pi}_n + \tau_n \mathbf{d}_n)$ occurs when

$$\left\langle \frac{\delta J(\boldsymbol{\pi}_{n+1})}{\delta \boldsymbol{\pi}}, \mathbf{d}_n \right\rangle = 0$$

that is,

$$\tau_n = -\left\langle \frac{\overline{\delta J}(\mathbf{d}_n)}{\delta \boldsymbol{\pi}}, \frac{\delta J(\boldsymbol{\pi}_n)}{\delta \boldsymbol{\pi}} \right\rangle \Big/ \left\langle \frac{\delta J(\boldsymbol{\pi}_n)}{\delta \boldsymbol{\pi}}, \frac{\delta J(\boldsymbol{\pi}_n)}{\delta \boldsymbol{\pi}} \right\rangle$$

This converts the Fletcher–Reeves algorithm to a direct iteration with no searches for τ_n . The idea of exploiting linearity in this way would seem to have more application to other minimization problems. It is especially valuable here.

The next section reports some experience with these methods.

4-6 NUMERICAL EXAMPLES

In this section several small examples will be treated by the methods of Secs. 4-4 and 4-5. Larger systems can be assembled from groups of smaller subsystems and with this in mind, these examples were developed as self-contained packages each capable of being a subproblem with arbitrary inputs and outputs for use later.

Since subscripts are less confusing than superscripts, *components of a subsystem* will be denoted by subscripts for the rest of this section.

Consider a basic second-order subsystem described by the following scalar equations:

$$\dot{x}_1 = x_2 \qquad x_1(0) = a_1$$

$$\dot{x}_2 = a_{21}x_1 + a_{22}x_2 + z_1 + z_2 \qquad x_2(0) = a_2$$

where x_1 and x_2 are the state variables, z_1 and z_2 are input variables (control or interaction), and a_{21} and a_{22} are arbitrary coefficients. The objective function is given as

$$J_i(z_1, z_2, \pi, \kappa) = \int_0^T \{\tfrac{1}{2}[(x_1 - r_1)^2 + z_1^2 + z_2^2] + (\kappa_1 x_1 + \kappa_2 x_2 - \pi_1 z_1 - \pi_2 z_2)\}\, dt$$

Here $r_1(t)$, $t \in [0, T]$ is a desired reference trajectory, and κ_1, κ_2, π_1, and π_2 are parameters chosen to manipulate J_i according to the way in which the inputs z_1 and z_2 are connected. This will be clear in the examples and completes the definition of the subproblem.

Example 4-2

Suppose that the simplest case is considered and a system of one subsystem is formed. Let z_1 be a control and choose z_2 using feedback

$$z_2 = x_1$$

This can be achieved by selecting κ_1, κ_2, π_1, and π_2 as follows:

$$\kappa_1 x_1 + \kappa_2 x_2 - \pi_1 z_1 - \pi_2 z_2 = \pi(x_1 - z_2)$$

That is,

$$\kappa_1 = \pi_2 = \pi$$

$$\kappa_2 = \pi_2 = 0$$

where π is a single coordinating function chosen to maximize

$$J(\pi) = J_1(z_1^*(\pi), z_2^*(\pi), \kappa, \pi)$$

where $z_1^*(\pi)$ and $z_2^*(\pi)$ are optimal solutions to the subproblem with κ and π chosen as shown.

A conjugate gradient method was used to solve the problem with $T = 10$ and π defined at 100 time increments with $h = \Delta t = 0.1$. Table 4-1 records $\| g_n \|$ versus n where

$$g_n = x_1^*(\pi) - z_2^*(\pi)$$

and

$$\| g_n \| = \int_0^T g_n'(t) \cdot g_n(t)\, dt/h$$

Table 4-1. Normed Interconnection Error versus Iteration for Example 4-2

$K = 100$, $h = 0.1$, $T = 10$ sec

n	0	1	2	3	4	5	6	7	8	9	10	11	12
$\| g_n \|$	91.3	7.82	4.05	0.49	0.09	0.03	0.016	0.0025	0.00055	0.00021	0.0002	0.00015	0.00014

and values of $x_1 - z_2$ are recorded in Table 4-2 for π chosen in 20 increments at $h = 0.01$ intervals. (The sudden change in the rate of convergence of $\| g_n \|$ in Table 4-1 is attributed to the computational noise masking the changes in π_n beyond iteration 8.)

Table 4-2. Interconnection Error versus Iteration for Example 4-2

$K = 20$, $h = 0.01$, $T = 0.2$ sec

n	0	1	2	3	4	5
$\| g_n \|$	4.03×10^{1}	4.69×10^{-2}	7.48×10^{-3}	2.4×10^{-3}	1.4×10^{-3}	10^{-4}
$g_n(h)$	9.0233	—0.0294	0.0012	0.0009	0.0005	$<10^{-4}$
$g_n(3h)$	9.0190	—0.0697	0.0093	0.0007	0.0004	$<10^{-4}$
$g_n(5h)$	9.0156	—0.1005	0.0154	0.0005	0.0003	$<10^{-4}$
$g_n(7h)$	9.0133	—0.1220	0.0198	0.0003	0.0002	$<10^{-4}$
$g_n(9h)$	9.0120	—0.1342	0.0223	0.0002	0.0001	$<10^{-4}$
$g_n(11h)$	9.0116	—0.1373	0.0229	0.0002	0.0001	$<10^{-4}$
$g_n(13h)$	9.0123	—0.1312	0.0217	0.0002	0.0001	$<10^{-4}$
$g_n(15h)$	9.0140	—0.1159	0.0186	0.0003	0.0002	$<10^{-4}$
$g_n(17h)$	9.0166	—0.0914	0.0137	0.0005	0.0003	$<10^{-4}$
$g_n(19h)$	9.0203	—0.0576	0.0069	0.0007	0.0004	$<10^{-4}$

Example 4-3

The utility of the subproblem package can now be exploited by interconnecting three of them in various configurations. The point of this is twofold: (1) it investigates the effect of structure on second-level coordination, and (2) it illustrates a technique for handling large systems composed of the same subsystems with an obvious saving in labor.

Figure 4-1 shows four configurations a, b, c, and d of the subsystems arranged in order of the number of loops. Thus a has no intrasubsystem

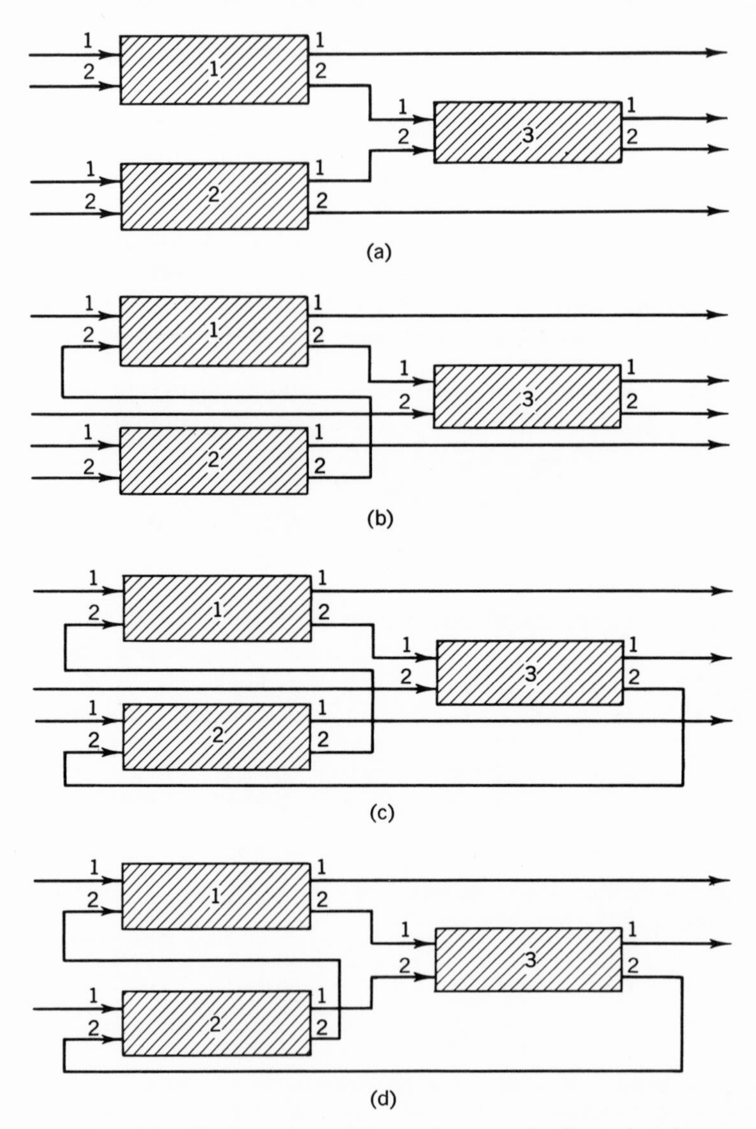

FIG. 4-1. Configurations of three subsystems for Example 4-3.

feedback, while d has the most. In order to explain the means of choosing the κ's and π's to give the coupling, we will consider the worst case as an example, using the connections of Fig. 4-1d.

The subsystems are numbered 1, 2, and 3, each having inputs 1 and 2 and outputs 1 and 2. Letting subscript ij denote the jth component of the ith subsystem variable, the interconnecting branches are as follows:

$$\begin{aligned} g_1(t) &= x_{22}(t) - z_{12}(t) \\ g_2(t) &= x_{12}(t) - z_{31}(t) \\ g_3(t) &= x_{21}(t) - z_{32}(t) \\ g_4(t) &= x_{32}(t) - z_{22}(t) \end{aligned} \qquad 0 \leqslant t \leqslant T$$

If g_1, g_2, g_3, and g_4 are identically zero, then the interconnections of Fig. 4-1d are enforced. Let multiplier π_i be associated with gradient g_i; then the components of κ and π for each subsystem must be chosen as follows:

$$\begin{aligned} \kappa_{22} &= \pi_{12} = \pi_1 \\ \kappa_{12} &= \pi_{31} = \pi_2 \\ \kappa_{21} &= \pi_{32} = \pi_3 \\ \kappa_{32} &= \pi_{22} = \pi_4 \end{aligned} \qquad 0 \leqslant t \leqslant T$$

As before, κ_{ij}, π_{ij} are the jth component of the ith subsystems κ, π, and $\kappa_{11} = \kappa_{31} = \pi_{11} = \pi_{21} \equiv 0$ for control terms. The three subproblems are solved for given $\pi_1, \ldots, \pi_4$ functions, and J_1, J_2, J_3, and $g_1, \ldots, g_4$ are evaluated. The second-level problem is to maximize $J(\pi)$ where

$$J(\pi) = \sum_{i=1}^{3} J_i(z_i(\kappa, \pi), z_2(\kappa, \pi), \kappa, \pi)$$

with κ and π defined above in terms of $\pi_1, \pi_2, \pi_3, \pi_4$. However, from the definitions of the g_i terms,

$$\frac{\delta J(\pi)}{\delta \pi} = (g_1, g_2, g_3, g_4)'$$

so that the conjugate gradient algorithm proceeds using the function $g_i(t)$. Figure 4-2 shows the effect of adding links by modifying the interconnections in Fig. 4-1a to d. It is interesting to see that rate of convergence is affected markedly although the actual system order remains at 6. This illustrates one criterion for choosing subsystems, namely, *that the subsystem should be chosen so that there are as few feedback loops as possible.* This principle is illustrated again in the next example.

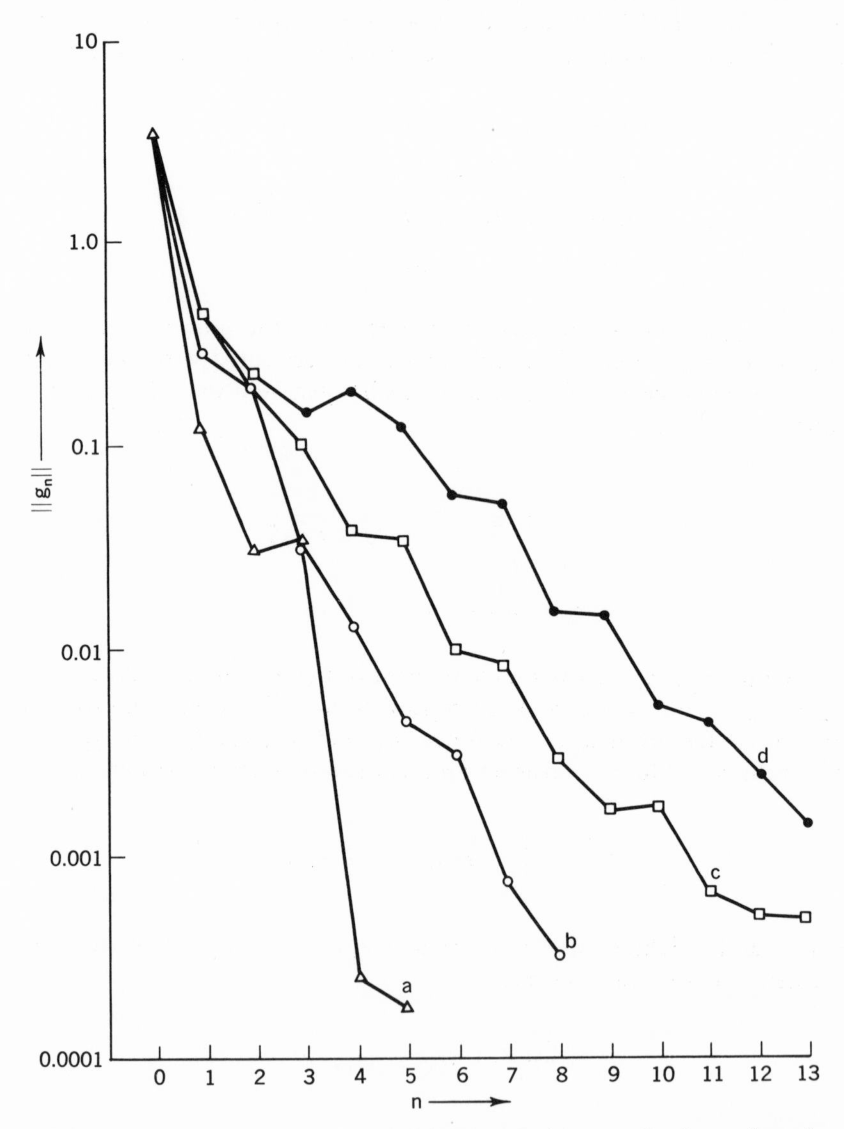

FIG. 4-2. Convergence rates of normed interconnection error for the configurations of Example 4-3.

Another sequence of calculations was performed with a third-order system defined as follows.

$$\dot{x}_1 = x_2 \qquad x_1(0) = a_1$$

$$\dot{x}_2 = x_3 \qquad x_2(0) = a_2$$

$$\dot{x}_3 = a_{31}x_1 + a_{32}x_2 + a_{33}x_3 + z_1 + z_2 \qquad x_3(0) = a_3$$

where z_1 and z_2 are chosen to minimize

$$\int_0^T [\tfrac{1}{2}(x_1^2 + x_2^2 + z_1^2 + z_2^2) + \kappa_1 x_1 + \kappa_2 x_2 - \pi_1 z_1 - \pi_2 z_2]\, dt$$

As before κ_1, κ_2, π_1, and π_2 are interconnected with other subproblems to enforce various subsystem connections.

Example 4-4

A large 45th-order subsystem can be synthesized by cascading 15 third-order subsystems of the type defined in the manner of Fig. 4-3.

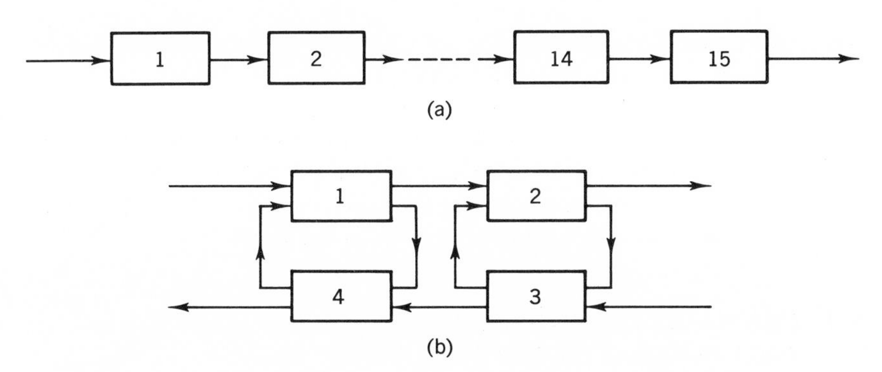

FIG. 4-3. (*a*) Cascade 45th-order system; (*b*) countercurrent 12th-order system.

Table 4-3 gives some idea of the rate of convergence of the conjugate gradient algorithm for $T = 1$ sec, using 10 increments to define each π component.

Example 4-5

A model of a countercurrent flow process such as a distillation column or a heat exchanger can be made by using the interconnections of Fig. 4-3*b*. Table 4-3 also shows the rate of convergence obtained for this 12th-order system.

In Examples 4-4 and 4-5 the number of π components caused storage difficulties on the Univac 1107 computer. Consequently, small optimiza-

Table 4-3. Normed Interconnection Error versus Iteration for Examples 4-4 and 4-5

$h = 0.1$, $K = 10$, $T = 1$ sec

	45th-order Cascade	12th-order Countercurrent
n	$\|\| \mathbf{g}_n \|\|$	$\|\| \mathbf{g}_n \|\|$
0	3.26	1.76
1	4.62×10^{-1}	7.96×10^{-1}
2	2.10×10^{-2}	6.93×10^{-2}
3	2.01×10^{-3}	2.12×10^{-2}
4	1.32×10^{-4}	8.52×10^{-3}
5	2.21×10^{-5} end	2.23×10^{-3}
6		8.36×10^{-4}
7		2.24×10^{-4}
8		4.40×10^{-5}
9		1.59×10^{-5}
10		

tion times had to be chosen and this contributes to the rate of convergence which varies roughly as exp (kT), for some small k.

Examples 4-4 and 4-5 confirm that feedback loops do materially slow the rate of convergence. The 45th-order cascade structure takes five iterations, while the 12th-order countercurrent takes nine iterations. In each case interconnections are matched to four decimal places. Further details and Algol programs are given by Reich.[18]

4-7 MODEL COORDINATION AND FEASIBLE DECOMPOSITION

In Sec. 4-2 it was indicated that there is another approach to decomposition in which the interconnection variables are fixed so that the system is not actually severed into subsystems. This is extremely useful if the decomposition method is to be used on-line with an actual operating system. In this section it is shown that this alternative approach is just the dual of the previous method of *goal coordination*, but suffers from a certain disadvantage.

Consider the problem formulated as an example at the end of Sec. 4-3. The primal and dual versions of this problem are repeated below.

PRIMAL PROBLEM:

$$\min J^P(\mathbf{m}, \mathbf{z}) = \tfrac{1}{2} \|\mathbf{x}(t_1)\|^2 \mathbf{P} + \tfrac{1}{2} \int_{t_0}^{t_1} (\|\mathbf{x}\|^2 \mathbf{Q}_1 + \|\mathbf{m}\|^2 \mathbf{Q}_2 + \|\mathbf{z}\|^2 \mathbf{Q}_3)\, dt \quad (64)$$

subject to $\dot{\mathbf{x}} = \mathbf{Ax} + \mathbf{Bm} + \mathbf{Cz} \qquad \mathbf{x}(t_0) = \mathbf{x}_0$ (65)

$$\mathbf{y} = \mathbf{Mx} + \mathbf{Nm} \tag{66}$$

$$\mathbf{z} = \mathbf{Ly} \tag{67}$$

$$\mathbf{R}_1\mathbf{x} + \mathbf{R}_2\mathbf{m} \geqslant 0 \tag{68}$$

DUAL PROBLEM:

$$\max J^d(\mathbf{x}, \mathbf{m}, \mathbf{z}, \boldsymbol{\mu}) = \mathbf{x}_0{}' \boldsymbol{\lambda}(t_0) - \tfrac{1}{2} \| \boldsymbol{\lambda}(t_2) \|^2 \mathbf{P}^{-1}$$

$$-\int_{t_0}^{t_1} \tfrac{1}{2}(\| \mathbf{x} \|^2 \mathbf{Q}_1 + \| \mathbf{B}'\boldsymbol{\lambda} + \mathbf{N}'\boldsymbol{\kappa} - \mathbf{R}_2{}'\boldsymbol{\mu} \| \mathbf{Q}_2^{-1} + \| \mathbf{z} \|^2 \mathbf{Q}_3)\, dt \tag{69}$$

subject to $\dot{\boldsymbol{\lambda}} + \mathbf{Q}_1\mathbf{x} + \mathbf{A}'\boldsymbol{\lambda} - \mathbf{R}_1{}'\boldsymbol{\mu} + \mathbf{M}'\boldsymbol{\kappa}$ (70)

$$\boldsymbol{\pi} = \mathbf{C}'\boldsymbol{\lambda} + \mathbf{Q}_3\mathbf{z} \tag{71}$$

$$\boldsymbol{\kappa} = \mathbf{L}'\boldsymbol{\pi} \tag{72}$$

$$\boldsymbol{\mu} \geqslant 0 \tag{73}$$

where $\mathbf{m}$ has been eliminated.

Suppose that the original primal system represents N coupled subsystems; then Eqs. (65), (66), and (68) represent N independent subsystems coupled by Eq. (67), in which case **A**, **B**, **C**, **M**, **N**, $\mathbf{R}_1$, $\mathbf{R}_2$, **P**, $\mathbf{Q}_1$, $\mathbf{Q}_2$, and $\mathbf{Q}_3$ will all be found from **N** appropriately sized block diagonal matrices.

Interestingly enough the dual system has the same structure of N dual subsystems coupled solely by Eq. (72). The static equation (70) separates into N subsystems in $\boldsymbol{\lambda}_i$, each of which has an output $\boldsymbol{\pi}_i$ which is coupled to the input $\boldsymbol{\kappa}_i$. The variables $\mathbf{x}_i$, $\boldsymbol{\mu}_i$, $\boldsymbol{\kappa}_i$, and $\mathbf{z}_i$ are the controllable inputs. It can now be seen that there are dual ways of decomposing a system depending upon whether one attacks the primal or the dual.

The original method of Sec. 4-2, using goal coordination, is to relax the coupling (67) by choosing the multipliers $\boldsymbol{\pi}$, and $\boldsymbol{\kappa} = \mathbf{L}'\boldsymbol{\pi}$. Equations (70) to (73) are satisfied while $\boldsymbol{\kappa}$ and $\boldsymbol{\pi}$ are adjusted until the coupling is enforced. The dual approach is to apply this method to the dual problem and relax Eq. (72) by defining the multiplier required for Eq. (72). To show that this is equivalent to choosing $\mathbf{y}$, and $\mathbf{z} = \mathbf{Ly}$, the dual problem will be solved.

Suppose the dual Lagrangian is $\mathscr{L}^d$ and the problem is to minimize $-J^d$ (in order to preserve the multiplier signs). Then

$$\mathscr{L}^d = \frac{\| \mathbf{x} \|^2 \mathbf{Q}_1 + \| \mathbf{B}'\boldsymbol{\lambda} + \mathbf{N}'\boldsymbol{\varkappa} - \mathbf{R}'_2\boldsymbol{\mu} \| \mathbf{Q}_2^{-1} + \| \mathbf{x} \|^2 \mathbf{Q}_3}{2}$$

$$+ (-\mathbf{Q}_1\mathbf{x} - \mathbf{A}'\boldsymbol{\lambda} + \mathbf{R}_1{}'\boldsymbol{\mu} - \mathbf{M}'\boldsymbol{\kappa} - \dot{\boldsymbol{\lambda}})' \boldsymbol{\lambda}^d$$

$$- \boldsymbol{\mu}'\boldsymbol{\mu}^d + (\mathbf{L}'\mathbf{C}'\boldsymbol{\lambda} + \mathbf{L}'\mathbf{Q}_3\mathbf{z} - \boldsymbol{\kappa})' \boldsymbol{\pi}^d \tag{74}$$

where $\boldsymbol{\lambda}^d$, $\boldsymbol{\mu}^d$, and $\boldsymbol{\pi}^d$ are multipliers taking the place of $\boldsymbol{\lambda}$, $\boldsymbol{\mu}$, and $\boldsymbol{\pi}$. The necessary conditions for this problem require that they should satisfy the following:

$$\dot{\boldsymbol{\lambda}}^d + \mathbf{B}\mathbf{Q}_2^{-1}(\mathbf{B}'\boldsymbol{\lambda} + \mathbf{N}'\boldsymbol{\kappa} - \mathbf{R}_2'\boldsymbol{\mu}) - \mathbf{A}\boldsymbol{\lambda}^d + \mathbf{C}\mathbf{L}\boldsymbol{\pi}^d = 0$$

$$\mathbf{Q}_1\mathbf{x} - \mathbf{Q}_1\boldsymbol{\lambda}^d = 0$$

$$\mathbf{Q}_3\mathbf{z} + \mathbf{Q}_3\mathbf{L}\boldsymbol{\pi}^d = 0$$

$$\boldsymbol{\mu}'\boldsymbol{\mu}^d = 0 \qquad \boldsymbol{\mu} \geqslant 0$$

$$\mathbf{R}_2\mathbf{Q}_2^{-1}(\mathbf{B}'\boldsymbol{\lambda} + \mathbf{N}'\boldsymbol{\kappa} - \mathbf{R}_2'\boldsymbol{\mu}) + \mathbf{R}_1\boldsymbol{\lambda}^d - \boldsymbol{\mu} = 0$$

$$\mathbf{N}\mathbf{Q}_2^{-1}(\mathbf{B}'\boldsymbol{\lambda} + \mathbf{N}'\boldsymbol{\kappa} - \mathbf{R}_2'\boldsymbol{\mu}) - \mathbf{M}\boldsymbol{\lambda}^d - \boldsymbol{\pi}^d = 0$$

Canceling $\mathbf{Q}_1$ and $\mathbf{Q}_3$ and substituting $\boldsymbol{\lambda}^d = -\mathbf{x}$ yields

$$\dot{\mathbf{x}} = \mathbf{A}\mathbf{x} + \mathbf{B}[-\mathbf{Q}_2^{-1}(\mathbf{B}'\boldsymbol{\lambda} + \mathbf{N}'\boldsymbol{\kappa} - \mathbf{R}_2'\boldsymbol{\mu})] + \mathbf{C}\mathbf{z} \qquad \mathbf{x}(t_0) = \mathbf{x}_0$$

$$(-\boldsymbol{\pi}^d) = \mathbf{M}\mathbf{x} + \mathbf{N}[-\mathbf{Q}_2^{-1}(\mathbf{B}'\boldsymbol{\lambda} + \mathbf{N}'\boldsymbol{\kappa} - \mathbf{R}_2'\boldsymbol{\mu})]$$

$$\mathbf{z} = \mathbf{L}(-\boldsymbol{\pi}^d)$$

$$\mathbf{R}_1\mathbf{x} + \mathbf{R}_2[-\mathbf{Q}_2^{-1}(\mathbf{B}'\boldsymbol{\lambda} + \mathbf{N}'\boldsymbol{\kappa} - \mathbf{R}_2'\boldsymbol{\mu})] \geqslant 0$$

These are recognizable as the optimized form of the system equations (65) to (68), with $\mathbf{m}$ chosen optimally and $\boldsymbol{\pi}^d = -\mathbf{y}$ as required.

Returning to the decomposition, relaxing the dual coupling restraint (72) requires that its multiplier $\boldsymbol{\pi}^d$ be chosen by Eq. (74). However $\boldsymbol{\pi}^d \equiv (-\mathbf{y})$ and this corresponds to defining the observed outputs in each primal subsystem while still preserving the coupling $\mathbf{z} = \mathbf{L}\mathbf{y}$. From Eqs. (69) and (74), the subproblem objective functions must have the form

$$J_i^d(\mathbf{x}_i, \mathbf{m}_i, \mathbf{z}_i, \boldsymbol{\mu}_i, \boldsymbol{\pi}^d) = \mathbf{x}_{0i}'\boldsymbol{\lambda}_i(t_0) - \tfrac{1}{2}\|\boldsymbol{\lambda}_i(t_1)\|^2\,\mathbf{P}_i^{-1}$$

$$- \int_{t_0}^{t_1} [\tfrac{1}{2}(\|\mathbf{x}_i\|^2\,\mathbf{Q}_{1i} + \|\mathbf{B}_i'\boldsymbol{\lambda}_i + \mathbf{N}'\boldsymbol{\kappa}_i - \mathbf{R}_{2i}'\boldsymbol{\mu}_i\|^2\,\mathbf{Q}_{2i}^{-1} + \|\mathbf{z}\|^2\,\mathbf{Q}_{3i})$$

$$- (\boldsymbol{\lambda}_i'\mathbf{C}_i + \mathbf{z}_i'\mathbf{Q}_{3i}) \sum_{j=1}^{N} \mathbf{L}_{ij}\boldsymbol{\pi}_j{}^d - \boldsymbol{\kappa}_i\boldsymbol{\pi}_i{}^d]\, dt$$

where $\mathbf{P}_i$, $\mathbf{Q}_{1i}$, $\mathbf{R}_{2i}$, $\mathbf{Q}_{2i}$, and $\mathbf{Q}_{3i}$ are the ith diagonal blocks of the block diagonal matrices $\mathbf{P}$, $\mathbf{Q}_1$, $\mathbf{R}_2$, $\mathbf{Q}_2$, and $\mathbf{Q}_3$.

The problem of maximizing Eq. (75) subject to Eqs. (70) to (73) will generate the necessary conditions for the integrated problem with the exception of the multiplier coupling equation (72), for otherwise $\boldsymbol{\pi}^d$ would be optimal. This suggests a *model coordination decomposition*, which consists of defining the coupling variables $\mathbf{y}$, $\mathbf{z} = L\mathbf{y}$, and solving for the subproblem solutions $\mathbf{x}_i^*(\mathbf{y})$, $\mathbf{m}_i^*(\mathbf{y})$, $\mathbf{z}_i$, $\boldsymbol{\lambda}_i^*(\mathbf{y})$, $\boldsymbol{\mu}_i^*(\mathbf{y})$, $\boldsymbol{\kappa}_i^*(\mathbf{y})$, and $\boldsymbol{\pi}_i^*(\mathbf{y})$ using the necessary conditions of the subproblems with the exception $\boldsymbol{\kappa} \neq \mathbf{L}'\boldsymbol{\pi}$; that is,

$$\boldsymbol{\kappa}_i^*(\mathbf{y}) \neq \sum_{j=1}^{N} \mathbf{L}'_{ji}\boldsymbol{\pi}_j^*(\mathbf{y}) \tag{76}$$

Since the dual has been effectively decomposed, it follows that choosing $\mathbf{y}$ to enforce Eq. (76) completes the necessary conditions for optimizing the dual-dual problem, i.e., the primal, since this process is involuntary. Thus $\mathbf{y}$ is chosen to *minimize*

$$J^d(\mathbf{y}) = \sum_{i=1}^{N} J_i^d(\mathbf{x}_i^*(\mathbf{y}), \mathbf{m}_i^*(\mathbf{y}), \mathbf{z}_i^*(\mathbf{y}), \boldsymbol{\mu}_i^*(\mathbf{y}); \boldsymbol{\pi}^d)$$

$$\equiv J_i(\mathbf{m}^*(\mathbf{y}), \mathbf{z}^*(\mathbf{y}); \mathbf{y})$$

with $\boldsymbol{\pi}^d = -y$. This follows from the way the dual is constructed from the primal by adding terms which vanish on the optimal trajectory.

Model coordination thus operates by adjusting the interconnections of an operating system of subsystems until its dual system of multipliers is coordinated. This is a very appealing property; however, as has been said, there is a snag, namely, given a system defined by $\dot{\mathbf{x}} = F(\mathbf{x}, \mathbf{m}, \mathbf{z}, t)$ and $\mathbf{z} = \mathbf{LG}(\mathbf{x}, \mathbf{m}, t)$, when $\mathbf{z} = \mathbf{Ly} = \mathbf{LG}$ is defined arbitrarily in advance it may not be possible to choose an $\mathbf{x}$, $\mathbf{m}$ which will satisfy the system equations. To do this requires, in effect, that each subsystem has a sufficient number of degrees of freedom to reproduce a desired set of coupling variables. Usually only very large systems have the property that the number of coupling variables is smaller than the number of control variables, which is a prerequisite.

An example of the use of this form of decomposition and model coordination arises in the electrical utility industry in a static sense. Independent utility companies negotiate contracts to borrow from each other amounts of energy at given power flows throughout each day. This takes advantage of cheap fuel supplies, shifting power demand due to time zones or excess capacity, etc. Almost the whole of the United States is linked in this way. There are two prices involved in each exchange. A generating price at the source company and a generating price at the consumer company. These are different and depend monotonically on the actual power levels in the two companies at the time of the power exchange. Clearly, if the generating price $\boldsymbol{\kappa}^*(\mathbf{y})$ for buying power level $\mathbf{y}$ is less than $\boldsymbol{\pi}^*(\mathbf{y})$, the cost of generating $\mathbf{y}$, then the consumer company buys more. This decreases the additional power he needs to generate and hence his own price $\boldsymbol{\pi}^*(\mathbf{y})$, while the power required and hence the source price $\boldsymbol{\kappa}^*(\mathbf{y})$ increases. At balance the interchange $\mathbf{y}^*$ minimizes the joint system costs. In practice, of course, the coordination is done at monthly intervals accompanied by a cash exchange, which one supposes is proportional to the difference in value.

$$\int_{t_0}^{t_1} [\boldsymbol{\pi}^*(\mathbf{y}) - \boldsymbol{\kappa}^*(\mathbf{y})]\, \mathbf{y}\, dt$$

where we have taken $\mathbf{L} = \mathbf{I}_s$ and $\mathbf{z} = \mathbf{y}$. This example represents a static version of the analysis presented here because each power company operates independently at its own instantaneous "least generating cost." The system output probably approximates the dynamic optimum because of the system's very short transient response. This example also illustrates how the risk of overdetermining the system can be completely avoided. In practice it is not possible to exchange exactly the prescribed amount of power and the solution actually serves as a guide line about which small fluctuations occur. It is probable that the negotiated exchange also includes penalties for prolonged deviations of the actual exchange from the agreed level. With this in mind, the following control analogy appears reasonable.

4-7-1 Pseudo–Model Coordination

It has been shown that model coordination may not be feasible because of the possibility of overdetermining the system constraints. It also has the disadvantage of prescribing outputs, which may not be desirable for a given choice of inputs.

An alternative approach is to define "desirable" interconnection variable magnitudes, and then penalize deviations from these.

Consider, for example, the problem

$$\min J_i^s(\mathbf{m}_i\,;\hat{\mathbf{y}}_i) = g_i + \int_{t_0}^{t_1} (f_i + \mathbf{k}\,\|\,\mathbf{y}_i - \hat{\mathbf{y}}_i\,\|^2\, \boldsymbol{\Lambda}_i)\, dt$$

$$\text{subject to } \dot{\mathbf{x}}_i = \mathbf{F}_i(\mathbf{x}_i\,, \mathbf{m}_i\,, \mathbf{z}_i\,, t) \qquad \mathbf{x}_i(t_0) = \mathbf{x}_{i0}$$

$$\mathbf{y}_i = \mathbf{G}_i(\mathbf{x}_i\,, \mathbf{m}_i\,, t) \qquad \mathbf{R}_i(\mathbf{x}_i\,, \mathbf{m}_i\,, \mathbf{z}_i t) \geqslant 0$$

$$\mathbf{z}_i = \sum_{j=1}^{N} \mathbf{L}_{ij}\mathbf{y}_j$$

where $\mathbf{\Lambda}_i$ is an $s \times s$ dimensional positive definite matrix. If $k \gg 0$, one expects that $\mathbf{y}_i^*(\hat{\mathbf{y}}_i) \simeq \hat{\mathbf{y}}_i$, the desired output, and $\mathbf{z}_i^*(\hat{\mathbf{y}}) \simeq \hat{\mathbf{z}}$, the desired input. Accordingly, the problem can be decomposed approximately by replacing the actual interaction $\mathbf{z}_i^*(\hat{\mathbf{y}}_1)$ with the desired value $\hat{\mathbf{z}}_i = \sum_{j=1}^{N} \mathbf{L}_{ij}\hat{\mathbf{y}}_j$, so that each subsystem is then independent.

Suppose that $\mathbf{y}$ is defined over $[t_0, t_1]$ and each subproblem undertakes to minimize

$$\min J_i^s(\mathbf{m}_i, \hat{\mathbf{y}}) = g_i + \int_{t_0}^{t_1} (f_i + \mathbf{k} \| \mathbf{y}_i - \hat{\mathbf{y}}_i \|^2 \mathbf{\Lambda}_i)\, dt \tag{77}$$

$$\begin{aligned} \text{subject to } \dot{\mathbf{x}}_i &= F_i(\mathbf{x}_i, \mathbf{m}_i, \hat{\mathbf{z}}_i, t) \qquad \mathbf{x}_i(t_0) = \mathbf{x}_0 \\ \mathbf{y}_i &= \mathbf{G}_i(\mathbf{x}_i, \mathbf{m}_i, \hat{\mathbf{z}}_i, t) \geqslant 0 \\ &\mathbf{R}_i(\mathbf{x}_i, \mathbf{m}_i, \hat{\mathbf{z}}_i, t) \geqslant 0 \\ \hat{\mathbf{z}}_i &= \sum_{j=1}^{N} \mathbf{L}_{ij}\hat{\mathbf{y}}_j \end{aligned} \tag{78}$$

If k is increased, the solutions $\mathbf{x}_i^*(\hat{\mathbf{y}})$, $\mathbf{m}_i^*(\hat{\mathbf{y}})$, $\mathbf{y}_i^*(\hat{\mathbf{y}})$ will be such that $\mathbf{y}_i^*(\hat{\mathbf{y}}) \to \hat{\mathbf{y}}_i$. Equality may not occur since, as previously discussed, $\hat{\mathbf{y}}_i$ may not be a feasible output.

Theorem 4-10: If J_i^s has a minimum for each $k\mathbf{\Lambda}_i \geqslant 0$ and the problem

$$\min \int_{t_0}^{t_1} (\| \mathbf{y}_i - \hat{\mathbf{y}}_i \|^2 \mathbf{\Lambda}_i)\, dt$$

subject to Eq. (78) has the unique solution $(\bar{\mathbf{x}}_i, \bar{\mathbf{y}}_i, \bar{\mathbf{m}}_i)$, then as $k \to \infty$ the solutions $\mathbf{x}_i^*(\hat{\mathbf{y}})$, $\mathbf{y}_i^*(\hat{\mathbf{y}})$, $\mathbf{m}_i^*(\hat{\mathbf{y}}) \to (\bar{\mathbf{x}}_i, \bar{\mathbf{y}}_i, \bar{\mathbf{m}}_i)$.

Proof: It follows from the assumptions that, for $k > 0$,

$$\begin{aligned} \min \left(g_i + \int_{t_0}^{t_1} f_i\, dt\right) &\leqslant \min \left[g_i + \int_{t_0}^{t_1} (f_i + k \| \mathbf{y}_i - \hat{\mathbf{y}}_i \|^2 \mathbf{\Lambda}_i)\, dt\right] \\ &\leqslant \bar{g}_i + \int_{t_0}^{t_1} (f_i + k \| \bar{\mathbf{y}}_i - \hat{\mathbf{y}}_i \|^2 \mathbf{\Lambda}_i)\, dt \end{aligned}$$

when $\bar{g}_i$, etc., denotes $g_i(\bar{\mathbf{x}}_i(t_1))$ and $\mathbf{x}$, $\mathbf{y}$, $\mathbf{m}$ satisfy Eq. (78). Thus $J_i^s(\mathbf{m}_i, \mathbf{z}_i, \hat{\mathbf{y}})$ is bounded below by a constant, and above by a linear function of k. Dividing by k and taking the limit

$$0 \leqslant \lim_{\mathbf{k} \to \infty} \left[\int_{t_0}^{t_1} (\| \mathbf{y}_i^*(\hat{\mathbf{y}}) - \hat{\mathbf{y}}_i \|^2 \mathbf{\Lambda}_i)\, dt\right] \leqslant \int_{t_0}^{t_1} (\| \bar{\mathbf{y}}_i - \hat{\mathbf{y}}_i \|^2 \mathbf{\Lambda}_i)\, dt$$

On the other hand since $\bar{\mathbf{x}}_i$, $\bar{\mathbf{y}}_i$, $\bar{\mathbf{m}}_i$ is also an optimal solution, for every $\mathbf{k} > 0$,

$$\int_{t_0}^{t_1} (\| \mathbf{y}_i^*(\hat{\mathbf{y}}) - \hat{\mathbf{y}}_i \|^2 \mathbf{\Lambda}_i)\, dt \geqslant \int_{t_0}^{t_1} (\| \bar{\mathbf{y}}_i - \hat{\mathbf{y}}_i \|^2 \mathbf{\Lambda}_i)\, dt$$

Thus in the limit we must have equality, which is attained at the unique point $\bar{\mathbf{x}}_i$, $\bar{\mathbf{y}}_i$, $\bar{\mathbf{m}}_i$

$$\lim_{\mathbf{k}\to\infty} [\mathbf{x}_i^*(\hat{\mathbf{y}}), \mathbf{y}_i^*(\hat{\mathbf{y}}), \mathbf{m}_i^*(\hat{\mathbf{y}})] = \bar{\mathbf{x}}_i, \bar{\mathbf{y}}_i, \bar{\mathbf{m}}_i$$

which completes the proof. Clearly if $\hat{\mathbf{y}}_i$ is an attainable output then $\bar{\mathbf{y}}_i = \hat{\mathbf{y}}_i$. Thus as $\mathbf{k}$ increases $\mathbf{y}_i^*(\hat{\mathbf{y}})$ approaches the value $\bar{\mathbf{y}}_i$, the nearest realizable $\hat{\mathbf{y}}_i$. Clearly if $\hat{\mathbf{y}}_i$ is chosen carefully, then $\bar{\mathbf{y}}_i \equiv \hat{\mathbf{y}}_i$. Since a control $\mathbf{m}_i^*(\hat{\mathbf{y}})$ is available for each subsystem independently, which makes $\mathbf{y}_i^*(\hat{\mathbf{y}}_i)$ approximate $\hat{\mathbf{y}}_i$ as closely as possible, the problem is to adjust $\hat{\mathbf{y}}_i$ so that $\mathbf{y}_i^*(\hat{\mathbf{y}}_i) \equiv \hat{\mathbf{y}}_i$, giving the *optimal coordination* $\mathbf{y}_i^*$. Theorem 4-10 shows that by penalizing sufficiently the deviation of $\mathbf{y}_i^*(\hat{\mathbf{y}})$ from $\hat{\mathbf{y}}_i$, $\mathbf{y}_i^*(\hat{\mathbf{y}})$ will approach the nearest realizable $\hat{\mathbf{y}}_i$, which is $\bar{\mathbf{y}}_i$.

Now, given a set of desired outputs $\hat{\mathbf{y}}$, each subsystem is optimized relative to J_i^s, the pseudo-objective function, so that it most closely reproduces $\hat{\mathbf{y}}_i$, with $\hat{\mathbf{z}}_i = \sum_{j=1}^{N} \mathbf{L}_{ij}\hat{\mathbf{y}}_j$ as the input. If each $\mathbf{y}_i^* \equiv \hat{\mathbf{y}}_i$, then the subsystems are directly connectable, for each subsystem reproduces the required output and receives the required input. We will call a $\hat{\mathbf{y}}$ with the property that $\mathbf{y}^*(\hat{\mathbf{y}}) = \hat{\mathbf{y}}$ the *optimal coordination*, denoted as $\mathbf{y}^*$. The next theorem indicates how $\mathbf{y}^*$ and $\mathbf{k}$ can be chosen. We will let $J^s(\hat{\mathbf{y}})$ be defined as

$$J^s(\hat{\mathbf{y}}) = \sum_{i=1}^{N} J_i^s(\mathbf{m}_i^*(\hat{\mathbf{y}}), \hat{\mathbf{y}})$$

Theorem 4-11: If each subsystem objective function J_s^i is minimized for $k > 0$, then the optimal coordination $\mathbf{y}^*$ minimizes $J^s(\mathbf{y})$ and gives the integrated optimal solution.

Proof: By direct calculation, the integrated optimum is given by

$$J_I^* = \min \sum_{i=1}^{N} \left[g_i + \int_{t_0}^{t_1} (f_i)\, dt \right]$$

where $(\mathbf{x}_i, \mathbf{y}_i, \mathbf{m}_i, \mathbf{z}_i)$, $i = 1, 2, \ldots, N$ satisfy the integrated system constraints. Now

$$J_I^* \equiv \min \sum_{i=1}^{N} \left[g_i + \int_{t_0}^{t_1} (f_i + k \| y_i - \mathbf{y}_i^* \| \mathbf{\Lambda}_i)\, dt \right]$$

where $\mathbf{y}_i^*$, $i = 1, 2, \ldots, N$ is optimal, so that

$$J_I^* \leqslant \min \sum_{i=1}^{N} \left[g_i + \int_{t_0}^{t_1} (f_i + k \, \| \, \hat{\mathbf{y}}_i - \mathbf{y}_i \, \| \, \mathbf{\Lambda}_i) \, dt \right]$$

subject to $\mathbf{z} = \hat{\mathbf{z}} = \mathbf{L}\hat{\mathbf{y}}$ with $\hat{\mathbf{y}}$ chosen arbitrarily. Thus,

$$J_I^* \leqslant \sum_{i=1}^{N} \min \left[g_i + \int_{t_0}^{t_1} (f_i + k \, \| \, \mathbf{y}_i - \hat{\mathbf{y}}_i \, \| \, \mathbf{\Lambda}_i) \, dt \right]$$

with $\mathbf{z} = \hat{\mathbf{z}} = \mathbf{L}\hat{\mathbf{y}}$, since this decouples the subsystems and

$$J_I^* \leqslant \sum_{i=1}^{N} J_i^s(\mathbf{m}_i^*(\hat{\mathbf{y}}), \hat{\mathbf{y}}) = J^s(\hat{\mathbf{y}})$$

as previously defined. Thus we have established $J_I^* \leqslant J^s(\hat{\mathbf{y}})$ and from the same argument $J_I^* = J^s(\mathbf{y}^*)$ when $\mathbf{y}^*$ is defined by the optimal solution $\mathbf{x}^*, \mathbf{y}^*, \mathbf{m}^*, \mathbf{z}^*$. This completes the proof.

The remarkable part of this result is that coordination can be effected independently of the magnitude of $\mathbf{k}$, as long as each subproblem has a minimum. The technique however requires that all subproblem solutions be obtained as parameters of $\hat{\mathbf{y}}$, the desired set of outputs, and has much in common with the technique of defining a multiplier $\boldsymbol{\pi}$.

4-8 APPLICATION TO STOCHASTIC PROBLEMS

4-8-1 Application to Smoothing Problems

As is well known, the problems of smoothing noisy data and of control system regulation are duals in a certain sense. The duality is deeper than the observation that they satisfy the same equations for they are in fact also duals in the programming sense of Sec. 4-3. Consequently, it is entirely reasonable that the methods developed earlier have direct application to the problem of smoothing.

The smoothing, filtering, and prediction problem can be introduced and formulated as follows. We are given $t_1 - t_0$ seconds of a statistical record of the vector functions $\mathbf{v}$ and $\mathbf{w}$. It is assumed that $\mathbf{v}$ and $\mathbf{w}$ are noise-contaminated observations of a linear dynamical process which is excited by noise in the following way.

The state $\mathbf{x}$ of the process satisfies a linear differential equation.

$$\begin{aligned} \dot{\mathbf{x}} &= \mathbf{A}\mathbf{x} + \mathbf{B}\mathbf{m} + \mathbf{C}\mathbf{z} \\ \mathbf{z} &= \mathbf{L}\mathbf{y} \\ \mathbf{y} &= \mathbf{M}\mathbf{x} + \mathbf{N}\mathbf{m} \end{aligned} \tag{79}$$

where (for the purposes of this section) $\mathbf{m}$ is a noise input and $\mathbf{z}$ is an interaction input. The observation $\mathbf{v}$ and $\mathbf{w}$ are assumed to be given by

$$\begin{aligned} \mathbf{v} &= \mathbf{Hx} + \boldsymbol{\eta} \\ \mathbf{w} &= \mathbf{z} + \boldsymbol{\zeta} \end{aligned} \tag{80}$$

where η and ζ are also noise inputs. The noise contaminations $\mathbf{m}$, $\boldsymbol{\eta}$, and $\boldsymbol{\zeta}$ are assumed to be drawn from Gaussian white noise processes GWN($\boldsymbol{\mu}$, $\boldsymbol{\sigma}$) with mean $\boldsymbol{\mu}$ and variance $\boldsymbol{\sigma}$, such that

$$\begin{aligned} \mathbf{m} &= GWN(0, \mathbf{S}) \\ \boldsymbol{\eta} &= GWN(0, \mathbf{Q}) \\ \boldsymbol{\zeta} &= GWN(0, \mathbf{R}) \end{aligned}$$

Finally, we have an estimate of the initial state of the process $\mathbf{x}(t_0)$ which is assumed to have a Gaussian density function of mean $\bar{\mathbf{x}}_0$ and variance $\mathbf{P}_0$. The matrices $\mathbf{S}$, $\mathbf{R}$, $\mathbf{Q}$, and $\mathbf{P}_0$ are all positive definite and symmetric.

This completes the description of the "message process" which generates the observations $\mathbf{v}$ and $\mathbf{w}$. In practice the model would represent a series of interacting subsystems whose state is partially observable but whose interaction is *completely* observable, both via noisy channels.

The problems are threefold.

1. *Smoothing*; i.e., obtain the best estimate of $\mathbf{x}(t)$ for $t_0 \leqslant t \leqslant t_1$ using all the data over the fixed interval $[t_0, t_1]$
2. *Filtering*; i.e., what is the best estimate of $\mathbf{x}(t_1)$ for given data, as t_1 increases
3. *Extrapolation*; i.e., what is the best estimate of $\mathbf{x}(T)$ for a fixed $T > t_1$, given the data over $[t_0, t_1]$

We are concerned only with part 1 but will make some comments on part 2.

The relationship between this problem and the previous problems is remarkably direct. The most probable estimate of $\mathbf{x}(t_0)$ is obtained by maximizing the Gaussian probability density function of $\mathbf{x}(t_0)$

$$[(2\pi)^n \mid \mathbf{P}_0 \mid]^{-1/2} [\exp[-\tfrac{1}{2} \| \mathbf{x}(t_0) - \bar{\mathbf{x}} \|^2 \mathbf{P}_0^{-1}]$$

where $|\mathbf{P}_0|$ is the determinant of $\mathbf{P}_0$. This is clearly the same as the deterministic problem of minimizing

$$\| \mathbf{x}(t_0) - \bar{\mathbf{x}} \|^2 \mathbf{P}_0^{-1}$$

especially since the variance-covariance matrix $\mathbf{P}_0$ is positive definite, by definition.

In order to obtain a deterministic equivalent problem for the complete system the coupling will be eliminated to put the system in standard form. Thus

$$\dot{\mathbf{x}} = (\mathbf{A} + \mathbf{CLM})\,\mathbf{x} + (\mathbf{B} + \mathbf{CLM})\,\mathbf{m}$$

$$\begin{pmatrix} \mathbf{v} \\ \mathbf{w} \end{pmatrix} = \begin{pmatrix} \mathbf{H} \\ \mathbf{LM} \end{pmatrix} \mathbf{x} + \begin{pmatrix} \eta \\ \mathbf{LNm} + \zeta \end{pmatrix}$$

that is,

$$\dot{\mathbf{x}} = \mathbf{A}^*\mathbf{x} + \mathbf{B}^*\mathbf{m}$$

$$\mathbf{v}^* = \mathbf{H}^*\mathbf{x} + \eta^*$$

where $\mathbf{m}$ and η^* are zero-mean correlated sources, with variance-covariance given by

$$\operatorname{var}\begin{pmatrix} \mathbf{m} \\ \eta^* \end{pmatrix} = \begin{pmatrix} \mathbf{S} & 0 & \mathbf{SN'L'} \\ 0 & \mathbf{Q} & 0 \\ \mathbf{LNS} & 0 & \mathbf{R} + \mathbf{LNSN'L'} \end{pmatrix} = \begin{pmatrix} \mathbf{S} & (\mathbf{S}^*)' \\ (\mathbf{S}^*) & \mathbf{Q}^* \end{pmatrix}$$

where we have defined

$$\begin{aligned} \mathbf{A}^* &= \mathbf{A} + \mathbf{CLM} \\ \mathbf{B}^* &= \mathbf{B} + \mathbf{CLN} \end{aligned} \qquad \mathbf{H}^* = \begin{pmatrix} \mathbf{H} \\ \mathbf{LM} \end{pmatrix} \qquad \eta^* = \begin{pmatrix} \eta \\ \mathbf{LNm} + \zeta \end{pmatrix} \qquad \mathbf{v}^* = \begin{pmatrix} \mathbf{v} \\ \mathbf{w} \end{pmatrix}$$

$$\mathbf{S}^* = [\mathbf{LNS} \quad 0] \qquad \mathbf{Q}^* = \begin{pmatrix} \mathbf{Q} & 0 \\ 0 & \mathbf{R} + \mathbf{LNSN'L'} \end{pmatrix}$$

Now, for this problem Cox[3] has shown that the a posteriori Gaussian density function has a negative exponential argument of

$$\tfrac{1}{2} \| \mathbf{x}(t_0) - \bar{\mathbf{x}}_0 \|^2_{\mathbf{P}_0^{-1}} + \tfrac{1}{2} \int_{t_0}^{t_1} \left[\left\| \begin{matrix} \mathbf{m} \\ \mathbf{v}^* - \mathbf{H}^* x \end{matrix} \right\|^2_{\begin{pmatrix} \mathbf{S} & (\mathbf{S}^*)' \\ (\mathbf{S}^*) & \mathbf{Q}^* \end{pmatrix}^{-1}} \right] dt \tag{81}$$

where $\mathbf{x}$ and $\mathbf{m}$ satisfy the integrated process model. Thus the maximum likelihood estimate of $\mathbf{x}$ can be obtained formally by minimizing the functional above. This problem is within the class of problems of the earlier sections.

The solution of the problem has been given by Cox and can be obtained by writing down the necessary conditions defining the multiplier $\boldsymbol{\lambda}$ and making a Riccati transformation of the form

$$\boldsymbol{\lambda}(t) = -\mathbf{K}^{-1}(t)[\mathbf{x}(t) - \mathbf{e}(t)]$$

where the defining equations for $\mathbf{K}$ and $\mathbf{e}$ are given by

$$\dot{\mathbf{K}} = (\mathbf{A}^*)\,\mathbf{K} + \mathbf{K}(\mathbf{A}^*)' + (\mathbf{B}^*)\,\mathbf{S}(\mathbf{B}^*)' - \boldsymbol{\Lambda}\mathbf{Q}^*\boldsymbol{\Lambda} \tag{82}$$

$$\mathbf{K}(t_0) = \mathbf{P}_0$$

where

$$\boldsymbol{\Lambda} = [\mathbf{K}(\mathbf{H}^*)' + \mathbf{B}^*\mathbf{S}^*](\mathbf{Q}^*)^{-1}$$

and

$$\dot{\mathbf{e}} = \mathbf{A}^*\mathbf{e} + \boldsymbol{\Lambda}(\mathbf{v}^* - \mathbf{H}^*\mathbf{e}) \qquad \mathbf{e}(t_0) = \mathbf{e}_0 \tag{83}$$

Equations similar to these are derived in the next section. In terms of the n-vector $\mathbf{e}$ and the $n \times n$ matrix $\mathbf{K}$, the functional takes on the minimal value

$$\tfrac{1}{2} \| \mathbf{x}(t_1) - \mathbf{e}(t_1)\|^2_{\mathbf{K}^{-1}(t_1)} + \text{const}$$

so that minimizing yields $\mathbf{x}(t_1) = \mathbf{e}(t_1)$ and this will be the *maximum likelihood estimate* of $\mathbf{x}(t_1) = \hat{\mathbf{x}}(t_1)$, with variance-covariance $\mathbf{K}(t_1)$. Thus the equation for the optimal filter $\hat{\mathbf{x}}$ is the equation for $\mathbf{e}$; namely,

$$\frac{d}{dt}[\hat{\mathbf{x}}(t_1)] = \mathbf{A}^*\hat{\mathbf{x}}(t_1) + \boldsymbol{\Lambda}(t_1)[\mathbf{v}^*(t_1) - \mathbf{H}^*\hat{\mathbf{x}}(t_1)] \tag{84}$$

$$\hat{\mathbf{x}}(t_0) = \bar{\mathbf{x}}_0$$

with $\boldsymbol{\Lambda}(t)$ is defined in terms of $\mathbf{K}$ as before.

The optimally smoothed trajectory can be found by inserting the minimizing $\bar{\mathbf{m}}$ into the message equation in a way which will be apparent after the next section.

4-8-2 Decomposition of the Smoothing Problem

Clearly the equations for $\mathbf{e}$ and $\mathbf{K}$ do not separate because of the arbitrary structure of $\mathbf{L}$. The separation can be achieved, however, by introducing the interaction explicitly. To do this, we first examine the quadratic form of the functional. By using Gaussian elimination it is straightforward to show that

$$\begin{pmatrix} \mathbf{S} & 0 & \mathbf{SN'L'} \\ 0 & \mathbf{Q} & 0 \\ \mathbf{LNS} & 0 & \mathbf{R} + \mathbf{LNSN'L'} \end{pmatrix}^{-1} = \begin{pmatrix} \mathbf{S}^{-1} + \mathbf{N'L'R}^{-1}\mathbf{LN} & 0 & -\mathbf{N'L'R}^{-1} \\ 0 & \mathbf{Q}^{-1} & 0 \\ -\mathbf{R}^{-1}\mathbf{LN} & 0 & \mathbf{R}^{-1} \end{pmatrix}$$

The corresponding change in the functional argument is as follows:

$$\left\| \begin{matrix} \mathbf{m} \\ \mathbf{v} - \mathbf{Hx} \\ \mathbf{w} - \mathbf{LMx} \end{matrix} \right\|^2 \begin{pmatrix} \mathbf{S} & 0 & \mathbf{SN'L'} \\ 0 & \mathbf{Q} & 0 \\ \mathbf{LNS} & 0 & \mathbf{R} + \mathbf{LNSN'L'} \end{pmatrix}^{-1}$$

$$\equiv (\| \mathbf{m} \|^2\, S^{-1} + \| \mathbf{v} - \mathbf{Hx} \|^2\, \mathbf{Q}^{-1} + \| \mathbf{w} - \mathbf{z} \|^2\, \mathbf{R}^{-1})$$

This introduces the coupling and since **S**, **Q**, and **R** are block diagonal, it separates into subproblems. Thus an equivalent deterministic problem to the statistical problem can be decomposed exactly as before and will generate a parametric problem as follows (for a given $\boldsymbol{\pi}$):

$$\min J(\mathbf{m}, \mathbf{z}; \boldsymbol{\pi}) = \tfrac{1}{2} \| \mathbf{x}(t_0) - \bar{\mathbf{x}}_0 \|^2 \mathbf{P}_0^{-1}$$

$$+ \int_{t_0}^{t_1} [\tfrac{1}{2} \| \mathbf{v} - \mathbf{Hx} \|^2 \mathbf{Q}^{-1} + \| \mathbf{w} - \mathbf{z} \|^2 \mathbf{R}^{-1} + \| \mathbf{m} \|^2 \mathbf{S}^{-1})$$

$$+ \boldsymbol{\pi}'(\mathbf{LMx} + \mathbf{LNm} - \mathbf{z})]\, dt \tag{85}$$

$$\text{subject to } \dot{\mathbf{x}} = \mathbf{Ax} + \mathbf{Bm} + \mathbf{Cz}$$

An appealing property of the decomposed problem is that it is obviously easier to manipulate, and the fact that $\mathbf{z}$ is observed introduces the strict convexity in a natural way. The functional $J(\mathbf{m}, \mathbf{z}; \boldsymbol{\pi})$ can be completely separated into **N** problems by introducing the multiplier $\boldsymbol{\kappa}$

$$\boldsymbol{\kappa} = \mathbf{L}'\boldsymbol{\pi}$$

where the block diagonal structure of **A**, **B**, **C**, $\mathbf{P}_0$, **Q**, **R**, **S**, **M**, and **N** implies that separation can be achieved.

Since the manipulations are simpler, the working development will be given in more detail, partly to explain the derivation of the previous solution.

The Hamiltonian function is $\mathscr{H}(\mathbf{x}, \mathbf{m}, \mathbf{v}, \mathbf{w}, \mathbf{z}, \boldsymbol{\lambda}, t)$.

$$\mathscr{H} = \tfrac{1}{2}(\| \mathbf{v} - \mathbf{Hx} \|^2 Q^{-1} + \| \mathbf{w} - \mathbf{z} \|^2 \mathbf{R}^{-1} + \| \mathbf{m} \|^2 \mathbf{S}^{-1})$$

$$+ \boldsymbol{\kappa}'(\mathbf{Mx} + \mathbf{Nm}) - \boldsymbol{\pi}'\mathbf{z} + \boldsymbol{\lambda}'(\mathbf{Ax} + \mathbf{Bm} + \mathbf{Cz})$$

The necessary conditions for a solution require that $\mathbf{m}$, $\mathbf{z}$ and $\boldsymbol{\lambda}$ must satisfy the following equations:

$$\mathscr{H}_{\mathbf{m}}' = \mathbf{m} + \mathbf{SN}'\boldsymbol{\kappa} + \mathbf{SB}'\boldsymbol{\lambda} = 0 \tag{86}$$

$$\mathscr{H}_{\mathbf{z}}' = \mathbf{z} - \mathbf{w} + \mathbf{R}(\mathbf{C}'\boldsymbol{\lambda} - \boldsymbol{\pi}) = 0 \tag{87}$$

$$\dot{\boldsymbol{\lambda}} + \mathscr{H}_{\mathbf{x}}' = \dot{\boldsymbol{\lambda}} + \mathbf{H}'\mathbf{Q}^{-1}(\mathbf{v} - \mathbf{Hx}) + \mathbf{M}'\boldsymbol{\kappa} + \mathbf{A}'\boldsymbol{\lambda} = 0$$

$$\boldsymbol{\lambda}(t_0) + \mathbf{P}_0^{-1}[\mathbf{x}(t_0) - \mathbf{x}_0] = 0 \qquad \boldsymbol{\lambda}(t_1) = 0$$

Eliminating $\mathbf{m}$ and $\mathbf{z}$ reduces the necessary conditions to the following pair

$$\dot{\mathbf{x}} = \mathbf{Ax} - \mathbf{BS}(\mathbf{B}'\boldsymbol{\lambda} + \mathbf{N}'\boldsymbol{\kappa}) - \mathbf{CR}(\mathbf{C}'\boldsymbol{\lambda} - \boldsymbol{\pi}) + \mathbf{Cw}$$

$$\dot{\boldsymbol{\lambda}} = \mathbf{A}'\boldsymbol{\lambda} - \mathbf{H}'\mathbf{Q}^{-1}(\mathbf{Hx} - \mathbf{v}) + \mathbf{M}'\boldsymbol{\kappa}$$

Now making a similar substitution as previously, $\boldsymbol{\lambda} = -\mathbf{P}^{-1}(\mathbf{x} - \mathbf{c})$, equations can be generated for $\mathbf{P}$ and $\mathbf{c}$.

$$\dot{\mathbf{P}} = \mathbf{AP} + \mathbf{PA}' + \mathbf{BSB}' + \mathbf{CRC}' - \mathbf{PH}'\mathbf{Q}^{-1}\mathbf{HP} \tag{88}$$

$$\mathbf{P}(t_0) = \mathbf{P}_0$$

$$\dot{\mathbf{c}} = \mathbf{Ac} + \mathbf{PH}'\mathbf{Q}^{-1}(\mathbf{v} - \mathbf{Hc}) + (\mathbf{CR}\boldsymbol{\pi} - \mathbf{BSN}'\boldsymbol{\kappa} - \mathbf{PM}'\boldsymbol{\kappa} + \mathbf{Cw}) \tag{89}$$

$$\mathbf{e}(t_0) = \bar{\mathbf{x}}_0$$

The boundary conditions on $\mathbf{c}$ and $\mathbf{P}$ are chosen to satisfy

$$-\mathbf{P}^{-1}[\mathbf{x}(t_0) - \mathbf{c}(t_0)] + \mathbf{P}_0^{-1}[\mathbf{x}(t_0) - \bar{\mathbf{x}}_0] = 0$$

Let $\mathbf{c}(t; \boldsymbol{\pi})$ be a solution to the second equation for a given $\boldsymbol{\pi}$, and $\boldsymbol{\kappa} = \mathbf{L}'\boldsymbol{\pi}$ function. Then the smoothed solutions are obtained by substituting $\bar{\mathbf{m}}$ and $\bar{\mathbf{z}}$ into the message process, where from the necessary conditions[1]

$$\bar{\mathbf{m}} = -\mathbf{S}[-\mathbf{B}'\mathbf{P}^{-1}(\bar{\mathbf{x}} - \mathbf{c}) + \mathbf{N}'\boldsymbol{\kappa}] \tag{90}$$

$$\bar{\mathbf{z}} = -\mathbf{R}[-\mathbf{C}'\mathbf{P}^{-1}(\bar{\mathbf{x}} - \mathbf{c}) - \boldsymbol{\pi}] + \mathbf{w} \tag{91}$$

The smoothing equation is integrated backward in time from $\mathbf{t}_1$ with boundary conditions given below.

$$\dot{\bar{\mathbf{x}}} = \mathbf{A}\bar{\mathbf{x}} + \mathbf{B}\bar{\mathbf{m}} + \mathbf{C}\bar{\mathbf{z}} \qquad \bar{\mathbf{x}}(t_1) = \mathbf{c}(t_1)$$

The boundary condition is derived from the fact that $\boldsymbol{\lambda}(t_1) = 0$. This procedure will generate $\bar{\mathbf{x}}(t; \boldsymbol{\pi})$ as smoothed solutions for the given $\boldsymbol{\pi}$ function.

From a decomposition point of view all the previous equations separate into $\mathbf{N}$ groups of equations, one per subsystem. Thus, the *smoothing* equation, the *filter* equation for $\mathbf{c}(\mathbf{t})$, and the *variance* equation for $\mathbf{P}(t)$ all represent $\mathbf{N}$ independent equations. This is because matrices $\mathbf{A}$, $\mathbf{B}$, $\mathbf{C}$, $\mathbf{Q}$, $\mathbf{R}$, $\mathbf{S}$, and $\mathbf{H}$ are all block diagonal. As a consequence $\mathbf{P}(t)$ is a block diagonal matrix for each t.

However $\boldsymbol{\pi}$ was chosen arbitrarily and as a consequence the subproblem solutions are not connected; hence

$$\mathbf{LM}\bar{\mathbf{x}}(\boldsymbol{\pi}) + \mathbf{LN}\bar{\mathbf{m}}(\boldsymbol{\pi}) \neq \bar{\mathbf{z}}(\boldsymbol{\pi})$$

[1] Note that $\mathbf{P}^{-1}$ satisfies a similar equation given below and this avoids a matrix inversion.

$$\frac{d}{dt}(\mathbf{P}^{-1}) + (\mathbf{P}^{-1})\mathbf{A} + \mathbf{A}'(\mathbf{P}^{-1}) + (\mathbf{P}^{-1})(\mathbf{BSB}' + \mathbf{CRC}')(\mathbf{P}^{-1}) - \mathbf{H}'\mathbf{QH} = 0$$

Applying the theory of Sec. 4-4, $\boldsymbol{\pi}$ can be found by maximizing the dual functional, which was shown to be the sum of the subproblem objective functions given by $J(\bar{\mathbf{m}}(\boldsymbol{\pi}), \bar{\mathbf{z}}(\boldsymbol{\pi}); \boldsymbol{\pi})$. Thus, the coordination problem, using goal coordination, is to maximize $J(\bar{\mathbf{m}}(\boldsymbol{\pi}), \bar{\mathbf{z}}(\boldsymbol{\pi}); \boldsymbol{\pi})$ where $\bar{\mathbf{x}}(\boldsymbol{\pi})$, $\bar{\mathbf{m}}(\boldsymbol{\pi})$, $\bar{\mathbf{z}}(\boldsymbol{\pi})$ are subproblem solutions for the given $\boldsymbol{\pi}$.

The gradient of $J(\bar{\mathbf{m}}(\boldsymbol{\pi}), \bar{\mathbf{z}}(\boldsymbol{\pi}); \boldsymbol{\pi})$ is the Fréchet derivative

$$\frac{\delta J(\bar{\mathbf{m}}(\boldsymbol{\pi}), \bar{\mathbf{z}}(\boldsymbol{\pi}); \boldsymbol{\pi})}{\delta \boldsymbol{\pi}} = \mathbf{LM}\bar{\mathbf{x}}(\boldsymbol{\pi}) + \mathbf{LN}\bar{\mathbf{m}}(\boldsymbol{\pi}) - \bar{\mathbf{z}}(\boldsymbol{\pi})$$

Numerical calculation of $\boldsymbol{\pi}^*$, the coordinating value, is simplified by the fact that $\bar{\mathbf{x}}(\boldsymbol{\pi})$ and $\mathbf{c}(\boldsymbol{\pi})$ are linear functions of $\boldsymbol{\pi}$ and $\mathbf{P}$ is independent of $\boldsymbol{\pi}$. Thus the simplified version of the conjugate gradient method may be used as outlined in Sec. 4-5. If only a numerical solution is required, there is no advantage in using the Riccati solution of $\mathbf{P}$ and $\mathbf{c}$ as opposed to the multipliers $\boldsymbol{\lambda}$, since the same number of equations is required at each iteration and in the latter case no $\mathbf{P}$ matrix need ever be computed.

Example 4-6

Consider a smoothing problem of finding the best estimate of temperature distribution over a rod, given measurements at certain points along it.

The temperature $\mathbf{x}(s, t)$ at position s and time t satisfies the one-dimensional heat conduction equation

$$\frac{\partial \mathbf{x}}{\partial t} = \mu \frac{\partial^2 \mathbf{x}}{\partial s^2}$$

where the diffusivity μ is assumed constant. Suppose the rod is divided into 16 equal segments of length Δ at positions s_i and let $\mathbf{x}_i(t) \equiv \mathbf{x}(s_i, t)$. A finite difference approximation to the heat equation can take the following form. Let

$$x_s = \frac{x_i - x_{i-1}}{\Delta}$$

$$x_{ss} \simeq \frac{x_{i+1} - 2x_i + x_{i-1}}{\Delta^2}$$

then

$$\frac{\partial x_i}{\partial t} = \frac{\mu}{\Delta^2}(x_{i+1} - 2x_i + x_{i-1})$$

or

$$\dot{x}_i = k(x_{i+1} - 2x_i + x_{i-1})$$

which is the simple finite difference approximation to be used for each segment.

Figure 4-4 shows the flow diagram of two elements, indicating the

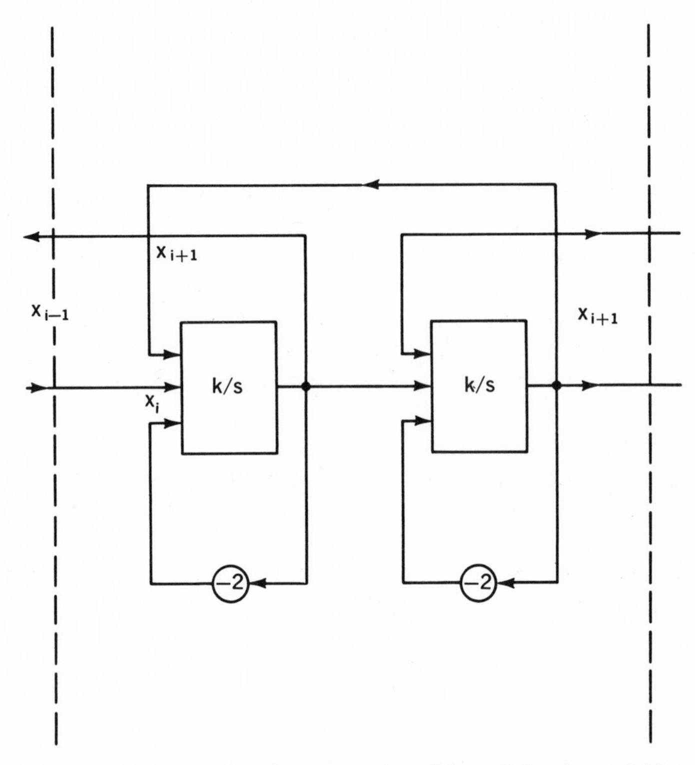

FIG. 4-4. A lumped approximation to a section of the rod showing variable origins.

countercurrent aspect of the model. Suppose 4 subsystems are composed of 4 elements each to form the 16-element discrete model of the rod. Figure 4-5 shows the configuration of interconnections of the four subsystems, each of which is modeled as follows.

Let *subscript ij* denote the *j*th element of *i*th-subsystem variables. The *i*th subsystem satisfies

$$\begin{pmatrix} \dot{x}_{i1} \\ \dot{x}_{i2} \\ \dot{x}_{i3} \\ \dot{x}_{i4} \end{pmatrix} = \begin{pmatrix} -2k & k & 0 & 0 \\ k & -2k & k & 0 \\ 0 & k & -2k & k \\ 0 & 0 & k & -2k \end{pmatrix} \begin{pmatrix} x_{i1} \\ x_{i2} \\ x_{i3} \\ x_{i4} \end{pmatrix} + \begin{pmatrix} k & 0 \\ 0 & 0 \\ 0 & 0 \\ 0 & k \end{pmatrix} \begin{pmatrix} z_{i1} \\ z_{i4} \end{pmatrix}$$

where z_{i1} and z_{i4} are interconnection inputs.

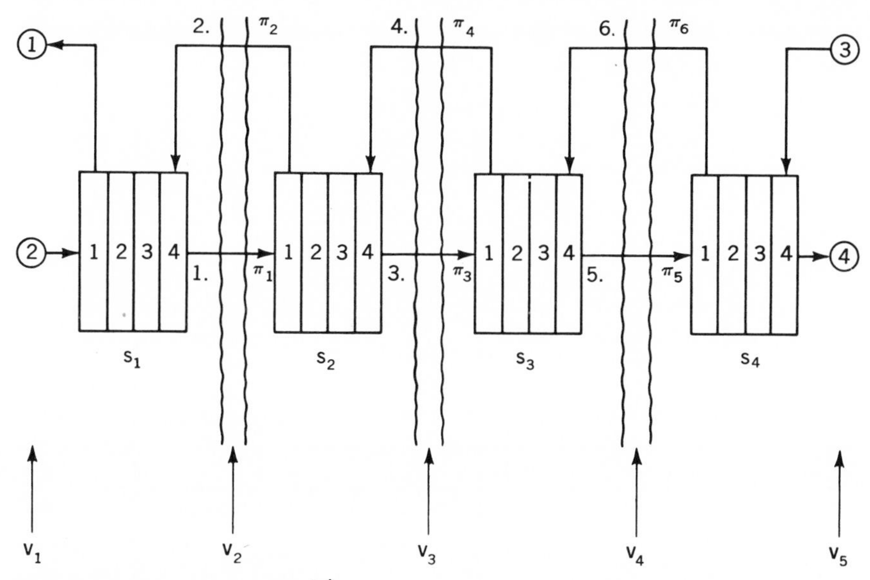

Interconnections are severed at)(

(1) Connection not needed

(2) Determined by x = 0 boundary condition

(3) Connection not needed

(4) Connection not needed

FIG. 4-5. Interconnections of subsystems s_1 to s_4 to form an approximation of the heat flow in the rod.

The subsystems are interconnected in the manner of Fig. 4-5, which has the following coupling relations for the links.

Link	Gradient	Parameter
link 1	gradient $g_1 = x_{21} - z_{14}$	parameter π_1
link 2	gradient $g_2 = x_{14} - z_{21}$	parameter π_2
link 3	gradient $g_3 = x_{31} - z_{24}$	parameter π_3
link 4	gradient $g_4 = x_{24} - z_{31}$	parameter π_4
link 5	gradient $g_5 = x_{41} - z_{34}$	parameter π_5
link 6	gradient $g_6 = x_{34} - z_{41}$	parameter π_6

Thus when all the gradients are zero, the subsystems are coupled to form the integrated model.

The reason that four subsystems were employed is that five observations were made at points separated by intervals of 4Δ. The observations are given below and are taken to be the average of the neighboring element temperatures. Thus

$$v_1 = \frac{x_{11} + b_{11}}{2} + \eta_1$$

$$v_2 = \frac{x_{21} + x_{14}}{2} + \eta_2$$

$$v_3 = \frac{x_{31} + x_{24}}{2} + \eta_3$$

$$v_4 = \frac{x_{41} + x_{34}}{2} + \eta_4$$

$$v_5 = x_{44} + \eta_5$$

where b_{11} is the boundary condition at $s = 0$ and η_i is a Gaussian white noise of zero mean and, for simplicity, unity variance. Similarly the initial temperature distribution is estimated at $t = t_0$ to have mean $\bar{x}_{ij}$ with unit variance at each point. Then using the equivalent deterministic model, the integrated problem can be set up and decomposed, using goal coordination.

Using the economic interpretation of the π_i's as prices, it is straightforward to write down the subsystem performance indices. These are, respectively,

$$J_1(\mathbf{z}_1\,;\,\boldsymbol{\pi}) = \tfrac{1}{2}\sum_{j=1}^{4}(x_{1j} - \bar{x}_{1j})^2 + \int_{t_0}^{t_1}\left\{\tfrac{1}{2}\left[v_1 - \frac{(x_{11} + b_{11})^2}{2}\right]^2 + \tfrac{1}{2}\left(v_2 - \frac{z_{14} + x_{14}}{2}\right)^2 + \pi_2 x_{14} - \pi_1 z_{14}\right\} dt$$

$$J_2(\mathbf{z}_2\,;\,\boldsymbol{\pi}) = \tfrac{1}{2}\sum_{j=1}^{4}(x_{2j} - \bar{x}_{2j})^2 + \int_{t_0}^{t_1}\left\{\tfrac{1}{2}\left[v_2 - \frac{(x_{21} + z_{21})^2}{2}\right]^2 + \tfrac{1}{2}\left(v_3 - \frac{z_{24} + x_{24}}{2}\right)^2 + \pi_1 z_{21} - \pi_2 z_{21} - \pi_3 z_{24} + \pi_4 x_{24}\right\} dt$$

$$J_3(\mathbf{z}_3\,;\,\boldsymbol{\pi}) = \tfrac{1}{2}\sum_{j=1}^{4}(x_{3j} - \bar{x}_{3j})^2 + \int_{t_0}^{t_1}\left[\tfrac{1}{2}\left(v_3 - \frac{x_{31} + z_{31}}{2}\right)^2 + \tfrac{1}{2}\left(v_4 - \frac{x_{34} + z_{34}}{2}\right)^2 + \pi_3 x_{31} - \pi_4 z_{31} - \pi_5 z_{34} + \pi_6 x_{34}\right] dt$$

$$J_4(\mathbf{z}_4\,;\,\boldsymbol{\pi}) = \tfrac{1}{2}\sum_{j=1}^{4}(x_{4j} - \bar{x}_{4j})^2 + \int_{t_0}^{t_1}\left[\tfrac{1}{2}\left(v_4 - \frac{x_{41} + z_{41}}{2}\right)^2 + \tfrac{1}{2}(v_5 - x_{44})^2 + \pi_5 x_{41} - \pi_6 z_{41}\right] dt$$

Given a set of π functions over the time interval $[t_0, t_1]$ and the observations v_i, subproblem solutions can be found and denoted as $x_{ij}^*(\pi)$, $z_{ij}^*(\pi)$.

The coordination problem is to find π^*. This is achieved by maximizing

$$J(\pi) = J_1(z_1^*(\pi); \pi) + J_2(z_2^*(\pi); \pi) + J_3(z_3^*(\pi); \pi) + J_4(z_4^*(\pi); \pi)$$

where $\mathbf{z}_i' \equiv (z_{i1}, z_{i2})'$ and each functional is evaluated along a solution to its subproblem for the given π. The gradient of $J(\pi)$, $\delta J(\pi)/\delta\pi$, is given by $\mathbf{g} = (g_1, g_2, g_3, g_4)$. A conjugate gradient method was used to maximize $J(\pi)$ and convergence was achieved for a variety of time intervals $[t_0, t_1]$, with results similar to those of Fig. 4-6, showing $\| \mathbf{g} \|$ versus the iteration number. Full details and the Algol program are reported by Tinkelman.[21]

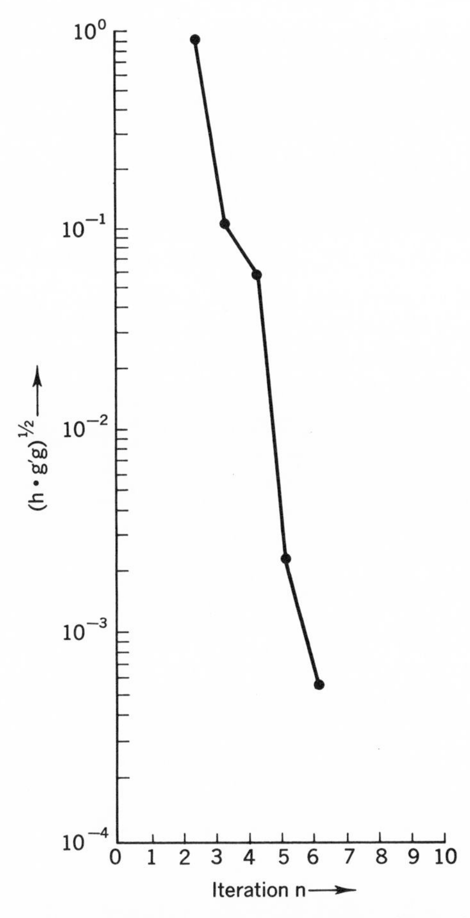

FIG. 4-6. Normed interconnection error versus iteration for the heat flow example.

An interesting aspect of this example is that lumped models of distributed systems naturally generate vast interconnections of fairly identical subsystems. Thus the "subsystem package" concept introduced in Sec. 4-6 becomes essential and produces great savings in the labor required to set up the subproblems since this need only be done once.

4-8-3 Decomposing the Optimal Filter

The most difficult and perhaps most interesting problem is to coordinate a series of optimal filters for the subproblems. Returning to the discussion at the end of Sec. 4-8-2 where the best estimate of $\mathbf{x}(t_1)$ is $\hat{\mathbf{x}}(t_1) = \mathbf{c}(t_1)$, the filter for $\mathbf{c}(t_1)$ and the error variance $\mathbf{P}(t_1)$ are given by

$$\begin{aligned}\dot{\mathbf{c}}(t_1) &= \mathbf{A}\mathbf{c}(t_1) + \mathbf{P}(t_1)\,\mathbf{H}'\mathbf{Q}^{-1}[\mathbf{v}(t_1) - \mathbf{H}\mathbf{c}(t_1)] \\ &\quad + [\mathbf{C}\mathbf{R}\boldsymbol{\pi}(t_1) - \mathbf{B}\mathbf{S}\mathbf{N}'\boldsymbol{\kappa}(t_1) - \mathbf{P}(t_1)\,\mathbf{M}'\boldsymbol{\kappa}(t_1) + \mathbf{C}\mathbf{w}(t_1)] \qquad (92)\\ \mathbf{c}(t_0) &= \bar{\mathbf{x}}_0 \\ \dot{\mathbf{P}}(t_1) &= \mathbf{A}\mathbf{P}(t_1) + \mathbf{P}(t_1)\,\mathbf{A}' + \mathbf{B}\mathbf{S}\mathbf{B}' + \mathbf{C}\mathbf{R}\mathbf{C}' - \mathbf{P}(t_1)\,\mathbf{H}'\mathbf{Q}^{-1}\mathbf{H}\mathbf{P}(t_1) \\ \mathbf{P}(t_0) &= \mathbf{P}_0\end{aligned}$$

Recall that the diagonal structure of the matrices $\mathbf{A}$, $\mathbf{B}$, $\mathbf{C}$, $\mathbf{M}$, $\mathbf{N}$, $\mathbf{P}_0$, $\mathbf{Q}$, $\mathbf{R}$, $\mathbf{S}$, and $\mathbf{H}$ implies that $\mathbf{P}(t_1)$ is diagonal, so that these equations represent N independent filters for given functions $\boldsymbol{\pi}(t)$ and $\boldsymbol{\kappa}(t) = \mathbf{L}'\boldsymbol{\pi}(t)$.

Suppose that the optimal coordination for the smoothing problem has been found, that is, $\boldsymbol{\pi}^*$ for an interval $[t_0, t_1]$. Since $\bar{\mathbf{x}}(t_1) = \hat{\mathbf{x}}(t_1) = \mathbf{c}(t_1)$ the same $\boldsymbol{\pi}^*$ function gives a filtered estimate which is coordinated. That is, since $\boldsymbol{\lambda}(t_1) = 0$ for each t_1,

$$\mathbf{m}^*(\boldsymbol{\pi}^*) = -\mathbf{S}\mathbf{N}'\boldsymbol{\kappa}^*$$

$$\mathbf{z}^*(\boldsymbol{\pi}^*) = \mathbf{w} + \mathbf{R}\boldsymbol{\pi}$$

from Eqs. (86) and (87) hence

$$\mathbf{L}\mathbf{M}\mathbf{c}(t_1\,;\,\boldsymbol{\pi}^*) + \mathbf{L}\mathbf{N}(-\mathbf{S}\mathbf{N}'\boldsymbol{\kappa}^*) = \mathbf{w} + \mathbf{R}\boldsymbol{\pi}^* \qquad (93)$$

Now from Eq. (92) $\mathbf{c}(\boldsymbol{\pi})$ is a continuous linear function of $\boldsymbol{\pi}$ defined over the whole interval $[t_0, t_1]$. Thus as t_1 advances with real time, Eq. (93) defines a linear integral equation for $\boldsymbol{\pi}^*$, the solution of which must *also* coordinate the smoothed system over each interval $[t_0, t_1]$. At present no simple solution to this problem is evident (that is, one which takes less work than solving the original problems).

4-8-4 Suboptimal Coordination

A reasonable alternative is to concentrate on choosing $\boldsymbol{\pi}$ only at t_1 so that the filter is coordinated only at t_1. This can be done by solving Eq. (93) for $\boldsymbol{\pi}$ as a function of $\mathbf{c}(\boldsymbol{\pi})$ and using this in the defining equation for $\mathbf{e}(t)$ to form a suboptimal filter.

From Eq. (93) and using $\boldsymbol{\kappa} = \mathbf{L}'\boldsymbol{\pi}$, let $\boldsymbol{\pi}^0$ be defined by

$$\boldsymbol{\pi}^0 = (\mathbf{R} + \mathbf{LNSN'L'})^{-1}(\mathbf{LMc} - \mathbf{w}) \tag{94}$$

where $\mathbf{LNSN'L} = (\mathbf{LN})\mathbf{S}(\mathbf{LN})'$ is at least positive semidefinite so that inverse exists for $\mathbf{R}$ positive definite. The suboptimal filter is now defined by an equation for $\mathbf{x}^0$, the suboptimal estimate obtained by substituting $\boldsymbol{\pi}^0$ into Eq. (92). Then

$$\frac{d\mathbf{x}^0}{dt} = \mathbf{Ax}^0 + \mathbf{PH'Q}^{-1}(\mathbf{v} - \mathbf{Hx}^0) + \mathbf{Cw}$$
$$+ (\mathbf{CR} - \mathbf{BSN'L'} - \mathbf{PM'L'})(\mathbf{R} + \mathbf{LNSN'L'})^{-1}(\mathbf{LMx}^0 - \mathbf{w})$$

Adding and subtracting $\mathbf{CLMx}^0$ from the right-hand side simplifies this expression to the final form

$$\begin{aligned}\frac{d\mathbf{x}^0(t_1)}{dt} &= (\mathbf{A} + \mathbf{CLM})\,\mathbf{x}^0(t_1) + \mathbf{P}(t_1)\,\mathbf{H'Q}^{-1}[\mathbf{v}(t_1) - \mathbf{Hx}^0(t_1)]\\ &\quad + [\mathbf{P}(t_1)\,\mathbf{M'L'} + (\mathbf{B} + \mathbf{CLN})\,\mathbf{SN'L'}](\mathbf{R} + \mathbf{LNSN'L'})^{-1}\\ &\quad \times [\mathbf{w}(t_1) - \mathbf{LMx}^0(t_1)]\\ \mathbf{x}^0(t_0) &= \bar{\mathbf{x}}_0\end{aligned} \tag{95}$$

with $\mathbf{P}(t_1)$ defined as the solution to the subsystem variance equation. This is the suboptimal filter on the basis of using a $\boldsymbol{\pi}$ which coordinates the smoothed trajectory only at t_1. On the other hand, the optimal filter for the integrated system is found by separating Eq. (84) into its component parts, namely,

$$\begin{aligned}\frac{d\hat{\mathbf{x}}(t_1)}{dt_1} &= (\mathbf{A} + \mathbf{CLM})\,\hat{\mathbf{x}}(t_1) + \boldsymbol{\kappa}(t_1)\,\mathbf{H'Q}^{-1}[\mathbf{v}(t_1) - \mathbf{H}\hat{\mathbf{x}}(t_1)]\\ &\quad + [\boldsymbol{\kappa}(t_1)\,\mathbf{M'L'} + (\mathbf{B} + \mathbf{CLN})\,\mathbf{S'N'L'}](\mathbf{R} + \mathbf{LNSN'L'})^{-1}\\ &\quad \times [\mathbf{w}(t_1) - \mathbf{LM}\hat{\mathbf{x}}(t_1)]\\ \hat{\mathbf{x}}(t_0) &= \bar{\mathbf{x}}_0\end{aligned}$$

with $\boldsymbol{\kappa}(t_1)$ defined by the solution of the integrated variance equation (82). It can be seen that these expressions are almost identical except for $\mathbf{P}$

and $\boldsymbol{\kappa}$. In fact, by subtracting Eq. (95) from the integrated message process model, an equation for the error $\tilde{\mathbf{x}}(t) = \mathbf{x}(t) - \mathbf{x}^0(t)$ can be obtained. The variance of this error can then be compared to the variance of the optimal estimation error.

$$\begin{aligned}\frac{d\tilde{\mathbf{x}}(t_1)}{dt} = \{&\mathbf{A} + \mathbf{CLM} - \mathbf{P}(t_1)\,\mathbf{H}'\mathbf{Q}^{-1}\mathbf{H} \\ &-[\mathbf{P}(t_1)\,\mathbf{M}'\mathbf{L}' + (\mathbf{B} + \mathbf{CLN})\,\mathbf{SN}'\mathbf{L}'](\mathbf{R} + \mathbf{LNSN}'\mathbf{L}')^{-1}\,\mathbf{LM}\}\,\tilde{\mathbf{x}}(t_1) \\ &+\{\mathbf{B} + \mathbf{CLN} - [\mathbf{P}(t_1)\,\mathbf{M}'\mathbf{L}' \\ &+ (\mathbf{B} + \mathbf{CLN})\,\mathbf{SN}'\mathbf{L}'](\mathbf{R} + \mathbf{LNSN}'\mathbf{L}')^{-1}\,\mathbf{LN}\}\,\mathbf{m}(t_1) \\ &-\{\mathbf{P}(t_1)\,\mathbf{H}'\mathbf{Q}^{-1}\}\,\boldsymbol{\eta}(t_1) - \{[\mathbf{P}(t_1)\,\mathbf{M}'\mathbf{L}' \\ &+ (\mathbf{B} + \mathbf{CLN})\,\mathbf{SN}'\mathbf{L}'](\mathbf{R} + \mathbf{LNSN}'\mathbf{L}')^{-1}\}\,\boldsymbol{\zeta}(t_1)\end{aligned} \tag{96}$$

This equation has the form

$$\frac{d\tilde{\mathbf{x}}(t_1)}{dt} = \mathscr{A}(t_1)\,\tilde{\mathbf{x}}(t_1) + \mathscr{B}(t_1)\,\mathbf{m}(t_1) - \mathscr{C}(t_1)\,\boldsymbol{\eta}(t_1) - \mathscr{D}(t_1)\,\boldsymbol{\zeta}(t_1)$$

where $\mathscr{A}$, $\mathscr{B}$, $\mathscr{C}$, and $\mathscr{D}$ are the expressions in braces multiplying $\mathbf{x}$, $\mathbf{m}$, $\boldsymbol{\eta}$, and $\boldsymbol{\zeta}$ in Eq. (96).

Now let $\Phi(t)$ be the solution to

$$\frac{d}{dt}[\Phi(t)] = \mathscr{A}(t)\,\Phi(t) \qquad \Phi(t_0) = \mathbf{I}_n$$

then the value of $\tilde{\mathbf{x}}(t_1)$ is given by

$$\tilde{\mathbf{x}}(t_1) = \Phi(t_1)\,\tilde{\mathbf{x}}(t_0) + \int_{t_0}^{t_1} \{\Phi(t_1)\,\Phi(-t)[\mathscr{B}\mathbf{m} - \mathscr{C}\boldsymbol{\eta} - \mathscr{D}\boldsymbol{\zeta}]\}\,dt$$

The errors at t_0, $\tilde{\mathbf{x}}(t_0)$ and the noise sources $\mathbf{m}$, $\boldsymbol{\eta}$, and $\boldsymbol{\zeta}$ are assumed independent. Letting $E(\mathbf{x})$ denote the expectation of $\mathbf{x}$, we get from the initial assumptions above

$$\begin{aligned}E(\boldsymbol{\eta}) = 0 \qquad E(\boldsymbol{\zeta}) &= 0 \qquad E(\mathbf{m}) = 0 \\ E[\boldsymbol{\eta}(s)\,\boldsymbol{\eta}'(\sigma)] &= \mathbf{Q}\delta(s-\sigma) \qquad E[\boldsymbol{\eta}(s)\,\boldsymbol{\zeta}'(\sigma)] = 0 \\ E[\boldsymbol{\zeta}(s)\,\boldsymbol{\eta}'(\sigma)] &= \mathbf{R}\delta(s-\sigma) \qquad E[\boldsymbol{\eta}(s)\,\mathbf{m}'(\sigma)] = 0 \\ E[\mathbf{m}(s)\,\mathbf{m}'(\sigma)] &= \mathbf{S}\delta(s-\sigma) \qquad E[\boldsymbol{\zeta}(s)\,\mathbf{m}'(\sigma)] = 0\end{aligned}$$

where $\delta(s)$ is the delta function. Using the independence yields

$$E[\tilde{\mathbf{x}}(t_1)\,\tilde{\mathbf{x}}'(t_1)] = \Phi(t_1)\,E[\tilde{\mathbf{x}}(t_0)\,\tilde{\mathbf{x}}'(t_0)]\,\Phi'(t_1)$$

$$+ \int_{t_0}^{t_1}\int_{t_0}^{t_1} (\Phi(t_1)\,\Phi(-t)\{\mathscr{B}E[\mathbf{m}(t)\,\mathbf{m}'(s)]\,\mathscr{B}'$$

$$+ \mathscr{C}E[\boldsymbol{\eta}(t)\,\boldsymbol{\eta}'(s)]\,\mathscr{C}' + \mathscr{D}E[\boldsymbol{\zeta}(t)\,\boldsymbol{\zeta}'(s)]\,\mathscr{D}'\}\,\Phi'(s)\,\Phi'(t_1))\,dt\,ds$$

where the time argument of the term in braces is t.

Substituting for the expectations and defining $E[\tilde{\mathbf{x}}(t)\tilde{\mathbf{x}}'(t)] = \mathbf{P}^0(t)$, the derivative of $\mathbf{P}^0(t_1)$ can be seen to satisfy

$$\frac{d}{dt_1}[\mathbf{P}^0(t_1)] = \mathscr{A}(t_1)\,\mathbf{P}_0(t_1) + \mathbf{P}^0(t_1)\,\mathscr{A}'(t_1)$$

$$+ \mathscr{B}(t_1)\,\mathbf{S}\mathscr{B}'(t_1) + \mathscr{C}(t_1)\,\mathbf{Q}\mathscr{C}'(t_1) + \mathscr{D}(t_1)\,\mathbf{R}\mathscr{D}'(t_1) \qquad (97)$$

and the initial error variance of $\tilde{\mathbf{x}}(t_0)$ requires

$$\mathbf{P}^0(t_0) = \mathbf{P}_0$$

This defines the suboptimal error variance-covariance matrix for each time t_1.

It is possible to show that $\mathbf{P}^0(t_1) > \mathbf{K}(t_1)$ as one obviously expects independent of $\mathbf{P}(t_1)$ for the general case. However, to avoid excessive algebra consider the special case $\mathbf{N} \equiv 0$, implying no disturbance coupling between subsystems. In this case $\mathbf{P}^0$ and $\mathbf{K}$ satisfy much simpler equations, namely,

$$\frac{d}{dt_1}[\mathbf{P}_0(t_1)] = [\mathscr{A}^* - \mathbf{P}(t_1)\,\Gamma]\,\mathbf{P}^0(t_1) + \mathbf{P}^0(t_1)[\mathscr{A}^* - \mathbf{P}(t_1)\,\Gamma]'$$

$$+ \mathbf{P}(t_1)\,\Gamma\mathbf{P}(t_1) + \mathbf{BSB}' \qquad (98)$$

$$\frac{d}{dt_1}[\mathbf{K}(t_1)] = [\mathscr{A}^* - \mathbf{K}(t_1)\,\Gamma]\,\mathbf{K}(t_1) + \mathbf{K}(t_1)[\mathscr{A}^* - \mathbf{K}(t_1)\,\Gamma]'$$

$$+ \mathbf{K}(t_1)\,\Gamma\mathbf{K}(t_1) + \mathbf{BSB}' \qquad (99)$$

where $\Gamma = (\mathbf{H}^*)'(\mathbf{Q}^*)^{-1}\mathbf{H}^*$ with $\mathbf{A}^*, \mathbf{H}^*$, and $\mathbf{Q}^*$ defined for the integrated problem in Sec. 3-8-1.

Letting $\Delta\mathbf{P} = \mathbf{P}^0 - \mathbf{K}$ and then subtracting these two equations leads directly to

$$\frac{d}{dt_1}[\Delta\mathbf{P}(t_1)] = [\mathscr{A}^* - \mathbf{P}(t_1)\,\Gamma]\,\Delta\mathbf{P}(t_1) + \Delta\mathbf{P}(t_1)[\mathscr{A}^* - \mathbf{P}(t_1)]'$$

$$+ [\mathbf{P}(t_1) - \mathbf{K}(t_1)]\,\Gamma[\mathbf{P}(t_1) - \mathbf{K}(t_1)]$$

which can be solved in terms of a matrix $\psi(t_1)$ satisfying

$$\frac{d}{dt}[\psi(t)] = [\mathscr{A}^* - \mathbf{P}(t)\,\Gamma]\,\psi(t) \qquad \psi(t_0) = \psi$$

and then

$$\Delta\mathbf{P}(t_1) = -\int_{t_0}^{t_1} \psi(t_1)\,\psi(-t)[\mathbf{P}(t) - \mathbf{K}(t)]\,\Gamma[\mathbf{P}(t) - \kappa(t)]\,\psi'(-t)\,\psi(t)\,dt$$

This shows that regardless of whether $\mathbf{P} > \mathbf{K}$ or $\mathbf{P} < \mathbf{K}$, $\Delta\mathbf{P}(t_1)$ is *negative* semidefinite and thus $\mathbf{P}^0(t_1) > \mathbf{K}(t_1)$. In principle it would be possible to choose the subsystems so that the error variance is minimized although the form of the equations does not presently encourage this.

This section has shown that a suboptimal rule can be used to simplify the coordination and has presented formulas for computing the error. The particular approximation is by no means unique, and so it is hoped that this will serve more as a stimulus to further work on this problem than as an answer.

4-9 COORDINATION OF LINEAR SYSTEMS BY CONTRACTION MAPPING

An entirely different, and in some respects more natural, approach has been proposed by Takahara.[19] It is based on the simple observation that if each subsystem knows the effects of the interaction from the other optimal subsystems, then it can be optimized independently of the integrated optimum. Of course this is ideal, but by exchanging optimal subsystem information in a sequence of iterations, the subsystem optima might be expected to converge to the integrated optimum. This question will be investigated in this section and will depend on the principle of contraction mapping.

Consider the linear formulation of the system subject to quadratic performance indices as used in Sec. 4-7. The problem will be considered as follows. Given the time-invariant linear system

$$\dot{\mathbf{x}} = \mathbf{Ax} + \mathbf{Bm} + \mathbf{Cz} \qquad \mathbf{x}(t_0) = \mathbf{x}_0 \tag{100}$$

$$\mathbf{y} = \mathbf{Mx} + \mathbf{Nm} \tag{101}$$

$$\mathbf{z} = \mathbf{Ly} \tag{102}$$

select $\mathbf{m}$ and $\mathbf{z}$ to minimize

$$J_I(\mathbf{m}, \mathbf{z}) = \tfrac{1}{2}\int_{t_0}^{t_1} (\|\mathbf{x}\|^2\mathbf{Q} + \|\mathbf{m}\|^2\mathbf{R} + \|\mathbf{z}\|^2\mathbf{S})\,dt$$

with **Q**, **R**, and **S** constant positive definite symmetric matrices. Also, **A**, **B**, **C**, **M**, **N**, **Q**, **R**, and **S** are block diagonal matrices representing **N** independent problems, and **L** is an arbitrary coupling matrix.

The necessary conditions for this problem use a Hamiltonian $\mathscr{H}(\mathbf{x}, \mathbf{m}, \mathbf{z}, \boldsymbol{\lambda}, \boldsymbol{\pi})$ defined by

$$\mathscr{H} = \tfrac{1}{2}(\| \mathbf{x} \|^2\mathbf{Q} + \| \mathbf{m} \|^2\mathbf{R} + \| \mathbf{z} \|^2\mathbf{S}) + \boldsymbol{\lambda}'(\mathbf{Ax} + \mathbf{Bm} + \mathbf{Cz})$$
$$+ \boldsymbol{\pi}'(\mathbf{LMx} + \mathbf{LNm} - \mathbf{z})$$

Then $\boldsymbol{\lambda}$ and $\boldsymbol{\pi}$ are defined by

$$\dot{\boldsymbol{\lambda}} + \mathscr{H}_{\mathbf{x}}' = \dot{\boldsymbol{\lambda}} + \mathbf{Qx} + \mathbf{A}'\boldsymbol{\lambda} + \mathbf{M}'\mathbf{L}'\boldsymbol{\pi} = 0 \tag{103}$$

$$\mathscr{H}_{\mathbf{z}}' = \mathbf{Sz} + \mathbf{C}'\boldsymbol{\lambda} - \boldsymbol{\pi} = 0 \tag{104}$$

$$\mathscr{H}_{\mathbf{m}}' = \mathbf{R}_{\mathbf{m}} + \mathbf{B}'\boldsymbol{\lambda} + \mathbf{N}'\mathbf{L}'\boldsymbol{\pi} = 0 \tag{105}$$

$$\boldsymbol{\lambda}(t_1) = 0 \tag{106}$$

The equations (100) to (106) define the solution to the integrated problem. Takahara proceeds by eliminating the interaction variables $\boldsymbol{\pi}$ and $\mathbf{z}$ by solving Eqs. (101), (102), and (104). As the preceding section shows, this can produce an excessive amount of algebra, and accordingly we will retain the structure.

Let $\boldsymbol{\kappa} = \mathbf{L}'\boldsymbol{\pi}$ and set up the following sequence of iterations. At cycle i let $(\mathbf{x}_i, \mathbf{y}_i, \mathbf{m}_i, \boldsymbol{\lambda}_i, \boldsymbol{\pi}_i)$ denote $(\mathbf{x}, \mathbf{y}, \mathbf{m}, \boldsymbol{\lambda}, \boldsymbol{\pi})$. Then (1) given functions $\boldsymbol{\kappa} = \boldsymbol{\kappa}_i$, $\mathbf{z} = \mathbf{z}_i$, solve the equations

$$\dot{\mathbf{x}} = \mathbf{Ax} + \mathbf{Bm} + \mathbf{Cz}_i \qquad \mathbf{x}(t_0) = \mathbf{x}_0 \tag{100a}$$

$$\mathbf{m} = -\mathbf{R}^{-1}(\mathbf{B}'\boldsymbol{\lambda} + \mathbf{N}'\boldsymbol{\kappa}_i) \tag{105a}$$

$$\dot{\boldsymbol{\lambda}} + \mathbf{Qx} + \mathbf{A}'\boldsymbol{\lambda} + \mathbf{M}'\boldsymbol{\kappa}_i = 0 \tag{103a}$$

$$\boldsymbol{\lambda}(t_1) = 0 \tag{106a}$$

Then $\boldsymbol{\pi}$ and $\mathbf{y}$ are given by

$$\boldsymbol{\pi} = \mathbf{Sz}_i + \mathbf{C}'\boldsymbol{\lambda} \tag{104a}$$

$$\mathbf{y} = \mathbf{Mx} + \mathbf{Nm} \tag{101a}$$

This gives $(\mathbf{x}_i, \mathbf{y}_i, \mathbf{m}_i, \boldsymbol{\lambda}_i, \boldsymbol{\pi}_i)$ as functions of the parameters $\boldsymbol{\kappa}_i$ and $\mathbf{z}_i$ at stage i. The fact that $\mathbf{x}_i$ and $\mathbf{z}_i$ are given severs the connections between subsystems and has the effect of decomposing the "primal" *and* "dual" systems as defined in Sec. 4-7. The variables $\boldsymbol{\pi}$ and $\mathbf{y}$ are outputs of the primal and dual subsystems, which are interconnected. (2) Compute $\boldsymbol{\kappa}_{i+1}$ and $\mathbf{z}_{i+1}$ as follows:

$$\boldsymbol{\kappa}_{i+1} = \mathbf{L}'\boldsymbol{\pi}_i$$

$$\mathbf{z}_{i+1} = \mathbf{L}\mathbf{y}_i \tag{102a}$$

and repeat (1) with i set to $i + 1$.

This coordinating rule distributes the solutions of the optimal subproblems which were based on the last estimate of the interaction variables $\boldsymbol{\kappa}_i$ and $\mathbf{z}_i$. It can be seen that Eqs. (100*a*) to (106*a*) duplicate Eqs. (100) to (106) apart from the order in which they are solved.

A remarkable property of this iteration is that it corresponds to solving a sequence of subproblems in which, at each stage i, a functional $J(\mathbf{m}; \boldsymbol{\kappa}_i, \mathbf{z}_i)$ is minimized by the subproblems where

$$J_c(\mathbf{m}; \boldsymbol{\kappa}_i, \mathbf{z}_i) = \tfrac{1}{2}\int_{t_0}^{t_1} (\| \mathbf{x} + \mathbf{Q}^{-1}\mathbf{M}'\boldsymbol{\kappa}_i \|^2\mathbf{Q} + \| \mathbf{m} + \mathbf{R}^{-1}\mathbf{N}'\boldsymbol{\kappa}_i \|^2\mathbf{R})\, dt$$

subject to

$$\dot{\mathbf{x}}_i = \mathbf{A}\mathbf{x} + \mathbf{B}\mathbf{m} + \mathbf{C}\mathbf{z}_i \qquad \mathbf{x}(t_0) = \mathbf{x}_0$$

Again, these subproblems separate into N independent problems. The necessary conditions for the problem generate Eqs. (100*a*), (105*a*), (103*a*), and (106*a*), as can be verified by inspection.

The subproblems are however, auxiliary to the argument and it can be seen that if the problem has merely a convex instead of a quadratic objective function then it would be more difficult, if not impossible, to construct them.

We will now consider the question of conditions under which this scheme will work. The algorithm arises from the following process. Let the n-dimensional vectors $\mathbf{x}_i$ be generated by the recursion

$$\mathbf{x}_{i+1} = \mathbf{A}\mathbf{x}_i$$

where $\mathbf{A}$ is an $n \times n$ matrix which constitutes a map of $\mathbf{x}_i$ into $\mathbf{x}_{i+1}$.

DEFINITION 1:

A mapping $\mathbf{A}$ of a metric space $R = \{\mathbf{x}\}$ into itself is a *contraction mapping* if there exists a number $\alpha < 1$ such that

$$\| \mathbf{A}\mathbf{x} - \mathbf{A}\mathbf{y} \| \leqslant \alpha \| \mathbf{x} - \mathbf{y} \|$$

for any two points $\mathbf{x}, \mathbf{y} \in R$.

Theorem 4-12: Every contraction mapping defined in a complete matric space R has one and only one fixed point ($\mathbf{Ax} = \mathbf{x}$ has one solution $\mathbf{x}^*$), and this can be generated by the sequence

$$\mathbf{x}_{i+1} = \mathbf{A}\mathbf{x}_i$$

such that

$$\lim_{i\to\infty}(\mathbf{x}_i) = \mathbf{x}^*$$

The proof is a standard one and may be found in Kolmogorov and Fomin.[10]

Now consider the algorithm given by Eqs. (100a) to (106a), which provide a linear map of $\mathbf{z}_i$, $\boldsymbol{\kappa}_i$ to the solution ($\mathbf{x}_i$, $\mathbf{m}_i$, $\mathbf{y}$, $\boldsymbol{\lambda}_i$, $\boldsymbol{\pi}_i$). Then Eq. (102a) provides a linear map of $\boldsymbol{\pi}_i$ and $\mathbf{y}_i$ to $\mathbf{z}_{i+1}$, $\boldsymbol{\kappa}_{i+1}$. Thus, in order to establish convergence to the optimum solution when $\mathbf{z}_i \equiv \mathbf{z}_{i+1}$, $\boldsymbol{\kappa}_i \equiv \boldsymbol{\kappa}_{i+1}$, we need to show that the combined map is a contraction mapping.

Eliminating $\mathbf{m}$, $\mathbf{y}$, and $\boldsymbol{\pi}$, the recursion requires the solution of the following equations.

$$\begin{aligned} \dot{\mathbf{x}} &= \mathbf{Ax} - (\mathbf{BR}^{-1}\mathbf{B}')\boldsymbol{\lambda} + (\mathbf{Cz}_i - \mathbf{R}^{-1}\mathbf{N}'\boldsymbol{\kappa}_i) \qquad & \mathbf{x}(t_0) &= \mathbf{x}_0 \\ \dot{\boldsymbol{\lambda}} &= -\mathbf{Qx} - \mathbf{A}'\boldsymbol{\lambda} - (\mathbf{M}'\boldsymbol{\kappa}_i) & \boldsymbol{\lambda}(t_1) &= 0 \end{aligned} \tag{107}$$

Then having found $\mathbf{x}_i$ and $\boldsymbol{\lambda}_i$, the variables $\boldsymbol{\kappa}_{i+1}$ and $\mathbf{z}_{i+1}$ are defined by

$$\begin{aligned} \boldsymbol{\kappa}_{i+1} &= \mathbf{L}'\mathbf{C}'\boldsymbol{\lambda}_i + \mathbf{L}'\mathbf{S}\mathbf{z}_i \\ \mathbf{z}_{i+1} &= \mathbf{LMx}_i - \mathbf{LNR}^{-1}\mathbf{B}'\boldsymbol{\lambda}_i - \mathbf{LNR}^{-1}\mathbf{N}'\boldsymbol{\kappa}_i \end{aligned} \tag{108}$$

Letting $\Phi(t, t_0)$ be the unique transition matrix of dimension $2n \times 2n$ which satisfies

$$\frac{d}{dt}[\Phi(t, t_0)] = \begin{pmatrix} \mathbf{A} & -\mathbf{BR}^{-1}\mathbf{B}' \\ -\mathbf{Q} & -\mathbf{A}' \end{pmatrix} \Phi(t, t_0) \tag{109}$$

with $\Phi(t_0, t_0) = \mathbf{I}_{2n}$, then the solution to Eq. (107) is given by the standard form

$$\begin{pmatrix} \mathbf{x}(t) \\ \boldsymbol{\lambda}(t) \end{pmatrix} = \Phi(t, t_0) \begin{pmatrix} \mathbf{x}(t_0) \\ \boldsymbol{\lambda}(t_0) \end{pmatrix} + \int_{t_0}^{t_1} \Phi(t, s)\,\mathbf{G} \begin{pmatrix} \boldsymbol{\kappa}_i(s) \\ \mathbf{z}_i(s) \end{pmatrix} ds$$

where $\mathbf{F}$ and $\mathbf{G}$ are defined as

$$\mathbf{F} = \begin{pmatrix} 0 & \mathbf{L}'\mathbf{C}' \\ \mathbf{LM} & -\mathbf{LNR}^{-1}\mathbf{B}' \end{pmatrix} \qquad \mathbf{G} = \begin{pmatrix} -\mathbf{R}^{-1}\mathbf{N}' & \mathbf{C} \\ -\mathbf{M}' & 0 \end{pmatrix}$$

Now since at t_1, $\boldsymbol{\lambda}(t_1) = 0$, $\mathbf{x}(t_0)$ and $\boldsymbol{\lambda}(t_0)$ must be related by

$$\boldsymbol{\lambda}(t_0) = -\Phi_{22}^{-1}(t_1, t_0)\left\{\Phi_{21}(t_1, t_0)\,\mathbf{x}(t_0) + \int_{t_0}^{t_1} [\Phi_{21}(t_1, s)\;\Phi_{22}(t_1, s)]\,\mathbf{G}\begin{pmatrix}\boldsymbol{\kappa}_i(s)\\ \mathbf{z}_i(s)\end{pmatrix} ds\right\}$$

where Φ_{11}, Φ_{12}, Φ_{21}, and Φ_{22} are the $n \times n$ partitioned blocks of Φ. We note that since $\Phi(t, t_0)$ is positive definite for all t, its principal minor Φ_{22} is also, and thus Φ_{22}^{-1} exists. Then the solution to the subproblems is given by the following:

$$\begin{bmatrix}\mathbf{x}(t)\\ \boldsymbol{\lambda}(t)\end{bmatrix} = \int_{t_0}^{t_1} \Phi(t, s)\,\mathbf{G}\begin{bmatrix}\boldsymbol{\kappa}_i(s)\\ \mathbf{z}_i(s)\end{bmatrix} ds - \int_{t_0}^{t_1} \psi(t, s, t_1, t_0)\,\mathbf{G}\begin{bmatrix}\boldsymbol{\kappa}_i(s)\\ \mathbf{z}_i(s)\end{bmatrix} ds \tag{110}$$

where for convenience

$$\psi(t, s, t_1, t_0) = \begin{bmatrix}\Phi_{12}(t, t_0)\\ \Phi_{21}(t, t_0)\end{bmatrix} [\Phi_{22}^{-1}(t_1, t_0)][\Phi_{21}(t_1, s)\;\Phi_{22}(t_1, s)] \tag{111}$$

and without loss of generality $\mathbf{x}(t_0) = \mathbf{x}_0 \equiv 0$, since the initial condition may be canceled by differencing consecutive iterations.

The final recursion for $\boldsymbol{\kappa}_i$, $\mathbf{z}_i$ can now be found by combining Eqs. (108) and (110).

$$\begin{bmatrix}\boldsymbol{\kappa}_{i+1}(t)\\ \mathbf{z}_{i+1}(t)\end{bmatrix} = \int_{t_0}^{t_1} \mathbf{F}\Phi(t, s)\,\mathbf{G}\begin{bmatrix}\boldsymbol{\kappa}_i(s)\\ \mathbf{z}_i(s)\end{bmatrix} ds - \int_{t_0}^{t_1} \mathbf{F}\psi(t, s, t_1, t_0)\,\mathbf{G}\begin{bmatrix}\boldsymbol{\kappa}_i(s)\\ \mathbf{z}_i(s)\end{bmatrix} ds$$

$$+ \begin{bmatrix}0 & \mathbf{L'S}\\ -\mathbf{LNR^{-1}N'} & 0\end{bmatrix}\begin{bmatrix}\boldsymbol{\kappa}_i(t)\\ \mathbf{z}_i(t)\end{bmatrix} \tag{112}$$

Since $\mathbf{x}_0 = 0$ without loss of generality, the optimal solution is $\mathbf{x}^* \equiv \boldsymbol{\lambda}^* \equiv 0$, and then $\mathbf{z}^* = \boldsymbol{\kappa}^* = 0$. Thus we need to show that $\boldsymbol{\kappa}_i \to 0$ $\mathbf{z}_i \to 0$ as $i \to \infty$.

Suppose that $t_1 = t_0$; then the recursion reduces to

$$\begin{bmatrix}\boldsymbol{\kappa}_{i+1}(t_0)\\ \mathbf{z}_{i+1}t(^0)\end{bmatrix} = \begin{bmatrix}0 & \mathbf{L'S}\\ -\mathbf{LNR^{-1}N} & 0\end{bmatrix}\begin{bmatrix}\boldsymbol{\kappa}_i(t_0)\\ \mathbf{z}_i(t_0)\end{bmatrix}$$

so that a sufficient condition for convergence is that

$$\left\|\begin{matrix}0 & \mathbf{L'S}\\ -\mathbf{LNR^{-1}N'} & 0\end{matrix}\right\| < 1$$

or equivalently using the Euclidean norm of the matrix, that *all* eigenvalues of $(\mathbf{LNR^{-1}N'L'S})$ are inside the unit circle, since if μ is an eigenvalue

$$\begin{vmatrix} \mu\mathbf{I}_n & \mathbf{L'S} \\ -\mathbf{LNR^{-1}N'} & \mu\mathbf{I}_n \end{vmatrix} = |\,\mu\mathbf{I}_n\,|^{-1}\,|\,\mathbf{I}_n + (\mathbf{LNR^{-1}N'})(\mu\mathbf{I}_n)^{-1}\,(\mathbf{L'S})|$$

$$= |\,\mu^2\mathbf{I}_n + \mathbf{LNR^{-1}N'L'S}\,|$$

and

$$\|\mathbf{A}\| = \max[\mathbf{u'Au};\ \mathbf{u'u} = 1] = \mu_{\max}[\mathbf{A}]$$

Suppose that $t_1 > t_0$; then upon examining $\Phi(t_1, t_0)$ and $\psi(t, s, t_1, t_0)$ we note that these are continuous functions of t and s for the time-invariant case. (This is still true if $\mathbf{A}$, $\mathbf{B}$, $\mathbf{C}$, $\mathbf{L}$, $\mathbf{M}$, and $\mathbf{N}$ and $\mathbf{Q}$, $\mathbf{R}$, and $\mathbf{S}$ are all continuous functions of time t.) Taking norms in Eq. (112) using the Euclidean norm throughout, and assuming bounded function $\boldsymbol{\kappa}_i$, $\mathbf{z}_i$ for each i,

$$\max_t \left[\left\| \begin{matrix} \boldsymbol{\kappa}_{i+1}(t) \\ \mathbf{z}_{i+1}(t) \end{matrix} \right\|\right] \leqslant \left\{ \int_{t_0}^{t_1} \max_{t,s}[\|\,\mathbf{F}\psi(t, s)\,\mathbf{G}\,\|]\,ds + \int_{t_0}^{t_1} \max_{t,s}[\|\,\mathbf{F}\psi(t, s, t_1, t_0)\mathbf{G}\,\|]\,ds \right.$$

$$\left. + \left\| \begin{matrix} 0 & \mathbf{L'S} \\ -\mathbf{LNR^{-1}N'} & 0 \end{matrix} \right\| \right\} \max_t \left[\left\| \begin{matrix} \boldsymbol{\kappa}_i(t) \\ \mathbf{z}_i(t) \end{matrix} \right\|\right]$$

$$= [k_1(t_1 - t_0) + k_2(t_1 - t_0) + k_3] \max_t \left[\left\| \begin{matrix} \boldsymbol{\kappa}_i(t) \\ \mathbf{z}_i(t) \end{matrix} \right\|\right]$$

where k_1, k_2, and k_3 are the maximum values of the three matrix norms. This is sufficient to establish the following theorem.

Theorem 4-13: If $\mu_{\max}[(\mathbf{LN})\mathbf{R}^{-1}(\mathbf{LN})'\mathbf{S}] < 1$, while there exists an interval $t_1^* - t_0$ such that for all t_1 in the open interval (t_0, t_1^*) the algorithm converges to the optimum solution.

Proof: Choose t_1^* so that

$$[k_1(t_1^* - t_0) + k_2(t_1 - t_0) + k_3)] = 1$$

which can be done since $k_3 < 1$. Then for $t_1 \in [t_0, t_1^*]$ and $t \in [t_0, t_1]$

$$\left[\max_t \left\| \begin{matrix} \boldsymbol{\kappa}_{i+1}(t) \\ \mathbf{z}_{i+1}(t) \end{matrix} \right\|\right] < \left[\max_t \left\| \begin{matrix} \boldsymbol{\kappa}_i(t) \\ \mathbf{z}_i(t) \end{matrix} \right\|\right]$$

Then for $\boldsymbol{\kappa}_i$, $\mathbf{z}_i \in C'[t_0, t_1]$ so that the maximum values are bounded, the iterates will converge to zero, the optimum solution. This completes the convergence proof.

Takahara reached a more general result which requires no bounds on $(\mathbf{LN})\mathbf{R}^{-1}(\mathbf{LN})'\mathbf{S}$. However, his performance functional does not allow interaction between $\mathbf{m}$ and $\mathbf{x}$, while in the case given here the integrated problem performance functional has the form

$$J_I(\mathbf{m}, \mathbf{LMx} + \mathbf{LNm}) = \tfrac{1}{2}\int_{t_0}^{t_1} [\|\mathbf{x}\|^2\mathbf{Q} + \|\mathbf{m}\|^2\mathbf{R} + \|\mathbf{Mx} + \mathbf{Nm}\|^2(\mathbf{L}'\mathbf{SL})]\, dt$$

which does allow interaction. More details, together with several examples, are given by Takahara.[19] The brevity of the interval $[t_0, t_1]$ over which convergence can be achieved can be overcome to some extent by a technique which might be called "smoothing." Instead of using $(\boldsymbol{\kappa}_{i+1}, \mathbf{z}_{i+1})$ as the next iterates, we use

$$\bar{\boldsymbol{\kappa}}_{i+1} = \epsilon\boldsymbol{\kappa}_{i+1} + (1-\epsilon)\,\boldsymbol{\kappa}_i$$
$$\bar{\mathbf{z}}_{i+1} = \epsilon\mathbf{z}_{i+1} + (1-\epsilon)\,\mathbf{z}_i \qquad 0 < \epsilon < 1$$

This trick generally achieves convergence for larger intervals $[t_0, t_1]$.

4-10 CONCLUSION AND SUMMARY

This chapter has surveyed the art of dynamic decomposition as it existed in 1968–1969. In general, all the methods depend on the same ideas exploited by Dantzig and Wolfe in their pioneering paper on linear programming decomposition.

In the early sections, the work by Macko showed the connection between the necessary variational properties of the integrated and decomposed problems. On the other hand, when dealing with sufficient conditions one is led to examine the primal and dual aspects of a restricted class of problems and to explore how these apply to decomposition. Having established decomposition methods, the solution techniques become of primary importance. It has been shown that, lacking any good way of dealing with "parametric subproblems," the dual problem can be maximized by conjugate gradient methods. Finally, examples of application to large problems and to statistical problems have illustrated the techniques.

In later sections the idea of model coordination has been discussed. The alternative method of iteration using contraction properties terminates the chapter.

It is hoped that a great deal of future work has been indicated to the reader. Lest it escape him, three areas are indicated again.

1. On a fundamental level, the decomposition method requires additional study of auxiliary problems or parametric problems along the lines of Falk and paraphrased in Sec. 4-4.

2. The computational requirement in solving parametric problems is to modify the solution efficiently as a function of the parameter. Computational techniques in this area need to be developed.
3. The discussion of model coordination is included only to present the problem. Techniques along these lines should be the most useful ones for controlling large systems in real time and merit further work.

REFERENCES:

1. Berkovitz, L. D.: Variational Methods in Problems of Control and Programming, *J. Math. Anal. Appl.*, vol. 3, no. 1, pp. 145–169, 1961.
2. Breakwell, J., J. Speyer, and A. E. Bryson: Optimization and Control of Nonlinear Systems Using the Second Variation, *SIAM J. Control*, vol. 1, no. 2, pp. 193–223, 1963.
3. Cox, H.: On the Estimation of State Variables and Parameters for Noisy Dynamical Systems, *IEEE Trans. Autom. Control*, vol. AC-7, pp. 5–12, January, 1964.
4. Dreyfus, S.: Variational Problems with Inequality Constraints, *J. Math. Anal. Appl.*, vol. 4, pp. 297–308, 1962.
5. Falk, J.: The Theory of Lagrange Multipliers with Application to Non-linear Programming, *Res. Anal. Corp. Rept.* TP 200, McLean, Va.
6. Fletcher, R., and M. J. D. Powell: A Rapidly Converging Descent Method for Minimization, *Computer J.*, vol. 6, pp. 163–168, 1963.
7. Fletcher, R., and R. Reeves: Function Minimization by Conjugate Gradients, *Computer J.*, vol. 7, pp. 149–154, 1964.
8. Kalman, R. E., Y. C. Ho, and K. S. Narendra: Controllability of Linear Dynamical Systems, *Contrib. Differential Equations*, vol. 1, no. 2, pp. 189–213, 1963.
9. Kelley, H., R. E. Kopp, and H. G. Moyer: A Trajectory Optimization Based on the Theory of the Second Variation, *Proc. AIAA Astrodyn. Conf.*, New Haven, Conn., 1963.
10. Kolmogorov, A., and S. Fomin: "Elements of the Theory of Functions and Functional Analysis," vol. 1, Graylock, 1957.
11. Lasdon, L., and J. Schoeffler: A Multilevel Technique for Optimization, *Proc. Joint Autom. Control Conf.*, 1965.
12. Lynch, D.: The Davidon Method in Function Space, *Proc. IEEE*, November, 1968.
13. Macko, D.: A Coordination Technique for Interacting Dynamic Systems, *Proc. Joint Autom. Control Conf.*, 1966.
14. Mitter, S. K.: "Function Space Methods in Optimal Control with Applications to Power Systems," doctoral dissertation, University of London, 1965.
15. Pearson, J. D.: On the Duality between Estimation and Control, *SIAM J. Control*, vol. 4, no. 4, pp. 594–600, 1966.
16. Pearson, J. D.: On Variable Metric Methods of Minimization, *Computer J.*, vol. 12, 1969.
17. Pearson, J. D.: Duality and a Decomposition Technique, *Proc. Intern. Conf. Programming Control*, 1965.
18. Reich, S.: "A Critique of a Decomposition Technique for Optimal Control Problems," master's thesis, Engineering Division, Case Western Reserve University, Cleveland, 1966.
19. Takahara, Y.: "A Multilevel Structure for a Class of Dynamic Optimization Problems," master's thesis, Engineering Division, Case Western Reserve University, Cleveland, 1965.

20. Takahashi, I.: A Note on the Conjugate Gradient Method, *Inform. Process. Japan*, vol. 5, pp. 45–49, 1965.
21. Tinkelman, M.: "Optimal Decomposition Applied to Optimal Smoothing and Filtering Problems," master's thesis, Case Western Reserve University, Cleveland, 1967.
22. Artosciewicz, H. A.: In John Todd (ed.), "Survey of Numerical Analysis," McGraw-Hill, 1962.
23. Warren, A., L. Lasdon, and S. K. Mitter: The Conjugate Gradient Method for Optimal Control Problems, *Proc. Ann. Allerton Conf. Circuit Syst. Theory*, 4*th*, 1966.
24. Wilde, D. J.: "Optimum Seeking Methods," Prentice-Hall, 1964.

5

AGGREGATION[†]

M. AOKI[‡]

We present in this chapter an approach to the problems associated with controls of large dynamic systems. It complements the multilevel approach, and is based on the concept of aggregation, which generalizes that of orthogonal or oblique projection[§] described in Chapter 2. This concept is also related to such concepts as the state vector partition and the modal controls.[19,28,43,46]

It will be shown in the course of discussions that the concept of aggregation is useful in deriving computationally efficient algorithms for suboptimal controls, state vector estimations, and system parameter identification. For example, a suboptimal control policy is derived by applying the concept of aggregation for a class of systems composed of weakly coupled subsystems.[1]

The concept of aggregation is defined and developed in the first part of the chapter and is applied to generate controls and estimate state vector and system parameters in the second part of the chapter.

5-1 INTRODUCTION

Consider two linear dynamic systems S_1 and S_2 where the dimension of S_1, n, is much larger than that of S_2, l. The system S_1 may be a dynamic system to be controlled (called the *plant*) and S_2 may be its model or

† The material of this chapter is based in part on Refs. 1 and 5.

‡ During the research reported herein, the author was supported in part by National Science Foundation Grant GK-2032. Certain preliminary aspects of the research were also aided by Intramural Research Grant 2391 of the University of California, Los Angeles.

§ The method of the paper applies, with suitable modifications, to certain stochastic systems as well as to deterministic systems. To present the basic idea most simply, the systems are assumed to be deterministic. See Refs. 35 and 42 for some simple examples of stochastic system aggregation.

S_1 and S_2 may be two models of the same plant of different complexity. The state vectors of these two systems, denoted by $\mathbf{x}$ and $\mathbf{z}$ respectively, are to satisfy under certain conditions the relationship

$$\mathbf{z} = \mathbf{C}\mathbf{x}$$

where $\mathbf{C}$ is an $l \times n$ constant matrix.

In many cases, it is desirable to choose the dynamic structure of S_2 to reflect a significant portion of the dynamics of S_1 in a sense to be specified later. When this is done, S_2 is called an aggregated model[20,22,41] of S_1.† For example, S_1 may describe the dynamics of a physical or a nonphysical object according to some classification of variables, and S_2 may give the description of the same object using a coarser grid of classification, hence of lesser dimension. This is a point of view which is appropriate for instance in mathematical models of economic systems.

There is another somewhat related point of view on the way such a relationship arises, where S_1 is taken to be a given system and S_2 is regarded as an observer of S_1. With this viewpoint, the linear transformation $\mathbf{C}$ is to be chosen, subject to certain constraints, such that the state vector $\mathbf{z}$ together with the originally available measurement $\mathbf{y}$ of S_1 is used to reconstruct $\mathbf{x}$ exactly or approximately.[6,14,33] Hence, the name "observer" for S_2.‡ Loosely speaking, the dynamic structure of S_2 is chosen to reflect only that part of the dynamics of S_1 which is not carried by the independent information contained in $\mathbf{y}$. This point of view is taken later in constructing a computationally efficient estimate of the state vector of S_1.

One of the important questions to be examined in this chapter is that of controlling S_1 via a control policy derived for S_2. The others are that of performance degradation suffered by S_1 and that of the stability of S_1 being controlled in this manner. It is shown in this chapter that the aggregation concept can be used to give bounds on the performance degradation due to suboptimal control in a computationally efficient manner. Discussions on the relation of the closed-loop stability of the original and the aggregated systems are also taken up later.

Control system designers usually have some freedom in choosing the aggregation matrix $\mathbf{C}$ subject to some constraints imposed by the problems. For a certain choice of $\mathbf{C}$, S_2 becomes an orthogonal projection of S_1 into the l-dimensional state space retaining the l most significant characteristic values and characteristic vectors of S_1 and showing that

† For the connection with dynamic controllers and cascade compensations see for example Refs. 4 (Sec. VII.4), 6, 13, and 51.

‡ We restrict our attention to a linear aggregation. Some isolated but stronger results are available with nonlinear aggregation, i.e., when $\mathbf{z}$ is a nonlinear function of $\mathbf{x}$.[12]

the model obtained by such a projection method is a special case of models constructed by aggregation.

5-2 AGGREGATION OF CONTROL SYSTEM

Consider a continuous-time constant coefficient dynamic system[†] described by the plant equation

$$\frac{d\mathbf{x}}{dt} = \mathbf{A}\mathbf{x} + \mathbf{B}\mathbf{u} \tag{1}$$

where $\mathbf{B}$ is an $n \times r$ matrix and $\mathbf{x}$ and $\mathbf{u}$ are respectively the state and control vectors of appropriate dimensions. Assume that $\mathbf{x}$ is directly observed and that S_1 is controllable (Ref. 53)[‡]. The system described by Eq. (1) will be referred to as S_1.

Consider an l-dimensional vector $\mathbf{z}$, called the aggregated state vector, defined by

$$\mathbf{z}(t) = \mathbf{C}\mathbf{x}(t) \tag{2}$$

where $\mathbf{C}$ is an $l \times n$ constant matrix, $l \leqslant n$. Assume that

$$\operatorname{rank} \mathbf{C} = l \tag{3}$$

The statement that $\mathbf{z}(t)$ satisfies the differential equation

$$\frac{d\mathbf{z}}{dt} = \mathbf{F}\mathbf{z} + \mathbf{G}\mathbf{u} \qquad \mathbf{z}(0) = \mathbf{C}\mathbf{x}(0) \tag{4}$$

is equivalent to the condition that $\mathbf{F}$ and $\mathbf{G}$ in Eq. (4) are related to $\mathbf{A}$ and $\mathbf{B}$ by

$$\mathbf{FC} = \mathbf{CA} \qquad \mathbf{G} = \mathbf{CB} \tag{5}$$

If $\mathbf{A}$ and $\mathbf{C}$ satisfy the matrix equation

$$\mathbf{CA} = \mathbf{CAC}'(\mathbf{CC}')^{-1}\,\mathbf{C} \tag{6}$$

then $\mathbf{F}$ in Eq. (5) is given by

$$\mathbf{F} = \mathbf{CAC}'(\mathbf{CC}')^{-1} \tag{7}$$

† Parallel developments can be made for discrete-time systems but are omitted since they are straightforward. See for example Ref. 27.

‡ When Eq. (1) is not controllable, the following discussions apply to the controllable and observable subsystem of Eq. (1).

The matrix $\mathbf{C}$ will be referred to as the aggregation matrix and $\mathbf{F}$ as the aggregated matrix or the aggregation of $\mathbf{A}$. Any polynomial in $\mathbf{A}$, $f(\mathbf{A})$, has $f(\mathbf{F})$ as its aggregation, i.e.,

$$\mathbf{C}f(\mathbf{A}) = f(\mathbf{F})\mathbf{C} \tag{6a}$$

For example, if $(\mathbf{I} - \mathbf{A})^{-1}$ exists, then $(\mathbf{I} - \mathbf{F})^{-1}$ is its aggregation. The aggregated state vector $\mathbf{z}$ satisfies the dynamic equation (4) with $\mathbf{F}$ given by Eq. (7).

If $\mathbf{A}$ is reducible[23] or if $\mathbf{A}$ is of rank l, then many aggregation matrices $\mathbf{C}$ satisfying Eq. (6) can be found. See Refs. 31, 44, and 45 for examples where the concept of aggregation has been used implicitly. If the system (1) is observed through $\mathbf{y} = \mathbf{Hx}$ and if the rank of the observability criterion matrix is $l < n$, then it can be shown[8] that there exists an l-dimensional vector $\mathbf{z}$ obtained by the projection of $\mathbf{x}$ onto the linear subspace spanned by the l linearly independent column vectors chosen from the observability criterion matrix. Here, the aggregation matrix takes the form of the orthogonal projection matrix. Then $\mathbf{y}$ is obtained by a suitable constant linear transformation of the projected vector $\mathbf{z}$. Thus, if the criterion function is an explicit function of $\mathbf{y}$ and $\mathbf{u}$ and if no unstable nonobservable modes exist, then the state vector equation can be taken to be l dimensional instead of n. Note also that linear control laws with incomplete state feedback[31,40] may be regarded as a special case of feedback using the aggregated state vector.

If it is assumed that the state vector of the plant governed by Eq. (1) is observed through

$$\mathbf{y} = \mathbf{Hx} \tag{1a}$$

where $\mathbf{H}$ is an $m \times n$ matrix, $m \leqslant n$, and if it is still desired to maintain the relation (2), then instead of Eq. (4), $\mathbf{z}$ obeys the dynamic equation[6,27,33]

$$\frac{d\mathbf{z}}{dt} = \mathbf{Fz} + \mathbf{Gu} + \mathbf{Dy} \tag{4a}$$

where $\mathbf{F}$, $\mathbf{G}$, and $\mathbf{D}$ are related to $\mathbf{A}$, $\mathbf{B}$, and $\mathbf{C}$ by

$$\begin{aligned} \mathbf{CA} - \mathbf{FC} &= \mathbf{DH} \\ \mathbf{G} &= \mathbf{CB} \end{aligned} \tag{5a}$$

The system described by Eqs. (1a) and (4a) is called an observer of the original system.

Then, any kth order polynomial in $\mathbf{A}$, $\phi(\mathbf{A})$, is related to $\phi(\mathbf{F})$ by

$$\mathbf{C}\phi(\mathbf{A}) - \phi(\mathbf{F})\mathbf{C} = \mathbf{DH}\phi_1(\mathbf{A}) + \mathbf{FDH}\phi_2(\mathbf{A}) + \cdots + \mathbf{F}^{k-1}\mathbf{DH}\phi_k(\mathbf{A}) \tag{6b}$$

where $\phi_i(\cdot)$, $1 \leqslant i \leqslant k$, are appropriate polynomials of order $<k$. Comparing Eqs. (5*a*) and (6*b*) with (5) and (6*a*), it is seen that the aggregation is the special case of observer with $\mathbf{D} = 0$. One way to satisfy Eq. (5*a*) is to choose **D** and **F** as[6]

$$\mathbf{D} = \mathbf{CAV}$$

$$\mathbf{F} = \mathbf{CAP}$$

where **P** and **V** are subject to the constraint

$$\mathbf{PC} + \mathbf{VH} = \mathbf{I}_n$$

where $\mathbf{I}_n$ is the $n \times n$ identity matrix. Another way is to choose **F** as a stable matrix with no common eigenvalues with **A** and choose ϕ in Eq. (6*b*) to be the characteristic polynomial of **A**. Then **C** can be solved for in Eq. (6*b*) since $\phi(\mathbf{A}) = 0$ and $\phi(\mathbf{F})$ is nonsingular, and **x** can be solved for from Eqs. (1*a*) and (2), provided the matrix

$$\mathbf{T} = \begin{pmatrix} \mathbf{C} \\ \mathbf{H} \end{pmatrix}$$

is nonsingular. See Ref. 14 for further discussion on this point. Also see Ref. 50.

In the rest of this chapter, **D** is taken to be zero, and $\mathbf{y} = \mathbf{x}$.

As mentioned in Sec. 5-1, **C** is the primary design parameter in constructing S_2 while **F** is given by Eq. (7). The choice of **C** is to be made in such a way that error in modeling the original system S_1 by S_2 is minimized in terms of the performance index for S_1. Thus the case of $l \ll n$ is of interest.†

In the next section we examine the relation among the modes of the original and aggregated systems and the question of how these modes are affected by the feedback controls.

5-3 AGGREGATION MATRIX AND MODES

Denote the n-dimensional row vectors of **C** by $\langle \mathbf{c}_i$, $1 \leqslant i \leqslant l$. Then the ith component of **z**, $\mathbf{z}_i$, is given by $\langle \mathbf{c}_i, \mathbf{x} \rangle$. Thus, $\mathbf{z}_i$ is the weighted sum of some components of **x**. If an additional condition is imposed on **C** that **C** has one entry in each column, then n components of **x** are grouped into at most l mutually exclusive sets. Then $\mathbf{z}_i$ is the weighted sum of the ith such class of components of **x**. Note that such vectors $\langle \mathbf{c}_i$ are

† Thus, we give up the hope of reconstructing the state vector of S_1 at time t exactly from the knowledge of $\mathbf{y}(t)$ and $\mathbf{z}(t)$ alone.

mutually orthogonal. By construction, all such **C**'s have maximal rank. The orthogonal projection of **x** into an l-dimensional subspace can be regarded as a special case of such **C** where after a suitable permutation **C** can be expressed as

$$\mathbf{C} = (\mathbf{I}_l \, , O_{l,n-l})$$

where I_l is the l-dimensional identity matrix and where $O_{l,n-l}$ is the $l \times (n - l)$ dimensional null matrix.

The matrix **F**, defined in Sec. 5-2, retains some of the characteristic values of **A** if **A** and **C** satisfy Eq. (5). To see this simply, assume that **A** has n distinct characteristic values denoted by $\lambda_1 , \lambda_2 ,..., \lambda_n$ and let $\mathbf{v}_1 ,..., \mathbf{v}_n$ be the associated normalized characteristic vectors. Then it follows from Eq. (5) that

$$\begin{aligned} \mathbf{CAv}_i &= \mathbf{FCv}_i \\ &= \lambda_i \mathbf{Cv}_i \end{aligned} \tag{8}$$

showing that if $\mathbf{Cv}_i \neq 0$, then it is a characteristic vector of **F** with the same characteristic value λ_i .

If the subspace spanned by the l row vectors of **C** is orthogonal to that spanned by $\mathbf{v}_{l+1} ,..., \mathbf{v}_n$, that is, if the row vectors of **C** spans the same subspace spanned by $\mathbf{w}_i$, $k \leqslant i \leqslant l$, where $\{\mathbf{w}_1 ,..., \mathbf{w}_n\}$ is the reciprocal basis of $\{\mathbf{v}_1 ,..., \mathbf{v}_n\}$, then

$$\mathbf{Cv}_i \neq 0 \qquad 1 \leqslant i \leqslant l \qquad \mathbf{Cv}_i = 0 \qquad l+1 \leqslant i \leqslant n \tag{9}$$

Thus, such a choice of **C** makes **F** inherit $\lambda_1 , \lambda_2 ,..., \lambda_l$ of **A** as its characteristic values. Physically, this means that the state vector of the n-dimensional dynamic system is aggregated in such a way that the l-dimensional aggregated dynamic system has l characteristic values of the original system.† Given such a **C**, it is easy to see which characteristic values of **A** are introduced into **F**.

The modes of $\mathbf{x}(t)$ and $\mathbf{z}(t)$ are next exhibited. These can be used to perform error analysis of such aggregated systems if desired. Using the spectral representation of **A**

$$\mathbf{A} = \sum_{i=1}^{n} \lambda_i \mathbf{v}_i \rangle\langle \mathbf{w}_i$$

† If the time response of S_1 is to be approximated by that of S_2 , then clearly **C** should be chosen in such a way that **F** inherits the dominant characteristic values of **A**. Generally it is more meaningful, however, to choose **C** so as to approximate the value of the criterion function for S_1 . See Sec. 5-5.

and assuming that the row vectors of $\mathbf{C}$ are mutually orthogonal and satisfy Eq. (9), one obtains from Eq. (7)

$$\mathbf{F} = \sum_{i=1}^{l} \lambda_i \begin{pmatrix} \langle \mathbf{c}_1, \mathbf{v}_i \rangle \\ \vdots \\ \langle \mathbf{c}_l, \mathbf{v}_i \rangle \end{pmatrix} \left(\frac{\langle \mathbf{w}_i, \mathbf{c}_1 \rangle}{\langle \mathbf{c}_1, \mathbf{c}_1 \rangle}, \ldots, \frac{\langle \mathbf{w}_i, \mathbf{c}_l \rangle}{\langle \mathbf{c}_l, \mathbf{c}_l \rangle} \right) \tag{10}$$

Denoting column vectors of $\mathbf{B}$ by $\mathbf{b}_i \rangle$, $1 \leqslant i \leqslant r$, one obtains from Eq. (1)

$$\mathbf{x}(t) = \sum_{i=1}^{l} \mathbf{e}^{\lambda_i t} \left[\langle \mathbf{w}_i, \mathbf{x}_1(0) \rangle + \sum_{j=1}^{r} \left(\int_0^t \mathbf{e}^{-\lambda_i \tau} \mathbf{u}_j(\tau)\, d\tau \right) \langle \mathbf{w}_i, \mathbf{b}_j \rangle \right] \mathbf{v}_i \rangle$$

$$+ \sum_{i=l+1}^{n} \mathbf{e}^{\lambda_i t} \left[\langle \mathbf{w}_i, \mathbf{x}_2(0) \rangle + \sum_{j=1}^{r} \left(\int_0^t \mathbf{e}^{-\lambda_i \tau} \mathbf{u}_j(\tau)\, d\tau \right) \langle \mathbf{w}_i, \mathbf{b}_j \rangle \right] \mathbf{v}_i \rangle \tag{11}$$

where $\mathbf{x}(0) = \mathbf{x}_1(0) + \mathbf{x}_2(0)$, $\mathbf{x}_1(0) \in \mathbf{L}(\mathbf{v}_1\rangle, \ldots, \mathbf{v}_l\rangle)$, $\mathbf{x}_2(0) \in \mathbf{L}(\mathbf{v}_{l+1}\rangle, \ldots, \mathbf{v}_n\rangle)$ and where $\mathbf{L}(\cdots)$ is a subspace spanned by the indicated vectors. With the choice of $\mathbf{C}$ satisfying Eq. (9),

$$\mathbf{z}(t) = \sum_{i=1}^{l} \mathbf{e}^{\lambda_i t} \left[\langle \mathbf{w}_i, \mathbf{x}_1(0) \rangle + \sum_{j=1}^{r} \left(\int_0^t \mathbf{e}^{-\lambda_i \tau} \mathbf{u}_j(\tau)\, d\tau \right) \langle \mathbf{w}_i, \mathbf{b}_j \rangle \right] \begin{pmatrix} \langle \mathbf{c}_1, \mathbf{v}_i \rangle \\ \vdots \\ \langle \mathbf{c}_l, \mathbf{v}_i \rangle \end{pmatrix} \tag{12}$$

Thus, from Eqs. (11) and (12) it is seen that the components of $\mathbf{x}(t)$ lying in $\mathbf{L}(\mathbf{v}_{l+1}\rangle, \ldots, \mathbf{v}_n\rangle)$ will be small if $|\lambda_i|$ is large with $\operatorname{Re} \lambda_i < 0$ and if $\langle \mathbf{w}_i, \mathbf{b}_j \rangle$ and $\langle \mathbf{w}_i, \mathbf{x}_2(0) \rangle$ are small for $l + 1 \leqslant i \leqslant n$, $1 \leqslant j \leqslant r$. The components of $\mathbf{x}(t)$ in $\mathbf{L}(\mathbf{v}_1, \ldots, \mathbf{v}_l)$ will be identical with $\mathbf{z}(t)$ if $\mathbf{C} = (\mathbf{I}_l, \mathbf{O}_{l,n+l})$. See Refs. 19, 38, 41, and 44 for the connection with modal analysis.

5-3-1 Modal Controls

As is evident from the previous discussions, it is possible to affect a subset of the characteristic values of $\mathbf{A}$ (i.e., the plant poles) by feedback controls where $\mathbf{Cx}$ instead of $\mathbf{x}$ is used to close the loop with an appropriately chosen $\mathbf{C}$ and a gain matrix. This is exactly the type of feedback controls employed in modal controls.[19,38,41,44] Since this point will be taken up again in connection with the stability analysis, we will merely illustrate this connection for a special case where the system dynamics are governed by

$$\frac{d\mathbf{x}}{dt} = \Lambda \mathbf{x} + \mathbf{Bu}$$

where

$$\Lambda \triangleq \operatorname{diag}(\lambda_1, \ldots, \lambda_n)$$

and where $\mathbf{B}$ is an $n \times r$ matrix such that $(\Lambda, \mathbf{B})$ is a controllable pair. Introduce the feedback control

$$\mathbf{u} = -\mathbf{KCx}$$

where $\mathbf{K}$ is a $r \times l$ gain matrix and where

$$\mathbf{C} = (\mathbf{I}_l, O_{l,n-l})$$

is the aggregation matrix. Note that only the first l modes of $\mathbf{x}$ are fedback. The dynamics of the closed-loop system are now governed by

$$\frac{d\mathbf{x}}{dt} = (\Lambda - \mathbf{BKC})\,\mathbf{x}$$

Then the modes of this system are given as the roots of the equation

$$\begin{aligned} O &= \det[\lambda\mathbf{I}_n - (\Lambda - \mathbf{BKC})] \\ &= \det\{\lambda\mathbf{I}_l - \mathbf{FGBK}\} \cdot \det\{\lambda\mathbf{I}_{n-l} - \operatorname{diag}(\lambda_{l+1}, \ldots, \lambda_n)\} \end{aligned} \tag{13}$$

where $\mathbf{F}$ and $\mathbf{G}$ are given by

$$\begin{aligned} \mathbf{F} &= \mathbf{C}\Lambda\mathbf{C}'(\mathbf{CC}')^{-1} = \operatorname{diag}(\lambda_1, \ldots, \lambda_l) \\ \mathbf{G} &= \mathbf{CB} \end{aligned}$$

This follows from the fact that[23]

$$\det \begin{pmatrix} \mathbf{A}_{11} & 0 \\ \mathbf{A}_{21} & \mathbf{A}_{22} \end{pmatrix} = \det \mathbf{A}_{11} \cdot \det \mathbf{A}_{22}$$

Equation (13) shows clearly that with this type of feedback, the modes $\lambda_{l+1}, \ldots, \lambda_n$ of the open-loop system are not affected at all by the feedback. Therefore, if it is desired to move $\lambda_1, \ldots, \lambda_l$ of the open-loop system, then it is only necessary to do so by feeding back only the corresponding modes, i.e., by considering the closed-loop control of the aggregated system

$$\frac{d\mathbf{z}}{dt} = \mathbf{Fz} + \mathbf{Gu}$$

where $\mathbf{CA} = \mathbf{FC}$ with $\mathbf{C} = (\mathbf{I}_l, \mathrm{O}_{l,n-l})$. See Ref. 50 for a general discussion on pole assignment. It is also to be noted that the aggregation matrix need not be the projection as assumed here. For example, take the matrices to be

$$\Lambda = \begin{pmatrix} \Lambda_1 & 0 \\ 0 & \Lambda_2 \end{pmatrix} \qquad \text{where } \Lambda_1 \text{ is } l \times l$$

and choose $\mathbf{C}$ to be $\mathbf{C} = (\mathbf{C}_1, \mathrm{O}_{l,n-l})$ where $\mathbf{C}_1$ is an $l \times l$ nonsingular matrix. Then, provided that

$$\mathbf{C}_1\Lambda_1 = \Lambda_1\mathbf{C}_1$$

is satisfied,

$$\det\{\lambda\mathbf{I}_n - (\Lambda - \mathbf{BKC})\} = \det\{\lambda\mathbf{I}_l - (\Lambda_1 - \mathbf{C}_1\mathbf{D}_1)\} \cdot \det(\lambda\mathbf{I}_{n-l} - \Lambda_2) \quad (14)$$

where $\mathbf{D}_1$ is the first $l \times l$ submatrix of $\mathbf{BK}$.

5-4 WEAKLY COUPLED SYSTEMS

In this section approximation is introduced in considering control policies for an n-dimensional linear time-invariant dynamic system S_1 that can be regarded as being composed of several subsystems of lower dimensions. According to the discussions so far presented in this chapter, it is seen that a proper choice of $\mathbf{C}$, for example one which makes $\mathbf{F}$ retain $\lambda_1, \ldots, \lambda_l$ of $\mathbf{A}$, requires the knowledge of the characteristic values of $\mathbf{A}$ (and characteristic vectors of $\mathbf{A}$). This is not very satisfactory since they may not be easily obtainable, especially for $\mathbf{A}$ with large n. If it is possible to choose $\mathbf{C}$ from the knowledge of the constituent subsystems disregarding intercouplings among them, then the construction of an aggregation system of a desired kind becomes much more feasible.

When S_1 is composed of several weakly coupled subsystems, in a sense to be made precise later, it is possible to replace the required a posteriori knowledge on $\mathbf{A}$, that is, λ_i and $\mathbf{v}_i\rangle$, $1 \leqslant i \leqslant n$ by a priori knowledge on the subsystems, i.e., knowledge of characteristic values and vectors of subsystems taken separately by disregarding the coupling. For the sake of presenting the essential ideas simply, discussions in this section will be based on S_1 consisting of two subsystems S_{11} and S_{12}. Note, however, that these subsystems need not exist as separate physical entities but can be merely mathematically convenient regroupings of the state vector components.

Consider two coupled dynamic systems described by

$$\frac{d}{dt}\begin{pmatrix} \mathbf{x}_1 \\ \mathbf{x}_2 \end{pmatrix} = \mathbf{A}\begin{pmatrix} \mathbf{x}_1 \\ \mathbf{x}_2 \end{pmatrix} \qquad \text{where} \quad \mathbf{A} = \begin{pmatrix} \Lambda_1 & \mathbf{A}_{12} \\ \mathbf{A}_{21} & \Lambda_2 \end{pmatrix} \quad (15)$$

where without loss of generality it is assumed that Λ_1 and Λ_2 are diagonal matrices, $\mathbf{x}_1$ is n_1-vector, and $\mathbf{x}_2$ is n_2-vector. $n_1 + n_2 = n$.† The characteristic values of Λ_1 are denoted by $\lambda_1, \ldots, \lambda_{n_1}$ and those of Λ_2 are denoted by $\mu_1, \ldots, \mu_{n_2}$. Assume also without loss of generality that characteristic values of S_{12} are larger than those of S_{11}.

The definition of weak coupling used in this paper is that proposed by Milne.[37] Define

$$r \triangleq \max_i |\lambda_i| \qquad \bar{\mathbf{R}} \triangleq \min_i |\mu_i|$$

The subsystems 1 and 2 are said to be weakly coupled if these conditions are met:

$$\frac{r}{\bar{\mathbf{R}}} \ll 1$$

and

$$\frac{n_1 \delta_{12} \delta_{21}}{\bar{\mathbf{R}}^2} \ll 1$$

where δ_{12} is the maximum of the moduli of the elements of $\mathbf{A}_{12}$ and where δ_{21} is similarly defined from the elements of $\mathbf{A}_{21}$. Thus, S_{11} and S_{12} are weakly coupled if their characteristic values are widely separated in modulus as given by the first condition and if the interaction terms are small as defined by the second condition.‡ Note that these conditions do not require the explicit knowledge of the characteristic values of the combined system.

Milne has also shown that the characteristic values of $\mathbf{A}$ are approximately given by $\hat{\lambda}_1, \ldots, \hat{\lambda}_{n_1}, \mu_1, \ldots, \mu_{n_2}$ where

$$|\hat{\lambda}_i \mathbf{I} - \Lambda_1 + \mathbf{A}_{12} \Lambda_2^{-1} \mathbf{A}_{21}| = 0 \qquad 1 \leqslant i \leqslant n_1 \tag{16a}$$

and μ_i is the eigenvalue of Λ_2.

The characteristic vectors are approximately given by

$$\mathbf{v}_i\rangle = \begin{pmatrix} \xi_i \\ -\Lambda_2^{-1} \mathbf{A}_{21} \xi_i \end{pmatrix} \qquad 1 \leqslant i \leqslant n_1$$

$$\mathbf{v}_{n_1+j}\rangle = \begin{pmatrix} \dfrac{\mathbf{A}_{12}}{\mu_j} \eta_j \\ \eta_j \end{pmatrix} \qquad 1 \leqslant j \leqslant n_2 \tag{16b}$$

† This is the simplest nontrivial example. Systems composed of subsystems connected in parallel, in cascade, or in cascade-parallel where each subsystem has its own control variables are trivial since the matrix **A** is either composed of diagonal submatrices or is in the reducible form.[23]

‡ Note that these conditions may be regarded as a slight generalization of the condition of Gersgorin's discs.[34]

where

$$(\Lambda_1 - \mathbf{A}_{12}\Lambda_2^{-1}\mathbf{A}_{21})\, \xi_i = \hat{\lambda}_i \xi_i \qquad \text{and} \qquad \Lambda_2 \eta_j = \mu_j \eta_j$$

Thus, for S_1 consisting of two weakly coupled subsystems, vectors given by Eq. (16*b*) are the approximate characteristic vectors of **A**. The appropriate aggregation matrix can now be chosen from the characteristic vectors of Eq. (16*b*). See also Ref. 16 for a related topic.

5-5 CONTROL OF A WEAKLY COUPLED SYSTEM VIA THE AGGREGATION MODEL

5-5-1 Control Law

Control of a weakly coupled system via control policies based on S_2 with an assumed aggregation matrix **C** is now considered. The criterion function for S_1 is assumed to be given by

$$\mathbf{J} = \int_0^\infty (\mathbf{x}'\mathbf{Q}\mathbf{x} + \mathbf{u}'\mathbf{R}\mathbf{u})\, dt$$

where $\mathbf{Q} \geqslant 0$ and $\mathbf{R} > 0$ so that $\mathbf{R}^{-1}$ exists.†

The dynamics of S_2 are given by

$$\frac{d\mathbf{z}}{dt} = \mathbf{F}\mathbf{z} + \mathbf{G}\mathbf{u} \qquad \mathbf{z}(0) = \mathbf{z}_0$$

Consider the behavior of S_2 where **u** is generated by

$$\mathbf{u} = -\mathbf{K}\mathbf{z} \tag{17}$$

and where **K** is a $r \times l$ gain matrix given by Eq. (18) below.

Now, if the system S_2 has a criterion function

$$\mathbf{J}_\mathbf{M} = \int_0^\infty (\mathbf{z}'\mathbf{Q}_\mathbf{M}\mathbf{z} + \mathbf{u}'\mathbf{R}\mathbf{u})\, dt$$

then it is well known [29] that for a controllable pair (**F**, **G**) the control law (17) results in a stable closed-loop system and is optimal when the control gain for S_2 is given by

$$\mathbf{K} = \mathbf{R}^{-1}\mathbf{G}'\mathbf{P} \tag{18}$$

where **P** is the solution of

$$O = \mathbf{F}'\mathbf{P} + \mathbf{P}\mathbf{F} - \mathbf{P}\mathbf{G}\mathbf{R}^{-1}\mathbf{G}'\mathbf{P} + \mathbf{Q}_\mathbf{M} \tag{19}$$

† $\mathbf{Q} \geqslant 0$ means that **Q** is positive semidefinite. With $\geqslant$ replaced by $>$, it means that the matrix is positive definite.

and the minimal value of $\mathbf{J_M}$, $\mathbf{J_M}^*$, is given by

$$\mathbf{J_M}^* = \mathbf{z}_0'\mathbf{P}\mathbf{z}_0$$

where

$$\mathbf{P} > 0$$

Now consider controlling S_1 by

$$\begin{aligned} \mathbf{u} &= -\mathbf{Kz} \\ &= -\mathbf{KCx} \end{aligned} \tag{20}$$

With the control given by Eq. (20), the state vector of S_1 is governed by the equation

$$\frac{d\mathbf{x}}{dt} = (\mathbf{A} - \mathbf{BKC})\,\mathbf{x}$$

The optimal state vector obeys the equation

$$\frac{d\mathbf{x}^*}{dt} = (\mathbf{A} - \mathbf{BK}^*)\,\mathbf{x}^* \tag{21}$$

with the optimal control

$$\mathbf{u}^* = -\mathbf{K}^*\mathbf{x}$$

where

$$\mathbf{K}^* = \mathbf{R}^{-1}\mathbf{B}'\mathbf{T}^*$$

and where $\mathbf{T}^*$ satisfies

$$O = \mathbf{A}'\mathbf{T}^* + \mathbf{T}^*\mathbf{A} - \mathbf{T}^*\mathbf{B}\mathbf{R}^{-1}\mathbf{B}'\mathbf{T}^* + \mathbf{Q} \tag{22}$$

When $\mathbf{C}$ is such that Eq. (6) is satisfied, then from Eqs. (5) and (7) and by pre- and postmultiplication of Eq. (19) by $\mathbf{C}'$ and $\mathbf{C}$, respectively,

$$O = \mathbf{A}'\mathbf{C}'\mathbf{PC} + \mathbf{C}'\mathbf{PCA} - \mathbf{C}'\mathbf{PCBR}^{-1}\mathbf{B}'\mathbf{C}'\mathbf{PC} + \mathbf{C}'\mathbf{Q_M}\mathbf{C} \tag{23}$$

Comparing Eq. (23) with Eq. (22), it is seen that $\mathbf{C}'\mathbf{PC}$ corresponds to $\mathbf{T}^*$ if $\mathbf{C}'\mathbf{Q_M}\mathbf{C}$ is made to correspond to $\mathbf{Q}$. Of course, they cannot be equated because $\mathbf{T}^*$ is of rank n while $\mathbf{C}'\mathbf{PC}$ is at most of rank l. The above arguments seem to indicate that the control given by Eq. (20) is of some interest as a suboptimal control for S_1 with a suitably chosen $\mathbf{Q_M}$, for example, if $\mathbf{Q_M}$ is chosen as

$$\mathbf{Q_M} = (\mathbf{CC}')^{-1}\,\mathbf{CQC}'(\mathbf{CC}')^{-1} \tag{24}$$

Example 5-1

Consider S_1 consisting of two weakly coupled subsystems with

$$\mathbf{A}_{11} = -0.1 \qquad \mathbf{A}_{22} = \text{diag}(-20, -30)\ \mathbf{A}_{12} = \mathbf{A}'_{21} = (1, 1) \qquad \mathbf{B}' = (1, 1, 1)$$

where $r/\mathbf{R} = 0.1/20$ and $n_1\, \delta_{12}\, \delta_{21}/(\mathbf{R})^2 = 1/400$ and the criterion function $\mathbf{J}$ with

$$\mathbf{Q} = \text{diag}(1, 1, 1) \qquad \mathbf{R} = 1$$

Take $\mathbf{C}$ to be a 2×3 matrix where $\langle \mathbf{c}_1 = (1, 0, 0)$ and $\langle \mathbf{c}_2 = (0, \frac{1}{2}, \frac{1}{2})$.†

Then from Eq. (7) the aggregated system S_2 has the differential equation (4) with

$$\mathbf{F} = \begin{pmatrix} -0.1 & 2 \\ 1 & -25 \end{pmatrix} \qquad \mathbf{G} = \begin{pmatrix} 1 \\ 1 \end{pmatrix}$$

and from Eq. (24) the criterion function $\mathbf{J_M}$

$$\mathbf{Q_M} = \text{diag}(1, 2)^{\ddagger}$$

The optimal and suboptimal control for S_1 are given respectively as

$$\mathbf{u}^* = -(0.97\mathbf{x}_1 + 0.07\mathbf{x}_2 + 0.05\mathbf{x}_3)$$

$$\mathbf{u} = -(0.97\mathbf{z}_1 + 0.12\mathbf{z}_2)$$

$$= -(0.98\mathbf{x}_1 + 0.06\mathbf{x}_2 + 0.06\mathbf{x}_3)$$

Note that the approximation is quite good, even though $\mathbf{FC} \neq \mathbf{CA}$ especially because $\mathbf{x}_2$ and $\mathbf{x}_3$ will decay to zero very rapidly.

With $\mathbf{C}$ given by $\langle \mathbf{c}_1 = (1, 0, 0)$ and $\langle \mathbf{c}_2 = (0, 5, 0)$, the suboptimal control is given by

$$\mathbf{u} = -(0.96\mathbf{x}_1 + 0.07\mathbf{x}_2)$$

5-5-2 Bounds on Performance Index

The use of the suboptimal control (20) in S_1 with $\mathbf{Q_M}$ given by Eq. (24) results in the value of the criterion function

$$\mathbf{J} = \mathbf{x}_0{}'\mathbf{T}\mathbf{x}_0 \tag{25}$$

† When the approximate eigenvalues of $\mathbf{A}$ computed by Eq. (16) are further approximated, $\mathbf{x}_i$, $i = 1, 2, 3$, are seen to correspond very approximately to the three canonical coordinates of $\mathbf{A}$. This $\mathbf{C}$ then takes $\mathbf{z}_1$ to be approximately the mode with the slowest decay and $\mathbf{z}_2$ to be the average of the other two modes with fast decay.

‡ It is possible to iteratively change $\mathbf{Q_M}$ to improve the suboptimal control. This topic is not pursued in this paper.[2]

where $\mathbf{T}$ satisfies the matrix equation

$$O = (\mathbf{A} - \mathbf{BKC})'\mathbf{T} + \mathbf{T}(\mathbf{A} - \mathbf{BKC}) + \mathbf{C}'\mathbf{K}'\mathbf{RKC} + \mathbf{Q} \tag{26}$$

provided that $(\mathbf{A} - \mathbf{BKC})$ is a stable matrix. Writing $\mathbf{T} = \mathbf{T}^* + \Delta$, matrix Δ satisfies

$$(\mathbf{A} - \mathbf{BKC})' \Delta + \Delta(\mathbf{A} - \mathbf{BKC}) = -\delta'\mathbf{BR}^{-1}\mathbf{B}'\delta \tag{27}$$

where $\delta \triangleq (\mathbf{T}^* - \mathbf{C}'\mathbf{PC})$ and is given by the solution of

$$\begin{aligned} 0 = (\mathbf{A} - \mathbf{BR}^{-1}\mathbf{B}'\mathbf{C}'\mathbf{PC})'\delta + \delta(\mathbf{A} - \mathbf{BR}^{-1}\mathbf{B}'\mathbf{C}'\mathbf{PC}) \\ -\delta\mathbf{BR}^{-1}\mathbf{B}'\delta + \mathbf{Q} - \mathbf{C}'(\mathbf{CC}')^{-1}\mathbf{CQC}'(\mathbf{CC}')^{-1}\mathbf{C} \end{aligned} \tag{28}$$

Therefore, if all characteristic values of $(\mathbf{A} - \mathbf{BKC})$ have negative real parts, i.e., if the suboptimal control (20) results in a stable S_1, then Δ is determined uniquely and is positive (semi-) definite since the right-hand side of Eq. (28) is either positive definite or positive semi-definite; i.e.,

$$\mathbf{T}^* \leqslant \mathbf{T} \tag{29a}$$

Note that $\mathbf{T}$ is far easier to obtain than $\mathbf{T}^*$. A lower bound on $\mathbf{T}^*$ can similarly be obtained as

$$0 \leqslant \mathbf{C}'\mathbf{PC} \leqslant \mathbf{T}^* \tag{29b}$$

It can be shown generally that aggregation matrices can be used to provide upper and lower bounds on solutions of matrix Riccati equations in a computationally efficient manner.[3] See Ref. 10 for a related topic.

Thus, an optimal aggregation matrix may be obtained by minimizing a suitable norm of Δ or more conveniently of $(\mathbf{T} - \mathbf{C}'\mathbf{PC})$ as a function of elements of $\mathbf{C}$ or of $\mathbf{Q}_M$ for a given $\mathbf{C}$. This is true whether S_1 is weakly coupled or not. Practically speaking, such optimization may be computationally inconvenient to perform. With the assumption of weak coupling, however, $\mathbf{C}$ obtained from a priori knowledge of S_1 may be made nearly optimal. Equation (28) or $(\mathbf{T} - \mathbf{C}'\mathbf{PC})$ can then be used to improve this initial choice interatively. Note that if $\mathbf{Q}$ is diag $(\mathbf{q}_1, \ldots, \mathbf{q}_n)$ and if $\mathbf{C}$ is the projection $\mathbf{C} = (\mathbf{I}_l, 0)$, then

$$\mathbf{Q} - \mathbf{C}'(\mathbf{CC}')^{-1}\mathbf{CQC}'(\mathbf{CC}')^{-1}\mathbf{C} = \text{diag}\,(\mathbf{q}_1, \ldots, \mathbf{q}_l, 0, \ldots, 0)$$

and Eq. (28) is easier to solve than solving for $\mathbf{T}^*$ from Eq. (22).

In designing a closed-loop control law via some quadratic performance index, it is known that the $\mathbf{Q}$ matrix can be put into an equivalent form to effect the simplifications of the optimal feedback gain computation.[44] This equivalence relation may be profitably used in choosing appropriate aggregation matrices. See Refs. 44 and 49 for the discussions on the effects of $\mathbf{Q}$ and $\mathbf{R}$ matrices on transient responses.

Performance degradation with several aggregation matrices can be discussed in a similar manner. Unlike the lower bound in Eq. (29*b*), a positive definite lower bound for $\mathbf{T}^*$ is obtained if at least two aggregation matrices are used. For example let $\mathbf{C}_i$ be an $l_i \times n$ aggregation matrix $i = 1, 2$. Assume for the sake of simplicity that $l_1 + l_2 = n$ and that

$$\text{Rank} \begin{bmatrix} \mathbf{C}_1 \\ \mathbf{C}_2 \end{bmatrix} = n$$

Then with $\mathbf{Q}_i \geqslant 0$ and $\mathbf{R}_i > 0$ as the symmetric matrices appearing in the performance index function for the ith aggregated system, it can be shown after some algebra that if $\mathbf{A} - \mathbf{BK}$ is a stable matrix where $\mathbf{K} \triangleq \mathbf{K}_1\mathbf{C}_1 + \mathbf{K}_2\mathbf{C}_2$ and where

$$\mathbf{K}_i = \mathbf{R}_i^{-1}\mathbf{B}_i'\mathbf{C}_i'\mathbf{P}_i$$

$$\mathbf{P}_i\mathbf{F}_i + \mathbf{F}_i'\mathbf{P}_i - \mathbf{P}_i\mathbf{G}_i\mathbf{R}_i^{-1}\mathbf{G}_i'\mathbf{P}_i + \mathbf{Q}_i = 0 \qquad i = 1, 2$$

then the solution of the matrix Riccati equation of S_1 can be expressed as

$$\mathbf{T} = \breve{\mathbf{P}} + \delta$$

where

$$\mathbf{P} = \mathbf{C}_1'\mathbf{P}_1\mathbf{C}_1 + \mathbf{C}_2'\mathbf{P}_2\mathbf{C}_2$$

and where

$$\delta = \int_0^\infty \mathbf{e}^{(\mathbf{A}-\mathbf{BK})'t}\psi\mathbf{e}^{(\mathbf{A}-\mathbf{BK})t}\,dt$$

$$\psi \triangleq \mathbf{Q} - \breve{\mathbf{Q}} + \breve{\mathbf{P}}\Lambda\breve{\mathbf{P}}$$

$$\breve{\mathbf{Q}} \triangleq \mathbf{C}_1'\mathbf{Q}_1\mathbf{C}_1 + \mathbf{C}_2'\mathbf{Q}_2\mathbf{C}_2$$

$$\Lambda \triangleq \begin{pmatrix} \mathbf{C}_1 \\ \mathbf{C}_2 \end{pmatrix}^{-1}$$

$$\times \begin{bmatrix} \mathbf{C}_1\mathbf{B}(\mathbf{R}_1^{-1}\mathbf{R}\mathbf{R}_1^{-1} - \mathbf{R}_1^{-1})\,\mathbf{B}'\mathbf{C}_1', & \mathbf{C}_1\mathbf{B}(\mathbf{R}_1^{-1}\mathbf{R}\mathbf{R}_2^{-1} - \mathbf{R}_1^{-1} - \mathbf{R}_2^{-1}) \\ \mathbf{C}_2\mathbf{B}(\mathbf{R}_2^{-1}\mathbf{R}\mathbf{R}_1^{-1} - \mathbf{R}_1^{-1} - \mathbf{R}_2^{-1})\,\mathbf{B}'\mathbf{C}_1', & \mathbf{C}_2\mathbf{B}(\mathbf{R}_2^{-1}\mathbf{R}\mathbf{R}_2^{-1} - \mathbf{R}_2^{-1})\,\mathbf{B}'\mathbf{C}_2' \end{bmatrix} (\mathbf{C}_1'\mathbf{C}_2')^{-1}$$

For example if the $\mathbf{Q}_i$ are chosen such that

$$\begin{pmatrix} \mathbf{Q}_1 & 0 \\ 0 & \mathbf{Q}_2 \end{pmatrix} < \begin{pmatrix} \mathbf{C}_1 \\ \mathbf{C}_2 \end{pmatrix}^{-1} \mathbf{Q}(\mathbf{C}_1'\mathbf{C}_2')^{-1}$$

and $\mathbf{R}_i$ are chosen such that Λ is positive definite, then δ is positive definite, and it can be shown that

$$0 < \mathbf{P} < \mathbf{T}^* \leqslant \mathbf{T} \tag{30}$$

Material in Secs. 5-5-1 and 5-5-2 can be extended to systems which are not weakly coupled provided that a suitable $\mathbf{C}$ can be found, e.g., if the characteristic values and vectors of $\mathbf{A}$ are approximately known. It is also straightforward to consider control over a finite time interval rather than $[0, \infty)$.

5-5-3 Error Analysis

Define the error vector

$$\mathbf{e} \triangleq \mathbf{z} - \mathbf{Cx}$$

Then

$$\dot{\mathbf{e}} = \mathbf{Fe} + (\mathbf{FC} - \mathbf{CA})\,\mathbf{x}$$

Thus, the discrepancy between $\mathbf{z}$ and $\mathbf{Cx}$ evolves with time as

$$\mathbf{e}(t) = e^{\mathbf{F}t}\mathbf{e}(0) + \int_0^t e^{\mathbf{F}(t-\tau)}(\mathbf{FC} - \mathbf{CA})\,\mathbf{x}(\tau)\,d\tau$$

Bounds on $\mathbf{e}(t)$ can be obtained from this. Sometimes $\mathbf{C}$ can be chosen in such a way that some components of $\mathbf{e}$ are zero. If $\mathbf{C}$ can be chosen such that the modes of $\mathbf{F}$ associated with nonzero components of $\mathbf{e}$ die out very quickly, then even if $\mathbf{FC} \neq \mathbf{CA}$, the suboptimal controls for S_1 derived from S_2 may still be satisfactory. The example given earlier is of this type. See Ref. 14 for a more detailed analysis.

Example 5-2

As another example of applications of the aggregation concept consider a control problem with a performance index independent of control and with a plant equation linear in the control variables. Such a problem has been discussed in Ref. 15 where the concept of aggregation has been used implicitly. The plant equation may depend nonlinearly on the state variables. This problem is special in the sense that the disaggregation operation can be performed without error; that is, $\mathbf{x}$ can be recovered without error. Consider a control system with the plant equation

$$\frac{d\mathbf{x}}{dt} = f(t, \mathbf{x}(t)) + \mathbf{Bu}(t)$$

and the performance index

$$J = \int_0^T W(t, \mathbf{x}(t))\, dt$$

where $\mathbf{B}$† is an $n \times r$ matrix with maximal rank $r \leqslant n$.

Since r n-dimensional column vectors of $\mathbf{B}$ are linearly independent by assumption, choose $n - r$ orthonormal vectors $\mathbf{c}_1\rangle, \ldots, \mathbf{c}_l\rangle$, $l = n - r$, such that each of them are orthogonal to the column vectors of $\mathbf{B}$.

$$\mathbf{C}' = (\mathbf{c}_1\rangle, \ldots, \mathbf{c}_l\rangle)$$

Now $\mathbf{c}_1\rangle, \ldots, \mathbf{c}_l\rangle$ together with the column vectors of $\mathbf{B}$ forms a basis

$$\mathbf{x}(t) = \sum_{i=1}^{l} \mathbf{z}_i(t)\, \mathbf{c}_i\rangle + \sum_{i=1}^{r} \mathbf{v}_i(t)\, \mathbf{b}_i\rangle$$

where $\mathbf{b}_i\rangle$ is the ith column vector of $\mathbf{B}$. In matrix notation

$$\mathbf{x}(t) = \mathbf{C}'\mathbf{z}(t) + \mathbf{B}\mathbf{v}(t)$$

where $\mathbf{z}(t)$ and $\mathbf{v}(t)$ are column vectors with $\mathbf{z}_i(t)$ and $\mathbf{v}_i(t)$ as their respective ith elements. The aggregation expression is

$$\mathbf{z}(t) = \mathbf{C}\mathbf{x}(t)$$

since

$$\mathbf{C}\mathbf{C}' = \mathbf{I}_l$$

$$\mathbf{C}\mathbf{B} = O_{l,r}$$

by construction.

The dynamic equation for $\mathbf{z}(t)$ is given by

$$\begin{aligned}\dot{\mathbf{z}} &= \mathbf{C}\dot{\mathbf{x}} \\ &= \mathbf{g}(t, \mathbf{z}(t), \mathbf{v}(t))\end{aligned}$$

where

$$\mathbf{g}(t, \mathbf{z}(t), \mathbf{v}(t)) \triangleq \mathbf{C}\mathbf{f}(t, \mathbf{C}'\mathbf{z} + \mathbf{B}\mathbf{v})$$

† In Ref. 15, a time-varying $\mathbf{B}$ is considered. This necessitates the use of a time-varying aggregation matrix. Although this does not introduce any difficulty provided that $\mathbf{B}$ satisfies certain regularity conditions, $\mathbf{B}$ is taken here to be constant for the sake of simplicity.

Its criterion function is given by

$$\mathbf{J} = \int_0^T \mathbf{w}(t, \mathbf{z}(t), \mathbf{v}(t))\, dt$$

where

$$\mathbf{w}(t, \mathbf{z}, \mathbf{v}) \triangleq \mathbf{W}(t, \mathbf{C}'\mathbf{z} + \mathbf{B}\mathbf{v})$$

When this optimal control problem for the aggregated system has been solved, the optimal control and the state vector for the original problem are related to those of the aggregated system by

$$\mathbf{x}^*(t) = \mathbf{C}'\mathbf{z}^*(t) + \mathbf{B}\mathbf{v}^*(t)$$

and

$$\mathbf{u}^*(t) = \dot{\mathbf{v}}^*(t) - (\mathbf{B}'\mathbf{B})^{-1}\mathbf{B}'f(t, \mathbf{x}^*(t))$$

if the indicated derivative exists for all t. Note that the optimal state vector $\mathbf{x}^*(t)$ can be reconstructed without error using $\mathbf{z}^*(t)$ and $\mathbf{v}^*(t)$.

5-6 STABILITY ANALYSIS

The closed-loop stability of S_1 when it is controlled by Eq. (20) is now investigated. From Eqs. (18) and (19)

$$(\mathbf{F} - \mathbf{GK})'\mathbf{P} + \mathbf{P}(\mathbf{F} - \mathbf{GK}) = -(\mathbf{K}'\mathbf{RK} + \mathbf{Q_M})$$

Let $(\mathbf{K}'\mathbf{RK} + \mathbf{Q_M})$ be positive definite. Then it is well known[23] that $[\mathbf{F} - \mathbf{GK}]$ is a stable matrix if and only if $\mathbf{P}$ is positive definite. A similar argument for $[\mathbf{A} - \mathbf{BKC}]$ breaks down because from Eq. (29) $\mathbf{T} \geqslant \mathbf{C}'\mathbf{PC}$ where $\mathbf{C}'\mathbf{PC}$ is only positive semidefinite. Thus, a stable $[\mathbf{F} - \mathbf{GK}]$ does not necessarily imply that $[\mathbf{A} - \mathbf{BKC}]$ is stable. This can be also seen from the fact that with an $l \times n$ aggregation matrix $\mathbf{C}$, at most l of the characteristic values of $\mathbf{A}$ can be moved by the feedback control. The remaining $n - l$ eigenvalues of $\mathbf{A}$ are unaffected by the feedback and remain the same; i.e., these are the uncontrollable modes of $\mathbf{A}$. This has been shown explicitly in Sec. 5-3 when $\mathbf{A}$ is a diagonal or a block diagonal matrix. The same conclusion holds if $\mathbf{A}$ is in a reduced form. Therefore, with the assumed forms of $\mathbf{A}$ and $\mathbf{C}$ if the number of unstable modes of $\mathbf{A}$ is not greater than l, $[\mathbf{A} - \mathbf{BKC}]$ will be a stable matrix when $[\mathbf{F} - \mathbf{CBK}]$ is. Another way of seeing this is by the s plane consideration when $\mathbf{A}$ is in the diagonal or reduced form. Then from Eqs. (1), (2), and (4),

$$\begin{aligned} \mathbf{A}(s) &= \mathbf{CX}(s) \\ &= \mathbf{C}(s\mathbf{I} - \mathbf{A})^{-1}\mathbf{BU}(s) \\ &= (s\mathbf{I} - \mathbf{F})^{-1}\mathbf{CBU}(s) \end{aligned} \tag{31}$$

where $\mathbf{X}(s)$, $\mathbf{U}(s)$, and $\mathbf{Z}(s)$ are the Laplace transforms of $\mathbf{x}(t)$, $\mathbf{u}(t)$, and $\mathbf{z}(t)$ respectively. Equation (31) shows that poles of S_1 corresponding to $\lambda_{l+1}, \ldots, \lambda_n$ are canceled to give the transfer function of S_2 and are not therefore controllable.

When $\mathbf{A}$ is not in a diagonal or reduced form, take the row vectors of $\mathbf{C}$ to be the row eigenvectors of $\mathbf{A}$ corresponding to $\lambda_1, \ldots, \lambda_l$; that is, $\langle \mathbf{w}_1, \ldots, \langle \mathbf{w}_l$ in the notation of Sec. 5-3. Then from Eq. (11) $\mathbf{F}$ is seen to inherit $\lambda_1, \ldots, \lambda_l$ as its eigenvalues. Hence, $[\mathbf{F} - \mathbf{GK}]$, with $\mathbf{K}$ given by Eq. (18), is a stable matrix; in other words, these l eigenvalues of $[\mathbf{A} - \mathbf{BKC}]$ have negative real parts. The remaining $n - l$ eigenvalues of $[\mathbf{A} - \mathbf{BKC}]$, however, assume new values with possibly positive real parts. Unless some simple test such as Gersgorin's theorem[34] implies that the remaining $n - l$ eigenvalues are stable,† it is necessary to test $[\mathbf{A} - \mathbf{BKC}]$ for stability. More generally, given an aggregated model with an arbitrary $\mathbf{C}$, it may be more convenient and faster to test $[\mathbf{A} - \mathbf{BKC}]$ directly, although it is possible to carry out the stability test for S_1 by first transforming $\mathbf{A}$ into a diagonal form. Note that this test of stability is computationally far easier to perform than solving for characteristic values and vectors.

5-7 IDENTIFICATION OF CONSTRAINED DYNAMIC SYSTEMS

5-7-1 Introduction

In Sec. 5-5, the aggregation concept has been applied to generate suboptimal controls. In this section, it is applied to identification problems. This is a problem of determining system parameter values in dynamic equations. It will be shown that the aggregation method can be applied to reduce the amount of computation associated with identification problems for a class of dynamic systems to be specified later. Consider an identification problem of the system S_1 described by

$$\mathbf{x}(i+1) = \mathbf{A}\mathbf{x}(i) + \tilde{\mathbf{f}}(i) \tag{32}$$

where $\mathbf{x}$ is an n-vector, and where $\tilde{\mathbf{f}}(i)$ represents a known forcing term.

Frequently one possesses some a priori knowledge of the system to be identified. It is assumed here that some of the elements of the matrix $\mathbf{A}$ are known constants. Some of the existing techniques of system identification do not take full advantage of the a priori information on $\mathbf{A}$ (beyond the mere quantitative requirements such as controllability and observability) and treat all elements of $\mathbf{A}$ (or n unknown elements in the

† Note that for weakly coupled systems, this test may suffice to determine whether or not $[\mathbf{A} - \mathbf{BKC}]$ is a stable matrix.

phase canonical form of **A**) equally as if they are all unknown;[25,26] i.e., these methods do not take advantage of the fact that some elements may be known. In other words structural information available a priori on dynamic system is not utilized.

Since the amount of computations in system identification grows as a nonlinear function of the dimension n, it is practically important to be able to reduce the amount of computations by taking advantage of the information on **A** which is available a priori, especially when n is large.

This section presents a way of incorporating the a priori information on the system identification algorithm. Since **A** is usually generated through a set of coupled equations written for individual subsystems comprising the total system and since the process of transforming **A** into the phase canonical or other canonical forms may result in having matrix elements which are in more complex expressions than the original system parameters to be identified, it is assumed that **A** is not necessarily in the phase canonical form. Express **A** as

$$\mathbf{A} = \mathbf{A}^0 + \mathbf{A}^1 \tag{33}$$

where all the known elements of **A** are grouped to form $\mathbf{A}^0$ and the elements of $\mathbf{A}^1$ are all unknown. Then Eq. (32) can be rewritten as

$$\mathbf{x}(i+1) = \mathbf{A}^1\mathbf{x}(i) + \mathbf{f}(i) \tag{34}$$

where the known term $\mathbf{A}^0\mathbf{x}(i)$ is lumped together with $\tilde{\mathbf{f}}$ to form a known forcing term **f**.

In many cases, $\mathbf{A}^1$ will be sparse; i.e., many elements of $\mathbf{A}^1$ will be zero, corresponding to the situations where **A** contains a few unknown elements and many known elements. Therefore, it is reasonable to assume that $\mathbf{A}^1$ satisfies certain linear constraints.

It is assumed that $\mathbf{A}^1$ satisfies the constraint equation

$$\mathbf{A}^1\mathbf{Q} = O_{n,k} \tag{35}^\dagger$$

where **Q** is a known $n \times k$ matrix of maximal rank k, $k \leqslant n$.

In the following development the superscript 1 will be dropped. As shown in Appendix A, Eq. (35) implies that **A** is at most of rank l, where $l = n - k$, and that **A** can be factored as

$$\mathbf{A} = \mathbf{A}_u\mathbf{A}_k \tag{36}$$

† It is also clear that the case where **A** itself satisfies a constraint of the type (35) can be treated similarly.

where $\mathbf{A}_u$ and $\mathbf{A}_k$ are $n \times l$ and $l \times n$ matrices respectively, and where $\mathbf{A}_u$ is unknown and $\mathbf{A}_k$ is a known matrix of rank l. Thus the identification of $\mathbf{A}$ reduces to that of identifying the $n \times l$ matrix $\mathbf{A}_u$. Consider now the problem of identifying $\mathbf{A}_u$ in Eq. (36).

5-7-2 Deterministic Systems with Constraints (1)

Least-squares Estimate†

Take the error of identification to be measured by

$$\mathbf{J}_\mathbf{N} = \sum_{i=0}^{\mathbf{N}-1} \| \mathbf{x}_{i+1} - \mathbf{A}\mathbf{x}_i - \mathbf{f}_i \|^2$$

$$= \mathrm{tr}\{-\mathbf{A}\mathbf{W}'_{\mathbf{N}-1,0} - \mathbf{W}_{\mathbf{N}-1,0}\mathbf{A}' + \mathbf{A}\mathbf{Z}_{\mathbf{N}-1,0}\mathbf{A}'\} + \text{const}$$

where

$$\begin{aligned} \mathbf{W}_{0,\mathbf{N}-1} &\triangleq \sum_{i=0}^{\mathbf{N}-1} (\mathbf{x}_{i+1} - \mathbf{f}_i)\, \mathbf{x}_i' \quad : n \times n \text{ matrix} \\ \mathbf{Z}_{0;\mathbf{N}-1} &\triangleq \sum_{i=0}^{\mathbf{N}-1} \mathbf{x}_i \mathbf{x}_i' \quad : n \times n \text{ matrix} \end{aligned} \tag{37}$$

and where $\mathbf{A}$ is given by Eq. (36). $\mathbf{J}_\mathbf{N}$ is to be minimized with respect to $\mathbf{A}_u$. The result of this minimization is the equation

$$\hat{\mathbf{A}}_u(\mathbf{N})\, \mathbf{A}_k \mathbf{Z}_{0,\mathbf{N}-1} \mathbf{A}_k' = \mathbf{W}_{0,\mathbf{N}-1} \mathbf{A}_k'$$

where $\hat{\mathbf{A}}_u(\mathbf{N})$ is the best estimate of $\mathbf{A}_u$ based on data $\mathbf{x}_0, \ldots, \mathbf{x}_{\mathbf{N}-1}$. If $\mathbf{N}$ is taken large enough so that $\mathbf{Z}_{0,\mathbf{N}-1}$ can be assumed to be positive definite, then

$$\hat{\mathbf{A}}_u(\mathbf{N}) = \mathbf{W}_{0,\mathbf{N}-1} \hat{\mathbf{A}}_k' (\mathbf{A}_k \mathbf{Z}_{0,\mathbf{N}-1} \mathbf{A}_k')^{-1} \tag{38}$$

and

$$\hat{\mathbf{A}}(\mathbf{N}) \triangleq \hat{\mathbf{A}}_u(\mathbf{N})\, \mathbf{A}_k$$

From Eq. (37), it is seen that

$$\mathbf{W}_{0,\mathbf{N}-1} = \mathbf{A}_u \mathbf{A}_k \mathbf{Z}_{0,\mathbf{N}-1}$$

Thus

$$\hat{\mathbf{A}}_u(\mathbf{N}) = \mathbf{A}_u \mathbf{A}_k \mathbf{Z}_{0,\mathbf{N}-1} \mathbf{A}_k' (\mathbf{A}_k \mathbf{Z}_{0,\mathbf{N}-1} \mathbf{A}_k')^{-1} = \mathbf{A}_u$$

† Only the method of least squares is considered in this chapter. For descriptions of the identification algorithm using the method of the maximum likelihood for systems for which only noisy measured data for x's are available see Ref. 52.

showing that for deterministic systems, $\mathbf{A}$ can be identified exactly, provided that $\mathbf{Z_N} > 0$. Note that in Eq. (38), the inversion of an $l \times l$ matrix is required rather than the inversion of an $n \times n$ matrix which would be required if $\mathbf{A}$ is identified directly.

Note also the fact that $\mathbf{J_N}$ defined by

$$\mathbf{J_N} = \sum_{i=0}^{\mathbf{N}-1} \| \mathbf{x}_{i+1} - \mathbf{A}\mathbf{x}_i - \mathbf{f}_i \|_{\mathbf{Q}}^2$$

leads to the same estimate of Eq. (38) as long as $\mathbf{Q} > 0$. See Appendix B for a derivation of Eq. (38) and for proof of the above statement.

In case the measurements are corrupted by noise, straightforward extensions of the results can be made to obtain $\hat{\mathbf{A}}_u$ as the least-squares estimate. This estimate is, however, generally unsatisfactory unless the measurement noise covariance matrices are quite small. For the sake of brevity, the noisy measurement case is not discussed here. It is clear, however, that a technique analogous to that of Sec. 5-7-3 can be used to reduce the computational work in generating some estimate of $\mathbf{A}_u$ such as the maximum likelihood estimate.[4,52]

Aggregation Method

Another scheme computationally simpler than that of Eq. (38) is next considered when $\mathbf{A}$ has the representation of Eq. (36). Regarding $\mathbf{A}_k$ in Eq. (36) as an aggregation matrix, define the aggregated state vector $\mathbf{z}$ at time i by

$$\mathbf{z}_i = \mathbf{A}_k \mathbf{x}_i \tag{39}$$

Then the $\mathbf{z}$ vector is governed by the dynamic equation given by

$$\mathbf{z}_{i+1} = \mathbf{F}\mathbf{z}_i + \mathbf{v}_i \tag{40}$$

where

$$\mathbf{v}_i \triangleq \mathbf{A}_k \mathbf{f}_i \qquad \mathbf{z}_0 \triangleq \mathbf{A}_k \mathbf{x}_0$$

and where $\mathbf{F}$ is the $l \times l$ matrix given by

$$\mathbf{F} = \mathbf{A}_k \mathbf{A}_u \tag{41†}$$

† More generally, if it is possible to find an $l \times n$ matrix $\mathbf{C}$ such that

$$\mathbf{C}\mathbf{A}_u\mathbf{A}_k = \mathbf{F}\mathbf{C}$$

then $\mathbf{z}_i$ is defined by

$$\mathbf{z}_i = \mathbf{C}\mathbf{x}_i$$

Take the criterion function of identification to be

$$\mathbf{J_N} = \sum_{i=0}^{\mathbf{N}-1} \| \mathbf{z}_{i+1} - \mathbf{F}\mathbf{z}_i - \mathbf{v}_i \|^2$$

The optimal estimate of $\mathbf{F}$, based on $\mathbf{N}$ observations, is given by

$$\hat{\mathbf{F}}(\mathbf{N}) = \mathbf{W_F}(\mathbf{N})\,\mathbf{Z_F}(\mathbf{N})^{-1}$$

provided $\mathbf{Z_F} > 0$ where

$$\mathbf{Z_F}(\mathbf{N}) \triangleq \sum_{0}^{\mathbf{N}-1} \mathbf{z}_i\mathbf{z}_i'$$

$$\mathbf{W_F}(\mathbf{N}) \triangleq \sum_{0}^{\mathbf{N}-1} (\mathbf{z}_{i+1} - \mathbf{v}_i)\,\mathbf{z}_i'$$

Note that $\mathbf{Z}_u$ and $\mathbf{M}_u$ are $l \times l$ matrices. Since

$$\mathbf{W_F}(\mathbf{N}) = \mathbf{F}\mathbf{Z_N}$$

$\mathbf{F}$ is identified exactly if no disturbances are present. Given an estimate of $\mathbf{F}$, Eq. (41) may be solved exactly for certain subclasses of $\mathbf{A}_u$.

Example 5-3

Consider a 3×3 matrix $\mathbf{A}$

$$\mathbf{A} = \begin{pmatrix} a & b & b \\ c & d & d \\ c & d & d \end{pmatrix}$$

It satisfies

$$\mathbf{AQ} = 0_{3,1}$$

where

$$\mathbf{Q}' = (0, 1, -1)$$

The matrix $\mathbf{A}$ has a factorization Eq. (36) where

$$\mathbf{A}_u = \begin{pmatrix} a & b \\ c & d \\ c & d \end{pmatrix} \qquad \text{and} \qquad \mathbf{A}_k = \begin{pmatrix} 1 & 0 & 0 \\ 0 & 1 & 1 \end{pmatrix}$$

This is the form discussed in Appendix A where

$$\mathbf{Q}_1 = \begin{pmatrix} 0 \\ 1 \end{pmatrix} \qquad \mathbf{Q}_2 = -1$$

then

$$\mathbf{A}_k = (\mathbf{I}_2, -\mathbf{Q}_1\mathbf{Q}_2^{-1})$$

Generally an estimate of $\mathbf{A}_u$ obtained as

$$\hat{\mathbf{A}}_u(\mathrm{N}) = \mathbf{A}_k{}^{+}\mathbf{W}_{\mathrm{F}}(\mathrm{N})\,\mathbf{Z}_{\mathrm{F}}(\mathrm{N})^{-1} \tag{42}$$ †

where $+$ denotes a pseudoinverse[4] is in error, since

$$\hat{\mathbf{A}}_u(\mathrm{N}) = \mathbf{A}_k{}^{+}\mathbf{A}_k\mathbf{A}_u$$

$$\mathbf{A} - \mathbf{A}(\mathrm{N}) = (\mathbf{A}_u - \hat{\mathbf{A}}_u(\mathrm{N}))\,\mathbf{A}_k = (\mathbf{I} - \mathbf{A}_k{}^{+}\mathbf{A}_k)\,\mathbf{A}_u\mathbf{A}_k$$

Thus, for a general $\mathbf{A}$ there will be an error in identification even if no random disturbances are present. For certain subclasses of $\mathbf{A}$, however, $\mathbf{A}$ will be identified exactly and the identification procedure of this section is computationally simpler than dealing with $\mathbf{A}$ directly, as the above example indicates. See the end of Sec. 5-7-3 for further examples.

5-7-3 Deterministic Systems with Constraints (2)

Next, consider the case where $\mathbf{A}$ is constrained by the equations

$$\begin{aligned} \mathbf{AQ} &= O_{n,k} \\ \mathbf{PA} &= O_{k,n} \end{aligned} \tag{43}$$

Then, as shown in Appendixes C and D, $\mathbf{A}$ has the representation

$$\mathbf{A} = \mathbf{DEC} \tag{44}$$

where $\mathbf{D}$, $\mathbf{E}$, and $\mathbf{C}$ are $n \times l$, $l \times l$, and $l \times n$ matrices respectively and where $\mathbf{E}$ is unknown with rank l. The matrices $\mathbf{C}$ and $\mathbf{D}$ respectively satisfy the constraint equations

$$\begin{aligned} \mathbf{CQ} &= 0_{(n,k)} \\ \mathbf{PD} &= 0_{(k,n)} \end{aligned} \tag{45}$$

Equation (43) is not as restrictive as it might appear at first sight.

If an $n \times n$ matrix $\mathbf{A}$ has n linear elementary divisors, it is known that there exists a complete set of left and right eigenvectors

$$\mathbf{A}\mathbf{u}_i = \lambda_i\mathbf{u}_i \qquad \mathbf{v}_i'\mathbf{A} = \lambda_i\mathbf{v}_i' \qquad 1 \leqslant i \leqslant n$$

† The same estimate of $\mathbf{A}_u$ results if Eq. (41) is substituted into the expression of $\mathbf{J}_\mathrm{N}$ and the resultant expression is minimized with respect to $\mathbf{A}_u$.

such that

$$\mathbf{v}_i{}'\mathbf{u}_j = 0 \qquad i \neq j$$

Normalize these eigenvectors by

$$\mathbf{u}_i{}'\mathbf{u}_i = \mathbf{v}_i{}'\mathbf{v}_i = 1 \qquad 1 \leqslant i \leqslant n$$

and denote

$$\mathbf{v}_i{}'\mathbf{u}_i = s_i \qquad 1 \leqslant i \leqslant n$$

then $\mathbf{A}$ can be represented as

$$\mathbf{A} = \sum_{i=1}^{n} \frac{\lambda_i}{s_i} \mathbf{u}_i \mathbf{v}_i{}'$$

$$= \mathbf{A}_1 + \mathbf{A}_2$$

where

$$\mathbf{A}_1 = \sum_{i=1}^{l} \frac{\lambda_i}{s_i} \mathbf{u}_i \mathbf{v}_i{}'$$

and

$$\mathbf{A}_2 = \sum_{l+1}^{n} \frac{\lambda_i}{s_i} \mathbf{u}_i \mathbf{v}_i{}'$$

If $\mathbf{A}_2$ is such that it can be treated as a perturbation term (for example if $\lambda_j \approx 0$, $l + 1 \leqslant j \leqslant n$) then

$$\mathbf{A} \approx \mathbf{A}_1$$

$$\mathbf{A}_1\mathbf{Q} = O_{n,k}$$

and

$$\mathbf{P}\mathbf{A}_1 = O_{n,k} \qquad k = n - l$$

where

$$\mathbf{Q} = [\mathbf{u}_{l+1}, \ldots, \mathbf{u}_n]$$

and

$$\mathbf{P}' = [\mathbf{v}_{l+1}, \ldots, \mathbf{v}_n]$$

Similar representation can be obtained from polar decomposition. For example, if rank $\mathbf{A} = r$, then

$$\mathbf{A} = \sum_{i=1}^{s} \lambda_i \mathbf{f}_i \mathbf{g}_i{}'$$

where λ_i^2, $\lambda \leqslant i \leqslant r$, are the positive eigenvalues of $\mathbf{A}'\mathbf{A}$ and

$$\mathbf{A}'\mathbf{A}\mathbf{g}_i = \lambda_i^2 \mathbf{g}_i$$

and

$$\mathbf{A}\mathbf{A}'\mathbf{f}_i = \lambda_i^2 \mathbf{f}_i$$

then

$$\mathbf{A} = \mathbf{F}_1 \Lambda_1 \mathbf{G}_1 + \mathbf{F}_2 \Lambda_2 \mathbf{G}_2$$

where

$$\mathbf{F}_1 = [\mathbf{f}_1, \ldots, \mathbf{f}_l] \qquad \mathbf{G}_1' = [\mathbf{g}_1, \ldots, \mathbf{g}_l] \qquad \Lambda_1 = \text{diag}(\lambda_1, \ldots, \lambda_l)$$

In this case, however, the eigenvalues of $\mathbf{F}_1 \Lambda_1 \mathbf{G}_1$ do not have any simple relation to $\lambda_1, \ldots, \lambda_l$ in general.

Now choose $\mathbf{C}$ subject to Eq. (45). This $\mathbf{C}$ is used to aggregate $\mathbf{x}$ into $\mathbf{z}$

$$\mathbf{z}_i = \mathbf{C}\mathbf{x}_i$$

which satisfies the dynamic equation

$$\mathbf{z}_{i+1} = \mathbf{F}\mathbf{z}_i + \mathbf{g}_i \tag{46}$$

where

$$\mathbf{g}_i \triangleq \mathbf{C}\mathbf{f}_i$$

and

$$\mathbf{F} \triangleq \mathbf{C}\mathbf{D}\mathbf{E} : l \times l \text{ matrix} \tag{47}$$

From the discussions in Sec. 5-7-2, it is seen that the matrix $\mathbf{F}$ can be identified exactly in the absence of random disturbances.

Having identified $\mathbf{F}$ exactly, Eq. (47) must be solved for $\mathbf{D}$ and $\mathbf{E}$ subject to Eq. (45). Choosing $\mathbf{C}$ and $\mathbf{D}$ so that $\mathbf{CD}$ is nonsingular yields

$$\mathbf{E} = (\mathbf{CD})^{-1} \mathbf{F}$$

and

$$\mathbf{A} = \mathbf{D}(\mathbf{CD})^{-1} \mathbf{F}\mathbf{C}$$

As will be shown in Sec. 5-8, there is some intuitive appeal for the choice of $\mathbf{C}$ and $\mathbf{D}$ subject to

$$\mathbf{CD} = \mathbf{I}_l \tag{48}$$

Then from Eq. (47) it is seen that $\mathbf{E}$ is identified directly as $\mathbf{F}$. For example, from Appendixes C and D, the matrices $\mathbf{C}$ and $\mathbf{D}$ can be represented as

$$\mathbf{C} = \mathbf{C}_2(-\mathbf{Q}_2\mathbf{Q}_1^{-1}, \mathbf{I}_l)$$

and

$$\mathbf{D} = \begin{pmatrix} -\mathbf{P}_1^{-1} & \mathbf{P}_2 \\ \mathbf{I}_l & \end{pmatrix} \mathbf{D}_2$$

where $\mathbf{C}_2$ and $\mathbf{D}_2$ are $l \times l$ nonsimilar matrices. If the matrix $\mathbf{D}_2$ is given by

$$\mathbf{D}_2 = (\mathbf{Q}_2\mathbf{Q}_1^{-1}\mathbf{P}_1^{-1}\mathbf{P}_2 + \mathbf{I}_l)^{-1}\, \mathbf{C}_2^{-1}$$

then Eq. (48) is satisfied.

Example 5-4

Consider a 3×3 matrix $\mathbf{A}$ constrained by $\mathbf{AQ} = 0$ where $\mathbf{Q}' = (1, 1, 1)$. Then $\mathbf{A} = \mathbf{A}_u\mathbf{A}_k$ where $\mathbf{A}_u = (a\rangle,\ b\rangle)$ where $\langle a = (a_1, a_2, a_3)$, $\langle b = (b_1, b_2, b_3)$ and where

$$\mathbf{A}_k = \begin{pmatrix} -1 & 1 & 0 \\ -1 & 0 & 1 \end{pmatrix}$$

Then $\mathbf{F}$ is identified exactly by $\mathbf{A}_k\mathbf{A}_u$. The null space of $\mathbf{A}_k$ is spanned by (1, 1, 1). Hence $\mathbf{A}$ is identified exactly from $\mathbf{F}$ if $\sum_i a_i = \sum_i b_i = 0$.

Example 5-5

The system of this example arises from a simplified model of an economic system where $\mathbf{A} = \mathbf{A}_1 + \mathbf{A}_2$ and where $\mathbf{A}_1$ is a known matrix, and $\mathbf{A}_2$ is the matrix containing five unknown parameters and is given by $\mathbf{A}_2 = [\mathrm{O}_{94}, a\rangle\langle 1, 1, 1, 1, 1]$ where O_{94} is a 9×4 null matrix and where $\langle a = (a_1, a_2, a_3, a_4, a_5, a_1, a_2, a_3, a_4)$. Then $\mathbf{A}_2\mathbf{Q} = 0$ where $\mathbf{Q}' = (\mathbf{I}_4, \mathrm{O}_{45})$. Thus $\mathbf{A}_2 = \mathbf{A}_u\mathbf{A}_k$ where $\mathbf{A}_u = a\rangle\langle 1, 1, 1, 1, 1$ and $\mathbf{A}_k = (\mathrm{O}_{54}, \mathbf{I}_5)$. Since $\mathbf{F}$ identifies $a_1 \times a_5$, the identification of $\mathbf{F}$ is enough to identify $\mathbf{A}_u$.

5-8 DISAGGREGATION: PROBLEM OF ESTIMATING STATE VECTORS OF CONSTRAINED DYNAMIC SYSTEMS

5-8-1 Exact Disaggregation for Constrained Systems

In this section we discuss the problem of estimating or reconstructing the state vector of dynamic systems from knowledge of aggregated system state vectors. This is called the problem of *disaggregation*.

The process of disaggregation is very simple and the state vector can be reconstructed exactly from the knowledge of the aggregated system state vector if the transition matrix $\mathbf{A}$ of the dynamic system has a special structure.

For example, given $\mathbf{x}_{i+1} = \mathbf{A}\mathbf{x}_i$ where the transition matrix $\mathbf{A}$ is assumed to be of the form $\mathbf{A} = \mathbf{DEC}$ where $\mathbf{D}$, $\mathbf{E}$, and $\mathbf{C}$ are matrices as given in Eq. (44), consider its aggregated dynamic system of dimension l

$$\mathbf{z}_{i+1} = \mathbf{F}\mathbf{z}_i$$

where

$$\begin{aligned} \mathbf{F} &= \mathbf{CDE} \\ \mathbf{z}_i &= \mathbf{C}\mathbf{x}_i \end{aligned} \tag{49}$$

We assume as in Sec. 5-7-3 that $\mathbf{CD}$ is nonsingular and that $\mathbf{z}_i$ is observed without error.

Then

$$\mathbf{x}_i = \tilde{\mathbf{C}}\mathbf{z}_i \qquad i = 1, 2, \ldots \tag{50}$$

where

$$\tilde{\mathbf{C}} = \mathbf{D}(\mathbf{CD})^{-1} \tag{51}$$

is the exact reconstruction of the state vector of the original dynamic system from that of the aggregated system.

The matrix $\tilde{\mathbf{C}}$ of Eq. (51) is called the disaggregation matrix. This is true since

$$\begin{aligned} \mathbf{x}_i - \tilde{\mathbf{C}}\mathbf{z}_i &= \mathbf{A}\mathbf{x}_{i-1} - \tilde{\mathbf{C}}\mathbf{F}\mathbf{z}_{i-1} \\ &= \mathbf{DEC}\mathbf{x}_{i-1} - \mathbf{D}(\mathbf{CD})^{-1}(\mathbf{CD})\mathbf{E}\mathbf{z}_{i-1} \cdot \\ &= (\mathbf{DEC} - \mathbf{DEC})\,\mathbf{x}_{i-1} \qquad i = 1, 2, \ldots \\ &= 0 \end{aligned}$$

More generally, given the control system

$$\mathbf{x}_{i+1} = \mathbf{A}\mathbf{x}_i + \mathbf{B}\mathbf{u}_i$$

where

$$\mathbf{A} = \mathbf{DEC}$$

consider the aggregated control system

$$\mathbf{z}_{i+1} = \mathbf{F}\mathbf{z}_i + \mathbf{G}\mathbf{u}_i$$

where

$$\begin{aligned} \mathbf{F} &= \mathbf{CDE} \\ \mathbf{G} &= \mathbf{CB} \end{aligned}$$

Then $\mathbf{x}_i$ is reconstructed exactly by

$$\mathbf{x}_i = \tilde{\mathbf{C}}\mathbf{z}_i + (\mathbf{I} - \tilde{\mathbf{C}}\mathbf{C})\,\mathbf{B}\mathbf{u}_{i-1} \qquad i = 1, 2, \ldots \tag{50a}$$

where $\tilde{\mathbf{C}}$ is given by Eq. (51).

Thus, for the class of systems satisfying the constraint given by Eq. (44), Eq. (50) or Eq. (50a) is the equation for the exact disaggregation. Note that $\mathbf{x}_0$ need not be known.

5-8-2 Disaggregation for General Systems†

We next consider a more general problem of reconstructing $\mathbf{x}_i$ of the control system governed by

$$\mathbf{x}_{i+1} = \mathbf{A}\mathbf{x}_i + \mathbf{B}\mathbf{u}_i \qquad i = 0, 1, \ldots \tag{52}$$

from the knowledge of the aggregated state vector given by

$$\mathbf{z}_i = \mathbf{C}\mathbf{x}_i \qquad i = 0, 1, \ldots \tag{53}$$

The matrix $\mathbf{A}$ is no longer assumed to be of the form given by Eq. (44). Suppose that $\mathbf{z}_j$, $j \leqslant k$, is the only available measurement on the dynamic system (51) to estimate $\mathbf{x}_k$. Since

$$\mathbf{x}_k = \phi_{k,j}\mathbf{x}_j + \sum_{l=j}^{k-1} \psi_{k,l}\mathbf{u}_l$$

where

$$\phi_{k,j} = \mathbf{A}^{k-j}$$

$$\psi_{k,l} = \mathbf{A}^{k-l-1}\mathbf{B}$$

we have

$$\mathbf{x}_j = \phi_{k,j}^{+}\left(\mathbf{x}_k - \sum_{l=j}^{k-1} \psi_{k,l}\mathbf{u}_l\right) + \mathbf{n}_j \tag{54}$$

where $\mathbf{n}_j$ is a vector in the null space of $\phi_{k,j}$, $\mathcal{N}(\phi_{k,j})$. From Eqs. (53) and (54),

$$\mathbf{z}_j = \mathbf{C}\phi_{k,j}^{+}\left(\mathbf{x}_k - \sum_{l=j}^{k-1} \psi_{k,l}\mathbf{u}_l\right) + \mathbf{C}\mathbf{n}_j$$

Let $\mathbf{P}_{k,j}$ be the orthogonal projection such that

$$\mathbf{P}_{k,j}\mathbf{z}_j = \mathbf{P}_{k,j}\mathbf{C}\phi_{k,j}^{+}\mathbf{x}_k \tag{55}$$

In other words, the null space of $\mathbf{P}_{k,j}$ is given by

$$\mathcal{N}(\mathbf{P}_{k,j}) = \mathbf{C}\mathcal{N}(\phi_{k,j}) + \mathcal{R}(\mathbf{C}\psi\mathbf{U})$$

† See Refs. 33 and 51 for related material.

where

$$\mathbf{U} \triangleq \mathbf{R}\left(\sum_{l=j}^{k-1} \psi_{k,l}\mathbf{u}_l\right)$$

and where

$$\psi = \phi_{k,j}^{+}$$

or dropping subscripts on $\mathbf{P}_{k,j}$,

$$\mathcal{N}(\mathbf{P}) = \mathcal{R}(\mathbf{CV}) + \mathcal{R}(\mathbf{C}\psi\mathbf{U})$$

where $\mathbf{V}$ is a matrix such that

$$\mathcal{R}(\mathbf{V}) = \mathcal{N}(\phi_{k,j})$$

From Eq. (55),

$$\mathbf{x}_k = \hat{\mathbf{x}}_k + \mathbf{L}_k \tag{56a}$$

where

$$\hat{\mathbf{x}}_k = (\mathbf{PC}\phi_{k,j}^{+})^{+}\mathbf{P}\mathbf{z}_j \in \mathcal{N}(\mathbf{PC}\phi_{k,j}^{+})^{\perp} \tag{56b}$$

and where $(\ )^{\perp}$ denotes the orthogonal complement, and $\mathbf{L}_k$ is a subspace of $\mathbf{R}^n$ given by

$$\mathbf{L}_k = \mathcal{N}(\mathbf{PC}\phi_{k,j}^{+}) \tag{56c}$$

Thus, $\mathbf{x}_k$ can be disaggregated exactly if $\mathbf{x}_k$ is in the subspace $\mathbf{L}_k^{\perp}$. When $\mathbf{x}_k$ is not in $\mathbf{L}_k^{\perp}$, only the component in $\mathbf{L}_k^{\perp}$ can be reconstructed from knowing only $\mathbf{z}_j$. The error vector of disaggregation lies in $\mathbf{L}_k$.

We next characterize the subspace $\mathbf{L}_k$, and then examine the effects of knowing $\mathbf{z}_{j_1}, \mathbf{z}_{j_2}, \ldots, \mathbf{z}_{j_k}$

$$j_1 < j_2 < \cdots < j_k \leqslant k$$

on $\mathbf{L}_k$.

PROPOSITION 1:

When only the observation $\mathbf{z}_j$ is available, then the state vector $\mathbf{x}_i$ of the dynamic system, $j \leqslant i$, is estimated as

$$\mathbf{x}_i \in \hat{\mathbf{x}}_i(\mathbf{z}_j) + \mathbf{L}_i(\mathbf{z}_j)^{\dagger}$$

where

$$\hat{\mathbf{x}}_i(\mathbf{z}_j) = (\mathbf{P}_{i,j}\mathbf{C}\phi_{i,j}^{+})^{+}\mathbf{P}_{i,j}\mathbf{z}_j$$

where

$$\mathbf{L}_i(\mathbf{z}_j) \supset \mathcal{N}(\mathbf{C}\phi_{i,j}) + \mathcal{R}\left(\sum_{l=j}^{i-1} \mathbf{A}^{i-1-l}\mathbf{B}\right)$$

and where $\mathbf{P}_{i,j}$ is the orthogonal projection defined earlier as in Eq. (55).

† It can also be shown that $\mathbf{x}_i \in \hat{\mathbf{x}}_i(\mathbf{z}_j) + \phi_{i,j}\mathcal{N}(\mathbf{C})$.

Proof: From Eq. (56*b*)

$$\mathbf{L}_k{}^{\perp} = \mathcal{N}(\mathbf{PC}\psi)^{\perp}$$

where

$$\psi = \phi_{k,j}^{+}$$

Now

$$\mathcal{N}(\mathbf{PC}\psi)^{\perp} = (\mathbf{C}\psi)^{\mathrm{T}}\,\mathcal{N}(\mathbf{P})^{\perp}$$

Since

$$\mathcal{N}(\mathbf{P})^{\perp} = \mathcal{N}(\mathbf{V}^{\mathrm{T}}\mathbf{C}^{\mathrm{T}}) \cap \mathcal{N}(\mathbf{U}^{\mathrm{T}}\psi^{\mathrm{T}}\mathbf{C}^{\mathrm{T}})$$

we have

$$\begin{aligned}\mathcal{N}(\mathbf{PC}\psi)^{\perp} &= (\mathbf{C}\psi)^{\mathrm{T}}\,[\mathcal{N}(\mathbf{V}^{\mathrm{T}}\mathbf{C}^{\mathrm{T}}) \cap \mathcal{N}(\mathbf{U}^{\mathrm{T}}\psi^{\mathrm{T}}\mathbf{C}^{\mathrm{T}})]\\ &= \psi^{\mathrm{T}}[\mathcal{R}(\mathbf{C}^{\mathrm{T}}) \cap \mathcal{N}(\mathbf{V}^{\mathrm{T}}) \cap \mathcal{N}(\mathbf{U}^{\mathrm{T}}\psi)^{\mathrm{T}}]\\ &\subset \psi^{\mathrm{T}}[\mathcal{R}(\mathbf{C}^{\mathrm{T}}) \cap \mathcal{R}(\phi_{k,j}^{\mathrm{T}})] \cap \mathcal{R}(\psi^{\mathrm{T}}) \cap \mathcal{N}(\mathbf{U}^{\mathrm{T}})\\ &= \psi^{\mathrm{T}}[\mathcal{R}(\mathbf{C}^{\mathrm{T}}) \cap \mathcal{R}(\phi_{k,j}^{\mathrm{T}})] \cap \mathcal{R}(\mathbf{U})^{\perp}\end{aligned}$$

where we use an identity

$$\mathbf{S}\mathcal{N}(\mathbf{TS}) = \mathcal{R}(\mathbf{S}) \cap \mathcal{N}(\mathbf{T})$$

for any matrices **S** and **T** such that **TS** is defined, where

$$\begin{aligned}\mathcal{R}(\mathbf{U}) &= \sum_{l=j}^{k-1} \mathcal{R}(\mathbf{A}^{k-l-1}\mathbf{B})\\ &= \mathbf{A}^{k-1-j}\mathcal{B} + \cdots + \mathcal{B}\\ \mathcal{B} &= \mathcal{R}(\mathbf{B})\end{aligned}$$

or

$$\begin{aligned}\mathcal{N}(\mathbf{PC}\psi) &\supset \mathcal{N}(\mathbf{C}\psi) + \mathcal{N}(\phi_{k,j}\psi) + \mathcal{R}(\mathbf{U})\\ &= \mathcal{N}(\mathbf{C}\psi) + \mathcal{R}(\mathbf{U})\end{aligned}$$

since

$$\mathcal{N}(\mathbf{A}^{+}) = \mathcal{N}(\mathbf{A}^{\mathrm{T}}) \subset \mathcal{N}(\mathbf{CA}^{+})$$

Proceeding in a similar manner we obtain:

PROPOSITION 2:

Given two observations $\mathbf{z}_j$ and $\mathbf{z}_k$, $j \leqslant k \leqslant i$, $\mathbf{x}_i$ can be reconstructed as

$$\mathbf{x}_i \in \hat{\mathbf{x}}_i(\mathbf{z}_j, \mathbf{z}_k) + \mathbf{L}_i(\mathbf{z}_j, \mathbf{z}_k)$$

where

$$\hat{\mathbf{x}}_i(\mathbf{z}_j, \mathbf{z}_k) = \begin{bmatrix} \mathbf{P}_{i,j}\mathbf{C}\phi_{i,j}^+ \\ \mathbf{P}_{i,k}\mathbf{C}\phi_{i,k}^+ \end{bmatrix}^+ \begin{pmatrix} \mathbf{P}_{i,j}\mathbf{z}_j \\ \mathbf{P}_{i,k}\mathbf{z}_k \end{pmatrix}$$

$$\mathbf{L}_i(\mathbf{z}_j, \mathbf{z}_k) = \mathcal{N}(\mathbf{P}_{i,j}\mathbf{C}\phi_{i,j}^+) \cap \mathcal{N}(\mathbf{P}_{i,k}\mathbf{C}\phi_{i,k}^+)$$

Using the results due to Ben-Israel and Katz,[11] we have

PROPOSITION 3:

$$\hat{\mathbf{x}}_i(\mathbf{z}_j, \mathbf{z}_k) = \hat{\mathbf{x}}_i(\mathbf{z}_j) + \mathbf{P}_j(\mathbf{P}_j + \mathbf{P}_k)^+ (\hat{\mathbf{x}}_i(\mathbf{z}_k) - \hat{\mathbf{x}}_i(\mathbf{z}_j))$$

where $\mathbf{P}_j$ is the orthogonal projection operator on $\mathcal{N}(\mathbf{P}_{i,j}\mathbf{C}\phi_{i,j}^+)$ and $\mathbf{P}_k$ is the orthogonal projection operator on $\mathcal{N}(\mathbf{P}_{i,k}\mathbf{C}\phi_{i,k}^+)$.

Proof: Apply their arguments to

$$\mathbf{P}_{i,j}\mathbf{C}\phi_{i,j}^+\mathbf{x}_i = \mathbf{P}_{i,j}\mathbf{z}_j$$

$$\mathbf{P}_{i,k}\mathbf{C}\phi_{i,k}^+\mathbf{x}_i = \mathbf{P}_{i,k}\mathbf{z}_k$$

since this set of equations possesses a common solution. Note that

$$\mathbf{P}_j = \mathbf{I} - (\mathbf{P}_{i,j}\mathbf{C}\phi_{i,j}^+)^+(\mathbf{P}_{i,j}\mathbf{C}\phi_{i,j}^+)$$

and the similar expression holds for $\mathbf{P}_k$.

By appropriately defining the null space of the orthogonal projection operator, we can derive expressions for $\hat{\mathbf{x}}_i$ given $\mathbf{z}_j$ and $\mathbf{u}_k$, j, $k \leqslant i$, and so on. In other words, the projection operator is chosen to annihilate the components of state vectors that are not available to the observer who knows only the past and current values of the aggregated state vectors.

5-8-3 Approximate Disaggregation

The above treatment in Secs. 5-8-1 and 5-8-2 can be extended to the state vector estimation with noisy measurements. To illustrate, consider the system

$$\mathbf{x}_{i+1} = \mathbf{A}\mathbf{x}_i$$

$$\mathbf{y}_i = \mathbf{H}\mathbf{x}_i + \eta_i$$

where

$$\mathbf{A} = \mathbf{DEC}$$

and where the means and the covariance matrices of $\mathbf{x}_0$ and noises are known. Since $\mathbf{x}_i$ can be reconstructed exactly by Eq. (50), it is not surprising that the Kalman filter estimate $\hat{\mathbf{x}}_i$ can be obtained as

$$\hat{\mathbf{x}}_i = \mathbf{G}\hat{\mathbf{z}}_i \qquad i > 0 \tag{50b}$$

where $\hat{\mathbf{z}}_i$ is the Kalman filter estimate of

$$\mathbf{z}_{i+1} = \mathbf{F}\mathbf{z}_i$$

$$\mathbf{y}_i = \mathbf{H}\mathbf{G}\mathbf{z}_i + \eta_i$$

provided only $\hat{\mathbf{x}}_0$ is generated from the original system by the optimal Kalman filter and $\hat{\mathbf{z}}_0 = \mathbf{C}\hat{\mathbf{x}}_0$ is used. It is quite straightforward to verify that the estimation error covariance matrices thus generated are equal to those associated with the optimal Kalman filter.

The computational advantage of the estimation scheme given by Eq. (50) or Eq. (50*b*) is obvious since the estimate can be propagated using the dynamics of the aggregated system, i.e., as $\mathbf{x}_{i+1} = \mathbf{G}\mathbf{F}\mathbf{z}_i$ rather than the original system.

Next consider the problem where **A** is only approximately represented by Eq. (44).

The matrices **D** and **C** in Eq. (44) are to be chosen to minimize the identification error in some appropriate sense, subject to the constraint equation so that the estimate of **A**, $\hat{\mathbf{A}}$, becomes a member of the class $\hat{\mathbf{A}}\mathbf{Q} = 0_{n,k}$, $\mathbf{P}\hat{\mathbf{A}} = 0_{k,n}$ where $\hat{\mathbf{A}} = \mathbf{DEC}$. This implies that the aggregation operation $\mathbf{z}_i = \mathbf{C}\mathbf{x}_i$ and the estimator of the form $\hat{\mathbf{x}}_i = \mathbf{D}\mathbf{z}_i$ are being used when **A** does not necessarily satisfy the structural constraint (45).

Practical significance of the above procedure is that a suboptimal estimate of $\mathbf{x}_j$ is obtained by (1) aggregating $\mathbf{x}_j$ into $\mathbf{z}_j$, (2) advancing time $\mathbf{z}_{j+1} = \mathbf{F}\mathbf{z}_j$, and (3) estimating $\mathbf{x}_{j+1}$ by $\hat{\mathbf{x}}_{j+1} = \mathbf{D}\mathbf{z}_{j+1} = \mathbf{DFC}\mathbf{x}_j$ where $\mathbf{PD} = 0_{k,l}$, $\mathbf{CQ} = 0_{l,k}$. This estimate is equivalent to that obtained by estimating $\hat{\mathbf{x}}_{j+1}$ by $\hat{\mathbf{x}}_j$ from $\hat{\mathbf{x}}_{j+1} = \hat{\mathbf{A}}\hat{\mathbf{x}}_j$ subject to the constraint on $\hat{\mathbf{A}}$ given by $\mathbf{P}\hat{\mathbf{A}} = 0$, $\hat{\mathbf{A}}\mathbf{Q} = 0$. Thus for **A** with certain structural properties, the aggregation and disaggregation operations can be performed with no loss of information and with a saving in the amount of computation. For other **A**'s, the operations involve certain loss of information which may be tolerable in view of the lesser computational requirements.

The type of estimation scheme indicated so far can be carried over to systems with forcing terms and/or with additive random disturbances in the plant and the observation equations. This, however, is not carried out in this chapter.

5-9 OTHER APPLICATIONS

Very often one is faced with the problem of inverting a matrix of large dimension, of evaluating its determinant and/or its characteristic values. It is known that the numbers of multiplications in inverting matrices grow as a nonlinear function of the dimensions of the matrices being inverted. Therefore, for large n, the matrix inversion is a nontrivial operation even with a modern digital computer. Similar comments can be made about evaluating determinants or characteristic values. For some special classes of matrices, however, the dimensions of matrices to be inverted can be reduced and similarly when the determinant must be evaluated.

A well-known example of such transformations is the following involving a matrix of rank 1, that is, a dyadic matrix.

$$\det(\mathbf{I}_n + \mathbf{a}\rangle\langle\mathbf{b}) = 1 + \langle\mathbf{b}, \mathbf{a}\rangle$$

where $\mathbf{a}\rangle$ and $\langle\mathbf{b}$ are n-dimensional column and row vectors, respectively.

More generally, an $n \times n$ matrix $\mathbf{S}$ with rank l, $l < n$, has a rank factorization $\mathbf{S} = \mathbf{BC}$ and

$$\det(\mathbf{A} - \mathbf{BC}) = \det \mathbf{A} \cdot \det(\mathbf{I}_l - \mathbf{CA}^{-1}\mathbf{B})$$

where $\mathbf{A}$ is a nonsingular $n \times n$ matrix, and $\mathbf{B}$ and $\mathbf{C}$ are respectively $n \times l$ and $l \times n$ matrices. Another well-known identity is

$$(\mathbf{A} - \mathbf{BC})^{-1} = \mathbf{A}^{-1} + \mathbf{A}^{-1}\mathbf{B}(\mathbf{I}_l - \mathbf{CA}^{-1}\mathbf{B})^{-1}\mathbf{CA}^{-1} \tag{57}$$

These and other matrix identity relations, useful in dealing with matrices of high dimension, can be regarded as examples of the aggregation concept as applied to matrix theory. Consider a special case where the aggregation matrix $\mathbf{C}$ is such that each column vector has one entry of unity. This is the type of aggregation used in input-output economic models.[20,21,41] Then, as discussed in Sec. 5-3, n components of the state vector $\mathbf{x}$ are grouped into l mutually exclusive subsets. Consider, therefore, a decomposition of $\mathbf{N} = \{1, 2, \ldots, n\}$ into

$$\mathbf{N} = \mathbf{N}_1 \cup \mathbf{N}_2 \cup \cdots \cup \mathbf{N}_l$$

where $\mathbf{N}_i$ is a subset of $\mathbf{N}$ with n_i elements such that $\mathbf{N}_i \cap \mathbf{N}_j = \phi$ and

$$\sum_{i=1}^{l} \mathbf{n}_i = \mathbf{n}$$

Then consider an aggregation matrix $\mathbf{C}$ with elements

$$\mathbf{C}_{ij} = \begin{cases} 1 & \text{if } j \in \mathbf{N}_i \\ 0 & \text{otherwise} \end{cases} \tag{58}$$

Consider a class of matrices

$$\mathscr{S} = \{\mathbf{S};\ \mathbf{S} = \mathbf{C}'\mathbf{FC} \text{ where } \mathbf{F} \text{ is } l \times l \text{ and where } \mathbf{C} \text{ is specified by Eq. (58)}\}$$

Then from Eq. (57)

$$(\mathbf{A} - \mathbf{S})^{-1} = [\mathbf{A} - \mathbf{C}'\mathbf{F}\mathbf{C}]^{-1}$$
$$= \mathbf{A}^{-1} + \mathbf{A}^{-1}\hat{\mathbf{S}}\mathbf{A}^{-1} \tag{59}$$

where

$$\hat{\mathbf{S}} \triangleq \mathbf{C}'\hat{\mathbf{F}}\mathbf{C} \in \mathbf{S}$$
$$\hat{\mathbf{F}} \triangleq \mathbf{F}(\mathbf{I}_l - \mathbf{C}\mathbf{A}^{-1}\mathbf{C}'\mathbf{F})^{-1}$$

Since the aggregation of $\mathbf{S}$, $\tilde{\mathbf{S}}$, is defined as

$$\tilde{\mathbf{S}} = \mathbf{C}\mathbf{S}\mathbf{C}'(\mathbf{C}\mathbf{C}')^{-1} = (\mathbf{C}\mathbf{C}')\,\mathbf{F}$$

and since

$$(\tilde{\mathbf{A}}^{-1}) = \mathbf{C}\mathbf{A}^{-1}\mathbf{C}'(\mathbf{C}\mathbf{C}')^{-1}$$

then

$$\hat{\mathbf{F}} = \mathbf{F}[\mathbf{I}_l - (\tilde{\mathbf{A}}^{-1})(\mathbf{C}\mathbf{C}')\,\mathbf{F}]^{-1}$$
$$= (\mathbf{C}\mathbf{C}')^{-1}\,\tilde{\mathbf{S}}[\mathbf{I}_l - (\tilde{\mathbf{A}}^{-1})\,\tilde{\mathbf{S}}]^{-1}$$

Similar but more complex identities are possible for more general aggregation matrices and their associated classes of matrices.

A possible application of Eq. (59) is to invert a matrix $[\mathbf{I} - \mathbf{T}]$ when $[\mathbf{I} - \mathbf{T}]$ is ill-conditioned[17] by rewriting it first as

$$[\mathbf{I} - \mathbf{S} - (\mathbf{T} - \mathbf{S})] \qquad \text{for some} \quad \mathbf{S} \in \mathbf{S}$$

if $\mathbf{S}$ can be chosen to make $\| \mathbf{T} - \mathbf{S} \|$ small. Even though $[\mathbf{I} - \mathbf{S}]$ may still be ill-conditioned, the dimension of the ill-conditioned matrix to be inverted is now lowered.

APPENDIX A

FACTORIZATION OF A

Assume without loss of generality that

$$\text{Rank } \mathbf{Q} = k$$

i.e., the column vectors of $\mathbf{Q}$, $\mathbf{q}_1, \ldots, \mathbf{q}_k$ are linearly independent. Writing

$$\mathbf{A} = \begin{bmatrix} \langle \mathbf{a}_1 \\ \vdots \\ \langle \mathbf{a}_n \end{bmatrix}$$

Eq. (35) implies that the vectors $\mathbf{a}_1, \ldots, \mathbf{a}_n$ are in a subspace perpendicular to that spanned by $\mathbf{q}_1, \ldots, \mathbf{q}_k$. Taking a basis of this subspace to be $\mathbf{q}_{k+1}, \ldots, \mathbf{q}_n$

$$\mathbf{A} = \mathbf{A}_u \mathbf{A}_k$$

where $\mathbf{A}_u$ is an unknown $n \times l$ matrix, and where

$$\mathbf{A}_k = \begin{bmatrix} \langle \mathbf{q}_{k+1} \\ \vdots \\ \langle \mathbf{q}_n \end{bmatrix}$$

is known. More specifically, write

$$\mathbf{Q}' = [\mathbf{Q}_1', \mathbf{Q}_2']$$

where $\mathbf{Q}_2$ is $k \times k$ nonsingular (after suitable permutations if necessary). Then we can take $\mathbf{A}_k$ to be

$$\mathbf{A}_k = [\mathbf{I}_l, -\mathbf{Q}_1 \mathbf{Q}_2^{-1}]$$

APPENDIX B

MINIMIZATION OF $\mathbf{J_N}$

Given

$$\mathbf{J}_N = \mathrm{tr}\{-\mathbf{A}\mathbf{W}_{0,N-1}^T - \mathbf{W}_{0,N-1}\mathbf{A}^T + \mathbf{A}\mathbf{Z}_{0,N-1}\mathbf{A}^T\} + \text{const}$$

where $\mathbf{w}_i = \mathbf{x}_{i+1} - \mathbf{f}_i$, it is desired to minimize $\mathbf{J}_N$ with respect to $\mathbf{A}_u$. The necessary condition for optimal $\mathbf{A}_u$ is given by

$$0 = \mathrm{tr}\{-\Delta\mathbf{A}_k\mathbf{W}_{0,N-1}^T - \mathbf{W}_{0,N-1}\mathbf{A}_k{}^T\Delta^T + \Delta\mathbf{A}_k\mathbf{Z}_{0,N-1}\mathbf{A}_k{}^T\hat{\mathbf{A}}_u + \hat{\mathbf{A}}_u\mathbf{A}_k\mathbf{Z}_{0,N-1}\hat{\mathbf{A}}_k{}^T\Delta^T\} \tag{B.1}$$

for arbitrary $n \times l$ matrix Δ, or

$$\hat{\mathbf{A}}_u\mathbf{A}_k\mathbf{Z}_{0,N-1}\mathbf{A}_k{}^T = \mathbf{W}_{0,N-1}\mathbf{A}_k{}^T \tag{B.2}$$

The same equation, Eq. (B.2), results if $\mathbf{J}_N$ is given by

$$\mathbf{J}_N = \sum_{0}^{N-1} (\mathbf{w}_i - \mathbf{A}\mathbf{x}_i)^T\,\mathbf{Q}(\mathbf{w}_i - \mathbf{A}\mathbf{x}_i) \qquad \text{where} \quad \mathbf{Q} > 0$$

APPENDIX C

Without loss of generality we can assume that rank

$$\text{Rank}\,\mathbf{Q} = \text{Rank}\,\mathbf{P} = k = n - l$$

Choose

$$\mathbf{C}' = [\mathbf{c}_1, \mathbf{c}_2, \ldots, \mathbf{c}_l] \qquad \text{and} \qquad \mathbf{D} = [\mathbf{d}_1, \ldots, \mathbf{d}_l]$$

such that the subspace spanned by $\mathbf{C}_i$, $1 \leqslant i \leqslant l$, is orthogonal to that of the column space of $\mathbf{Q}$ and that the subspace spanned by $\mathbf{d}_i$, $1 \leqslant i \leqslant l$, is orthogonal to the row space of $\mathbf{P}$. Then

$$\mathbf{A} = \mathbf{DEC}$$

where $\mathbf{C}$ and $\mathbf{D}$ satisfy Eq. (45).

The following representation of $\mathbf{A}$ satisfying the constraint equations in Eq. (43) is shown in a paper by Fischer[22] and included here for the sake of completeness.

Write $\mathbf{A}$ as

$$\begin{pmatrix} \mathbf{A}_{11} & \mathbf{A}_{12} \\ \mathbf{A}_{21} & \mathbf{A}_{22} \end{pmatrix}$$

where $\mathbf{A}_{11}$ is a $k \times k$ and $\mathbf{A}_{22}$ is a $l \times l$ matrix. By suitable permutations one can write

$$\mathbf{P} = (\mathbf{P}_1 \,\vdots\, \mathbf{P}_2)$$

where $\mathbf{P}_1$ is a $k \times k$ and $\mathbf{P}_2$ is a $k \times l$ matrix and where $\mathbf{P}_1$ is non-singular. Similarly, write $\mathbf{Q}$ as

$$\mathbf{Q} = \left(\frac{\mathbf{Q}_1}{\mathbf{Q}_2}\right)$$

where $\mathbf{Q}_1$ is a $k \times k$ and $\mathbf{Q}_2$ is an $l \times k$ matrix and where $\mathbf{Q}_1$ is non-singular. The constraint equations when written out give

$$\begin{aligned} \mathbf{P}_1\mathbf{A}_{11} + \mathbf{P}_2\mathbf{A}_{21} &= 0 \\ \mathbf{P}_1\mathbf{A}_{12} + \mathbf{P}_2\mathbf{A}_{22} &= 0 \\ \mathbf{A}_{11}\mathbf{Q}_1 + \mathbf{A}_{12}\mathbf{Q}_2 &= 0 \\ \mathbf{A}_{21}\mathbf{Q}_1 + \mathbf{A}_{22}\mathbf{Q}_2 &= 0 \end{aligned}$$

Solving for submatrices of $\mathbf{A}$,

$$\mathbf{A} = \begin{bmatrix} \mathbf{P}_1^{-1}\mathbf{P}_2\mathbf{A}_{22}\mathbf{Q}_2\mathbf{Q}_1^{-1} & -\mathbf{P}_1^{-1}\mathbf{P}_2\mathbf{A}_{22} \\ -\mathbf{A}_{22}\mathbf{Q}_2\mathbf{Q}_1^{-1} & \mathbf{A}_{22} \end{bmatrix}$$

$$= \begin{bmatrix} \mathbf{P}_1^{-1} & \mathbf{P}_2 \\ -\mathbf{I}_l & \end{bmatrix} \mathbf{A}_{22}(\mathbf{Q}_2\mathbf{Q}_1^{-1}, -\mathbf{I}_l)$$

which is in a form given by Eq. (44) upon identifying $\mathbf{D}$ with

$$\begin{bmatrix} \mathbf{P}_1^{-1} & \mathbf{P}_2 \\ -\mathbf{I}_l & \end{bmatrix}$$

$\mathbf{F}$ with $\mathbf{A}_{22}$ and $\mathbf{C}$ with $(\mathbf{Q}_2\mathbf{Q}_{-}^{1}, -\mathbf{I}_l)$.

The fact that $\mathbf{A}$ of Eq. (44) satisfies the constraints of Eq. (43) is clear from the matrices $\mathbf{D}$ and $\mathbf{C}$.

APPENDIX D

EXISTENCE OF C AND D MATRICES

Write

$$\mathbf{C} = (\mathbf{C}_1 \vdots \mathbf{C}_2)$$

where $\mathbf{C}_1$ and $\mathbf{C}_2$ are $l \times k$ and $l \times l$ matrices respectively, and where $\mathbf{C}_2$ is chosen to be nonsingular.

From Eq. (45) and Appendix C,

$$\mathbf{C}_1 = -\mathbf{C}_2\mathbf{Q}_2\mathbf{Q}_1^{-1} \tag{D.1}$$

Write

$$\mathbf{D} = \begin{pmatrix} \mathbf{D}_1 \\ \hline \mathbf{D}_2 \end{pmatrix}$$

where $\mathbf{D}_1$ and $\mathbf{D}_2$ are $k \times l$ and $l \times l$ matrices respectively. From the constraint Eq. (45)

$$\mathbf{P}_1\mathbf{D}_1 + \mathbf{P}_2\mathbf{D}_2 = 0$$

or

$$\mathbf{D}_1 = -\mathbf{P}_1^{-1}\mathbf{P}_2\mathbf{D}_2 \tag{D.2}$$

since $\mathbf{P}_1$ is chosen to be nonsingular.

REFERENCES:

1. Aoki, M.: Control of Large Scale Dynamic Systems by Aggregation, *IEEE Trans. Autom. Control*, vol. AC-13, no. 3, pp. 246–253, June, 1968.
2. Aoki, M.: Control of Dynamic Systems Containing a Small Parameter by Aggregation, *Proc. IFAC Symp. Sensitivity*, 2*d*, Dubrovnik, Yugoslavia, 1968.
3. Aoki, M.: Note on Aggregation and Bounds for the Solution of the Matrix Riccati Equations, *J. Math. Anal. Appl.*, February, 1968.
4. Aoki, M.: "Optimization of Stochastic Systems," Academic, 1967.
5. Aoki, M.: On Identification of Constrained Dynamic Systems with High Dimensions, *Proc. Ann. Allerton Conf. Circuit Syst. Theory*, 5*th*, 1967, pp. 191–200.
6. Aoki, M., and J. R. Huddle: Estimation of State Vector of a Linear Stochastic System with a Constrained Estimation, *IEEE Trans. Autom. Control*, vol. AC-12, no. 4, pp. 432–433, August, 1967.
7. Aoki, M., and P. C. Yue: On Utilization of Structural Information to Improve

Identification Accuracy, *Proc. U. S.–Japan Seminar on Learning Process in Control Systems, Univ. Nagoya,* Nagoya, Japan, 1970.
8. Balakrishnan, A. V.: "Foundation of System Theory and State Space," Engineering 228A class notes, University of California, Los Angeles, fall, 1966.
9. Bass, R. W., and I. Gura: High Order System Design via State-Space Considerations, *Proc. Joint Autom. Control Conf.*, 1965, pp. 311–318.
10. Bellman, R., and R. Kalaba: "Quasilinearization and Nonlinear Boundary Value Problems," RAND Corporation, Santa Monica, Calif., 1965.
11. Ben-Israel, A.: On the Geometry of Subspaces in Euclidean n-spaces, *SIAM J. Appl. Math.*, vol. 15, no. 5, pp. 1184–1198, September, 1967.
12. Bellman, R. E., and W. Karush: On the Maximum Transform and Semi-group of Transformations, *Syst. Devel. Corp. Rept.* SP-719, Santa Monica, Calif., March, 1962.
13. Brasch, F. M., Jr., and J. B. Pearson: Pole Placement Using Dynamic Compensators, *IEEE Trans. Autom. Control*, vol. AC-15, no. 1, pp. 34–43, Feb., 1970.
14. Bongiorno, J. J., Jr., and D. C. Youla: On Observers in Multivariable Control Systems, *Polytechnic Inst. Brooklyn Rept.* P1BMR1-1383-67, October, 1967.
15. Butman, S.: A Method for Optimizing Control for Costs in Discrete Systems with Linear Controllers, *Proc. Hawaii Intern. Conf.*, University of Hawaii, 1968.
16. Chao, Huang: Uncoupling Method for Diagonal Band Matrix Equation, *J. Eng. Mech. Div., Am. Soc. Civil Engrs.*, vol. 93, pp. 138–147, August, 1967.
17. Davison, E. J.: Method for Simplifying Linear Dynamic Systems, *IEEE Trans. Autom. Control*, vol. AC-11, pp. 93–101, January, 1966.
18. Dvoretzky, A.: On Stochastic Approximations, *Proc. Symp. Math. Statist. Probability*, 3*d*, Berkeley, col. 1, pp. 39–55, 1956.
19. Ellis, J. K., and G. W. T. White: An Introduction to Modal Analysis and Control, *Control*, April, May, June, 1965, p. 193.
20. Fei, John Ching-Han: A Fundamental Theorem for the Aggregation Problem of Input-Output Analysis, *Econometrica*, vol. 24, no. 4, pp. 400–412, October, 1956.
21. Fielder, M., and V. Ptàk: On Aggregation in Matrix Theory and Its Application to Numerical Inverting of Large Matrices, *Bull. Acad. Polon. Sci., Ser. Sci. Math., Astr. Phys.*, vol. 11, no. 12, pp. 757–759, 1963.
22. Fischer, W. D.: Optimal Aggregation in Multi-equation Prediction Models, *Econometrica*, vol. 30, no. 4, pp. 744–769, October, 1962.
23. Gantmacher, F. R.: "The Theory of Matrices," vol. I, Chelsea, 1959.
24. Ho, B. L., and R. E. Kalman: *Regelungstechnik*, vol. 14, no. 12, pp. 545–548, 1960.
25. Ho, Y. C., and B. H. Whalen: An Approach to the Identification and Control of Linear Dynamic Systems with Unknown Parameters, *IEEE Trans. Autom. Control*, vol. AC-8, no. 3, pp. 255–256, July, 1963.
26. Ho, Y. C., and R. C. K. Lee: Identification of Linear Dynamic Systems, *Inform. Control*, vol. 8, pp. 93–110, 1965.
27. Huddle, J. R.: "Suboptimal Control of Linear Discrete-time Stochastic Systems using Memory Element, doctoral dissertation", Department of Engineering, University of California, Los Angeles, Calif., 1966.
28. Joseph, P. D.: Suboptimal Linear Filtering, *TRW Space Technol. Labs, Inc., Rept.* 10C 9321.4653, Redondo Beach, Calif., December, 1963.
29. Kalman, R. E.: Contribution to the Theory of Optimal Control, *Bol. Soc. Mat. Mexicana*, pp. 102–119, 1960.
30. Kalman, R. E., and R. S. Bucy: New Results in Linear Filtering and Prediction Theory, *Trans. Am. Soc. Mech. Engrs.*, ser. D, vol. 83, no. 1, pp. 95–108, March, 1961.

31. Larson, R. E.: Optimum Combined Control and Estimation for Partially Controlled Systems, *Proc. Ann. Allerton Conf. Circuit Syst. Theory, 4th*, 1966, pp. 600–610.
32. Lee, R. C. K.: "Optimal Estimation, Identification and Control," M.I.T., 1964.
33. Luenberger, D. G.: "Determining the State of a Linear System with Observer of Low Dynamic Order," doctoral dissertation, Department of Electrical Engineering, Stanford University, Stanford, Calif., 1963.
34. Marcus, M., and H. Minc: "A Survey of Matrix Theory Inequalities," Allyn and Bacon, 1964.
35. Manetsch, T. J.: Transfer Function Representation of the Aggregate Behavior of a Class of Economic Processes, *IEEE Trans. Autom. Control*, vol. AC-4, no. 4, pp. 693–698, October, 1966.
36. Meditch, J. S.: A Class of Suboptimal Linear Controls, *IEEE Trans. Autom. Control*, vol. AC-11, no. 3, pp. 433–439, July, 1966.
37. Milne, R. D.: The Analysis of Weakly Coupled Dynamical Systems, *Intern. J. Control*, vol 2, pp. 171–199, August, 1965.
38. Mitra, D.: "On the Reduction of Complexity of Linear Dynamical Models," Atomic Energy Establishment, Rept. AEEW-R520, Dorset, England, 1967.
39. Rao, C. R.: Information and Accuracy Attainable in the Estimation of Statistical Parameters, *Bull. Calcutta Math. Soc.*, vol. 37, no. 3, pp. 81–91, September, 1945.
40. Rekasius, Z. V.: Optimal Linear Regulators with Incomplete State Feedback, *IEEE Trans. Autom. Control*, vol. AC-12, no. 3, pp. 296–299, June. 1967.
41. Rosenblatt, D.: "On Aggregation and Consolidation in Linear Systems," Statistics Department, Technical Report C, American University, Washington, D.C., August, 1956.
42. Rosenblatt, M.: Functions of Markov Processes, *Z. Wahrscheinlichkeits Verv. Geb.* 5, pp. 232–243, 1966.
43. Rosenbrock, H. H.: Distinctive Problems of Process Control, *Chemical Eng. Progr.*, vol. 58, no. 9, pp. 43–50, 1962.
44. Rynaski, E. G., and R. F. Whitbeck: "Theory and Application of Linear Optimal Control," Cornell Aeronautical Lab., Inc., Department IH-(94)-F-1, Buffalo, N. Y., October, 1965.
45. Sain, M.: On the Control Applications of a Determinant Equality Related to Eigenvalue Computation, *IEEE Trans. Autom. Control*, vol. AC-11, pp. 109–111, January, 1966.
46. Simon, J. D.: Theory and Application of Modal Control, *Case Western Reserve Univ. Syst. Res. Cen. Rept.* SRC 104-A-67-46, 1967.
47. Sworder, D. D.: "Optimal Adaptive Control Systems," Academic, 1966.
48. Tuel, W. G., Jr.: An Improved Algorithm for the Solution of Discrete Regulation Problems, *IEEE Trans. Autom. Control*, vol. AC-12, no. 5, pp. 527–528, October, 1967.
49. Tyler, J. S., and F. B. Tuteur: The Use of a Quadratic Performance Index to Design Multivariable Control Systems, *IEEE Trans. Autom. Control*, vol. AC-11, no. 1, pp. 84–92, January, 1966.
50. Wonham, W. M.: On Pole Assignment in Multi-input Controllable Linear Systems, *IEEE Trans. Autom. Control*, vol. AC-12, no. 6, pp. 660–665, December, 1967.
51. Wonham, W. M.: Dynamic Observers—Geometric Theory, *IEEE Trans. Autom. Control*, vol. AC-15, no. 2, pp. 258–259, April, 1970.
52. Yue, P. C.: "The Identification of Constrained Systems with High Dimension," doctoral dissertation, Department of System Science, University of California, Los Angeles, Calif., 1970.
53. Zadeh, L. A., and C. A. Desoer: "Linear System Theory," McGraw-Hill, 1963.

6

DISTRIBUTED MULTILEVEL SYSTEMS

DAVID A. WISMER

6-1 INTRODUCTION

Previous chapters have indicated a formalism for decomposing high-dimensional optimization problems into several lower-dimensional subsystems and then applying a second-level coordination algorithm to achieve an optimum for the overall system. This theory was developed for the optimization of both static and dynamic systems. Examples were given in Chapters 1 and 4 to illustrate two particular coordination schemes called model coordination and goal coordination and a gradient-type algorithm was presented to solve the second-level problem.

It seems clear that the efficient application of these or any other coordination methods depends upon the following two properties of the system decomposition: (1) each subsystem optimization problem must converge rather quickly for a wide range of initial estimates (assuming that an iterative technique is employed), and (2) the coupling constraints for each subsystem should be formulated such that the second-level coordination algorithm converges rapidly. If one selects at random fairly low-order systems on which to test his knowledge of multilevel methods, it will become apparent that the assumptions of Chapters 1 and 4 may not always hold. In particular, the subsystems in goal coordination may not possess minima with respect to the coordinating variables but rather saddle points or maxima which are unacceptable. In model coordination sufficient degrees of freedom may not be available in each subsystem to satisfy the additional constraints which are imposed. Hence, an alternative approach is to seek a broad class of high-dimensional optimization problems for which multilevel optimization is particularly well suited.

This chapter reviews the problems of static and dynamic optimization when the model equality constraints are difference equations or differential-difference equations arising from the discretization of partial differential equations. It will be shown that a wide class of these problems is particularly well suited to solution by multilevel methods. The dimensionality of these systems is proportional to the number of mesh points in the grid used in forming the difference approximations and this number approaches infinity as the grid is refined for increased accuracy. Thus there is no question that the magnitude of the computational problem is large.

Certainly other methods have been investigated for the optimization of distributed parameter systems which do not require discretization of the partial differential equations. These methods are well documented and will not be reviewed here except to say that they are generally limited to a very restricted class of partial differential equations (e.g., linear equations with quadratic cost functionals). In addition to their limited application, such methods also impose a sizable computational burden in order to obtain numerical solutions. However, if the computations for the discrete approach can be done efficiently, a more general theory and a large amount of computational experience can be employed. In this way, it is expected that solutions can be obtained for nonlinear systems, time- and space-varying systems, and systems with nonhomogeneous boundary conditions. Key questions, of course, are whether or not (1) the approximate problem is *consistent* with the continuous problem, (2) the approximate solution is *convergent* to the continuous solution, and (3) the solution procedure is *stable*. These questions are treated in the next section.

6-2 DISCRETE AND SEMIDISCRETE PROBLEM FORMULATION

One important reason for using a discrete or semidiscrete approach in the solution of optimization problems involving partial differential equations is that the continuous problem may not be capable of solution given the present state of theory and computation. Thus we wish to show that virtually any optimization problem involving partial differential equations can be posed as a tractable problem using some lumped approximation. For reasons of clarity, however, it would be unwise to present the discussion of multilevel optimization in such great generality. Thus a fairly general class of problems will be posed in this section and we will show how these problems can be approximated in various ways to make them tractable for multilevel optimization methods. Having dispensed with this aspect of the problem formulation, a somewhat simpler system will then be employed.

Consider now the system equation

$$\frac{\gamma \mathbf{u}}{\gamma t}(\mathbf{X}, t) = \mathscr{G}(\mathbf{u}(\mathbf{X}, t), \mathbf{m}(\mathbf{X}, t), \mathbf{X}, t) \qquad \mathbf{X} \in \Omega \quad t \geqslant t_0 \tag{1}$$

with boundary conditions

$$\alpha(\mathbf{X}, t)\, u_i(\mathbf{X}, t) + \beta(\mathbf{X}, t)\, \frac{\gamma u_i}{\gamma n}(\mathbf{X}, t) = f_i(\mathbf{X}, t) \qquad \mathbf{X} \in \Omega_b \quad t \geqslant t_0 \tag{2}$$

for each element i of the vector $\mathbf{u}$. The initial conditions are given by

$$\mathbf{u}(\mathbf{X}, t_0) = \mathbf{u}_0(\mathbf{X}) \qquad \mathbf{X} \in \Omega \tag{3}$$

The distributed parameter system described by Eqs. (1) to (3) is a multivariable system with a state vector $\mathbf{u}$ defined over an n-dimensional spatial domain Ω and time. The region Ω is assumed to be a given finite connected region in Euclidean n space and Ω_b is the boundary of Ω. The control vector $\mathbf{m}$ is similarly defined. The operator $\mathscr{G}$ is a spatially varying vector differential operator on $\mathbf{u}$ which may include parameters which are functions of $\mathbf{u}$, $\mathbf{m}$, $\mathbf{X}$, or t. The functions α, β, and f_i are real-valued functions, piecewise C^1 on Ω_b and C^2 on t which satisfy the inequalities

$$\begin{aligned} \alpha(\mathbf{X}, t) &\geqslant 0 \\ \beta(\mathbf{X}, t) &\geqslant 0 \qquad \mathbf{X} \in \Omega_b \quad t \geqslant t_0 \\ \alpha(\mathbf{X}, t) + \beta(\mathbf{X}, t) &> 0 \end{aligned} \tag{4}$$

The symbol n indicates a direction normal to the boundary. In case the distributed control vector $\mathbf{m}$ is not present, the only means of controlling the system may be through the functions f_i defined on the boundary only. Such a problem is called the boundary control problem for partial differential equations and must be accounted for in our subsequent treatment. We note that there is no analog to boundary control for systems described by ordinary differential equations.

The performance functional to be minimized in our optimal control problem is denoted by

$$J(\mathbf{m}) = \int_{\bar{\Omega}} P_0(\mathbf{u}(\mathbf{X}, t_1), \mathbf{X})\, \mathbf{dX} + \int_{\bar{\Omega}} \int_{t_0}^{t_1} P_1(\mathbf{u}(\mathbf{X}, t), \mathbf{m}(\mathbf{X}, t), t)\, dt\, \mathbf{dX} \tag{5}$$

where P_0, P_1 are real-valued functions of class C^2 on t and piecewise C^1 on Ω. The symbol $\bar{\Omega}$ represents the closure of Ω.

The control $\mathbf{m}$ may also be required to satisfy inequality constraints of the form

$$\mathbf{R}(\mathbf{u}(\mathbf{X}, t), \mathbf{m}(\mathbf{X}, t), \mathbf{X}, t) \geqslant \mathbf{0} \qquad \mathbf{X} \in \Omega \quad t \geqslant t_0 \tag{6}$$

and terminal constraints of the form

$$\psi_0(t_1) = 0$$
$$\boldsymbol{\psi}(\mathbf{u}(\mathbf{X}, t_1), \mathbf{X}) = \mathbf{0} \qquad \mathbf{X} \in \Omega \tag{7}$$

where $\mathbf{R}$ is a vector-valued function of dimension r with components R_i which are of class C^2 on t and piecewise C^1 on Ω. The vector-valued function $\boldsymbol{\psi}$ is of dimension q with components ψ_i having the same properties as R_i. The scalar function $\psi_0(t_1)$ specifies the final time t_1. We assume that the functions $\mathbf{R}$ and $\boldsymbol{\psi}$ are consistent with the boundary conditions in Eq. (2).

For the boundary control problem we will consider minimizing the functional

$$J(\mathbf{f}) = \int_{\Omega} \int_{t_0}^{t_1} P_1(\mathbf{u}(\mathbf{X}, t), \mathbf{X}, t)\, dt\, d\mathbf{X} \tag{8}$$

subject to the partial differential equation side constraints

$$\frac{\gamma \mathbf{u}}{\gamma t}(\mathbf{X}, t) = \mathscr{G}(\mathbf{u}(\mathbf{X}, t), \mathbf{X}, t) \qquad \mathbf{X} \in \Omega \quad t \geqslant t_0 \tag{9}$$

and boundary and initial conditions given by Eqs. (2) and (3). The inequality constraints become

$$\mathbf{R}(\mathbf{f}(\mathbf{X}, t), \mathbf{X}, t) \qquad \mathbf{X} \in \Omega_b \quad t \geqslant t_0 \tag{10}$$

and the terminal conditions are given by Eq. (7). Cases where the boundary control $\mathbf{f}$ appears explicitly in the criterion functional can also be treated if the spatial integration of $\mathbf{f}$ is taken only over the boundary domain.

The system described by Eq. (1) is sufficiently general to include all the standard types of partial differential equations. For example, the hyperbolic and biharmonic classes of equations which involve higher-order time derivatives can be formulated in the notation of Eq. (1) by employing the notion of state variables.[8] The single exception is the class of elliptic partial differential equations which describes the steady-state behavior of the systems described by Eq. (1). These equations are time independent and can be represented by

$$\mathscr{G}(\mathbf{u}(\mathbf{X}), \mathbf{m}(\mathbf{X}), \mathbf{X}) = \mathbf{0} \qquad \mathbf{X} \in \Omega \tag{11}$$

with boundary conditions given by

$$\alpha_i(\mathbf{X})\, u_i(\mathbf{X}) + \beta_i(\mathbf{X}) \frac{\gamma u_i}{\gamma n}(\mathbf{X}) = f_i(\mathbf{X}) \qquad \mathbf{X} \in \Omega \tag{12}$$

The criterion functional and inequality constraints then become

$$J(\mathbf{m}) = \int_{\bar{\Omega}} P_1(\mathbf{u}(\mathbf{X})\,\mathbf{m}(\mathbf{X}), \mathbf{X})\,d\mathbf{X} \tag{13}$$

$$\mathbf{R}(\mathbf{u}(\mathbf{X}), \mathbf{m}(\mathbf{X}), \mathbf{X}) \geqslant \mathbf{0} \qquad \mathbf{X} \in \Omega \tag{14}$$

These continuous-space, continuous-time partial differential equations can now be approximated in several ways. By a discrete approximation we will mean replacing all derivatives by differences and all integrals by sums, thereby yielding a mathematical programming problem having a large number of algebraic or transcendental equality constraints. Alternatively, a semidiscrete approximation can be used on partial differential equations of the form of Eq. (1) where only the spatial derivatives are replaced by differences and the equality constraints then become sets of ordinary differential equations. Likewise, the criterion functional retains a single integral with respect to time and we have a problem in the calculus of variations. In this chapter we will employ discretization in the space domain only, thus resulting in semidiscrete approximations of partial differential equations of the form of Eq. (1) and discrete approximations of elliptic partial differential equations.

We discretize the space variables by defining a column vector

$$\mathbf{X}_i = [i_1(\Delta x_1), i_2(\Delta x_2), \ldots, i_j(\Delta x_j), \ldots, i_n(\Delta x_n)]' \tag{15}$$

which in effect places a grid on the region Ω. The prime denotes the vector transpose. Here the elements of

$$\mathbf{i} = [i_1, i_2, \ldots, i_j, \ldots, i_n]'$$

are integers defined by $i_j = 0, 1, \ldots, N_j$ where

$$N_j = \frac{1}{\Delta x_j}[(x_j)_{\max} - (x_j)_{\min}]$$

Denoting the set of points defined by Eq. (15) by #, the following terms can be defined: *mesh point*—a point in #; *interior point*—a point in $\# \cap \Omega$; *boundary point*—a point in $\# \cap \Omega_b$; *exterior point*—a point belonging to $\# \cap C(\bar{\Omega})$ where $C(\cdot)$ represents the complement operator; *regular point*—a point belonging to $\# \cap \Omega$ and such that all adjacent points also belong to $\# \cap \Omega$; *irregular point*—a point belonging to # which is not a regular point.

For the case when $\beta_i = 0$ in Eq. (2), irregular points can be treated as regular points by appropriately defining a boundary (pseudo) mesh point at the intersection of the boundary and the line segment connecting the irregular point with each exterior point. In case $\beta_i > 0$, the boundary can be approximated by orthogonal line segments or by other suitable devices as discussed by Varga[11] and Young.[14] For our purposes, it is sufficient to see that methods exist for handling any irregular boundary. For simplicity in the subsequent discussion, we will assume that $\beta_i = 0$ in Eq. (2) and that all mesh points are regular points.

The outcome of discretizing the operator $\mathscr{G}$ depends upon the specific form of $\mathscr{G}$ and upon the accuracy of the approximation. For example, if $\mathscr{G}$ contained no derivative terms, the discrete operator $\mathbf{G}_\mathbf{i}$ at $\mathbf{X}_i$ would be a function of variables at $\mathbf{X}_i$ only. If first-derivative terms are present, at least one additional mesh point must be used in the approximation. However, which additional point and indeed how many additional points are brought into the operator $\mathbf{G}_\mathbf{i}$ depends on the specific approximation employed. Obviously, additional accuracy can be obtained by using more points in the approximation. It is not so obvious perhaps that the choice of the difference operator may affect the numerical stability of the solution. It is advisable in these applications to employ difference operators which are unconditionally stable.

In the approximation of second derivatives, at least three mesh points are needed. Here again higher numbers of points will give greater accuracy but at considerable expense in computational complexity. Higher-order derivatives can be treated in an analogous manner with third-order derivatives requiring at least five points, etc. One additional aspect of discretization which greatly affects the number of points brought into $\mathbf{G}_\mathbf{i}$ is n, the number of space variables. For example, the simplest approximation of two second-order partial derivatives with respect to different space variables requires five points, three such derivatives require seven points, etc.

For clarity of presentation, the operator $\mathscr{G}$ will be assumed to contain at most second-order space derivatives and the simplest possible approximation will be used. In this case we can express

$$\mathscr{G}(\mathbf{u}(\mathbf{X}_\mathbf{i}, t), \mathbf{m}(\mathbf{X}_\mathbf{i}, t), \mathbf{X}_\mathbf{i}, t) \cong \mathbf{G}_\mathbf{i}(\mathbf{u}_\mathbf{i}(t), \mathbf{u}_{\mathbf{i}\pm\mathbf{I}_k}(t), \mathbf{m}_\mathbf{i}(t), t) \qquad k = 1, 2, \dots\ n \quad (16)$$

where $\mathbf{I}_k = \{\mathbf{i} \mid \mathbf{i} = \mathbf{0}$ except for the kth element which equals $1\}$, $\mathbf{i}$ ranges over all regular points, and the functions $\mathbf{G}_\mathbf{i}$ are assumed to be real valued and of class C^2.

A few examples will help to clarify the discretization and at the same time serve as a vehicle for demonstrating the formulation of the multilevel problem. We will return to the general problem later in the chapter.

Example 6-1

Consider the following coupled system involving two states u_1 and u_2 and two controls m_1 and m_2 distributed over one space variable and time:

$$\begin{aligned} \frac{\gamma u_1}{\gamma t} &= \frac{\gamma^2 u_1}{\gamma x^2} + u_2 + m_1 \\ \frac{\gamma u_2}{\gamma t} &= \frac{\gamma^2 u_2}{\gamma x^2} + u_1 + m_2 \end{aligned} \tag{17}$$

The initial and boundary conditions are

$$\mathbf{u}(x, 0) = \mathbf{u}_0(x)$$

and

$$\mathbf{u}(0, t) = \mathbf{u}(1, t) = \mathbf{0}$$

respectively. Employing the simple three-point formula to approximate the spatial derivatives yields

$$\begin{aligned} \dot{u}_1^{\,i} &= h^{-2}[u_1^{i+1} + u_1^{i-1} - 2u_1^{\,i}] + u_2^{\,i} + m_1^{\,i} \\ \dot{u}_2^{\,i} &= h^{-2}[u_2^{i+1} + u_2^{i-1} - 2u_2^{\,i}] + u_1^{\,i} + m_2^{\,i} \end{aligned} \tag{18}$$

where

$$\begin{aligned} h &= x_{i+1} - x_i = \Delta x \\ u_j^{\,i} &= u_j(x_i) \\ i &= 1, 2, \ldots, N_1 \end{aligned}$$

Since the operator in Eq. (17) is linear, we can use vector-matrix notation to represent the entire system of equations. It is convenient to define a vector of all possible variables as follows

$$\mathbf{U} = [u_1^{\,1}, u_1^{\,2}, \ldots, u_1^{N_1}, u_2^{\,1}, u_2^{\,2}, \ldots, u_2^{N_1}]'$$

In order to reduce the dimensionality of the problem by multilevel methods, we will define subsystems by partitioning $\mathbf{U}$ into

$$\mathbf{U} = [\mathbf{U}_1, \mathbf{U}_2, \ldots, \mathbf{U}_j, \ldots, \mathbf{U}_N]'$$

where $\mathbf{U}_j$ is the state vector for the jth subsystem and has dimension n_j. A control vector $\mathbf{M}$ is similarly defined and partitioned. Depending on the total dimension of $\mathbf{U}$, we can determine how it should best be partitioned. For example, if we take $\mathbf{U}_1 = [u_1^1, \ldots, u_1^{N_1}]$ and $\mathbf{U}_2 = [u_2^1, \ldots, u_2^{N_1}]$, Eq. (18) can be written

$$\begin{bmatrix} \dot{\mathbf{U}}_1 \\ \dot{\mathbf{U}}_2 \end{bmatrix} = h^{-2} \begin{bmatrix} \mathbf{A}_{11} & \mathbf{A}_{12} \\ \mathbf{A}_{21} & \mathbf{A}_{22} \end{bmatrix} \begin{bmatrix} \mathbf{U}_1 \\ \mathbf{U}_2 \end{bmatrix} + \begin{bmatrix} \mathbf{M}_1 \\ \mathbf{M}_2 \end{bmatrix}$$

where $\mathbf{A}_{12} = \mathbf{A}_{21} = \mathbf{I}$ is the identity matrix and $\mathbf{A}_{11} = \mathbf{A}_{22}$ are called Jacobi or tridiagonal matrices having -2 on the main diagonal and 1 on either side of the main diagonal with all other elements zero. The dimension of all matrices $\mathbf{A}_{ij}$ is $N_1 \times N_1$. In order to decompose Eq. (18) into independent subsystems, we want to expose the coupling terms. Hence, we write Eq. (18) as

$$\dot{\mathbf{U}}_j = h^{-2}\mathbf{A}_{jj}\mathbf{U}_j + \mathbf{M}_j + h^{-2} \sum_{\substack{k=1 \\ (k \neq j)}}^{N} \mathbf{A}_{jk}\mathbf{U}_k \tag{19}$$

In later sections we will show that this formulation satisfies the two requirements for the efficient use of multilevel techniques which we stated earlier, namely, (1) that the subsystems converge rapidly and (2) that a second-level controller can be found which converges rapidly for this class of problems.

We should recognize here that a further partitioning of $\mathbf{U}$ will also result in subsystems of the form given by Eq. (19). In general, the $\mathbf{A}_{jj}$ will be block diagonal Jacobi matrices of different dimensions and the $\mathbf{A}_{jk}$ will be sparse in that they have few nonzero elements. The sparsity of $\mathbf{A}_{jk}$ indicates a low degree of coupling between the subsystems. It is these two properties which make it possible to satisfy the desired convergence criteria listed above. Not all partial differential equations of the form of Eq.(1) can be represented in this way and hence not all such partial differential equations are amenable to multilevel solution as discussed here.

Example 6-2

We will demonstrate now an example of a partial differential equation which when discretized does not have the form mentioned above and which cannot be treated by the methods of this chapter. Consider the simplest hyperbolic equation, the wave equation

$$\nabla_t^2 u = \nabla_x^2 u \tag{20}$$

where

$$\nabla_\alpha^2 = \frac{\gamma^2}{\gamma\alpha^2}$$

When written in normal form, Eq. (20) becomes

$$\nabla_t \begin{bmatrix} u \\ v \end{bmatrix} = \begin{bmatrix} 0 & \nabla_x \\ \nabla_x & 0 \end{bmatrix} \begin{bmatrix} u \\ v \end{bmatrix} \tag{21}$$

The block diagonal submatrices are now zero and all nonzero entries correspond to coupling terms. This decomposed system does not have either of the desirable properties mentioned above and indeed the class of hyperbolic equations in general is not well suited to solution by multilevel techniques. The same is true for the class of biharmonic equations illustrated by the beam equation

$$\nabla_t^2 u = -\nabla_x^4 u \tag{22}$$

Putting Eq. (22) in normal form yields

$$\nabla_t \begin{bmatrix} u \\ v \end{bmatrix} = \begin{bmatrix} 0 & -\nabla_x^2 \\ \nabla_x^2 & 0 \end{bmatrix} \begin{bmatrix} u \\ v \end{bmatrix} \tag{23}$$

which again does not have the properties desired.

One might think from Example 6-1 that the application of multilevel techniques to distributed parameter systems is limited to linear partial differential equations. This is not the case although linearity is a convenient property especially for notational purposes. The next example shows that a decomposed form analogous to Eq. (19) can be obtained for more general problems.

Example 6-3

Consider now the general nonlinear system in two states and one space variable

$$\begin{aligned} \frac{\gamma u_1}{\gamma t} &= g_1(u_1, u_2, m_1, m_2) \\ \frac{\gamma u_2}{\gamma t} &= g_2(u_1, u_2, m_1, m_2) \\ \mathbf{u}(x, 0) = \mathbf{u}_0(x) &\qquad \mathbf{u}(0, t) = \mathbf{u}(1, t) = \mathbf{0} \end{aligned} \tag{24}$$

From our experience in Example 6-2, we will assume that g_i contains no derivatives with respect to u_j if i does not equal j. For convenience we also assume no space-varying derivatives in the control variables.

Discretization now yields

$$\begin{aligned} \dot{u}_1^{\,i} &= g_1(u_1^{\,i}, u_1^{i\pm1}, u_2^{\,i}, m_1^{\,i}, m_2^{\,i}) \\ \dot{u}_2^{\,i} &= g_2(u_2^{\,i}, u_2^{i\pm1}, u_1^{\,i}, m_1^{\,i}, m_2^{\,i}) \end{aligned} \tag{25}$$

which can be written as

$$\begin{aligned}\dot{\mathbf{U}}_1 &= \mathbf{G}_1(\mathbf{U}_1, \mathbf{U}_2, \mathbf{M}_1, \mathbf{M}_2)\\ \dot{\mathbf{U}}_2 &= \mathbf{G}_2(\mathbf{U}_1, \mathbf{U}_2, \mathbf{M}_1, \mathbf{M}_2)\end{aligned} \tag{26}$$

Although it is not readily apparent from the notation, the class of problems represented by this general example is amenable to solution by multilevel methods. In particular, Eq. (26) has all the desirable properties of Eq. (19) except linearity. These facts will be clarified by a nonlinear example to be considered in a later section.

Before proceeding with a discussion of the criterion functionals, we will demonstrate by an example the formulation of a boundary control problem.

Example 6-4

The boundary control problem as indicated by Eq. (19) has no control variable distributed over space but rather is controlled only on the boundary domain. An example is given by the scalar one-dimensional system

$$\frac{\gamma u}{\gamma t} = \frac{\gamma^2 u}{\gamma x^2} \tag{27}$$

with initial and boundary conditions

$$u(x, 0) = u_0(x) \tag{28}$$

and

$$u(0, t) = f_0(t) \qquad u(1, t) = f_1(t) \tag{29}$$

respectively. Clearly the only controls are the functions $f_0(t)$ and $f_1(t)$ on the boundary. The difference equations are

$$\dot{u}^i = h^{-2}[u^{i+1} + u^{i-1} - 2u^i] \tag{30}$$

for all interior points $i = 1, 2, \ldots, N_1 - 1$. The boundary points are given by Eq. (29) as

$$\begin{aligned}u(0, t) &= u^0 = f_0(t)\\ u(1, t) &= u^{N_1} = f_1(t)\end{aligned} \tag{31}$$

Thus the boundary control function is absorbed into the system equation in a manner exactly like the distributed control of Example 6-1. In this case

$$\dot{\mathbf{U}} = h^{-2}\mathbf{A}\mathbf{U} + h^{-2}\mathbf{F} \tag{32}$$

where $\mathbf{A}$ is a Jacobi matrix and $\mathbf{F}$ is a column vector of the form

$$\mathbf{F} = [f_0, 0, 0, \ldots, f_1]'$$

It is particularly convenient that boundary control problems can be treated within the framework established for distributed controls. We note here that the $\mathbf{U}$ vector contains only internal points of the space region, the boundary points being treated separately. In the case where nonzero boundary conditions are specified in the distributed control problem, a result of the form

$$\dot{\mathbf{U}} = h^{-2}(\mathbf{AU} + \mathbf{F}) + \mathbf{M} \tag{33}$$

occurs where the elements of $\mathbf{F}$ are known constants. This is a slightly more general form than Eq. (18) in Example 6-1.

Next we consider the formation of difference equations corresponding to the remaining system equations (2) to (7). Assuming that the functions P_0, P_1, and $\mathbf{R}$ do not contain spatial derivatives, Eq. (5) can be written

$$J(\mathbf{m}_i) = \sum_{i \in \bar{\Omega}} \left[P_{0i}(\mathbf{u}_i) + \int_{t_0}^{t_1} P_{1i}(\mathbf{u}_i, \mathbf{m}_i, t)\, dt \right] \tag{34}$$

where we have divided out the $\Delta\mathbf{X}$ terms since they do not influence the optimal solution. The remaining equations are

$$\alpha_i(t)\, \mathbf{u}_i(t) = \mathbf{f}_i(t) \qquad \mathbf{X}_i \in \Omega_b \quad t \geqslant t_0 \tag{35}$$

$$\mathbf{u}_i(t_0) = \mathbf{u}_{0i} \qquad \mathbf{X}_i \in \Omega \tag{36}$$

$$\mathbf{R}_i(\mathbf{u}_i(t), \mathbf{m}_i(t), t) \geqslant \mathbf{0} \qquad \mathbf{X}_i \in \bar{\Omega} \quad t \geqslant t_0 \tag{37}$$

$$\begin{aligned} \psi_0(t_1) &= 0 \\ \boldsymbol{\psi}_i(\mathbf{u}_i(t_1)) &= \mathbf{0} \qquad \mathbf{X}_i \in \Omega \end{aligned} \tag{38}$$

In forming Eq. (34), the spatial integrals become sums over all the interior and boundary points. However, it is easily seen that the optimal values of $\mathbf{m}_i$ are specified on the boundary by the boundary conditions. Thus, if $\mathbf{R}_i > \mathbf{0}$, a necessary condition for a minimum of Eq. (34) is that

$$\frac{\gamma P_{1i}}{\gamma \mathbf{m}_i}(\alpha_i^{-1}\mathbf{f}_i, \mathbf{m}_i, t) = \mathbf{0} \qquad \mathbf{X}_i \in \Omega_b \tag{39}$$

where $\gamma P_{1i}/\gamma \mathbf{m}_i$ is the Fréchet derivative of P_{1i} at $\mathbf{m}_i$. If the values of $\mathbf{m}_i$ obtained in Eq. (39) violate Eq. (37), then one or more elements of $\mathbf{m}_i$ is obtained from

$$\mathbf{R}_i(\alpha_i^{-1}\mathbf{f}_i, \mathbf{m}_i, t) = \mathbf{0} \qquad \mathbf{X}_i \in \Omega_b \tag{40}$$

It is possible that all values of $\mathbf{m}_i$ may not be unique on the boundary. However, we now observe that the sum in Eq. (34) need be taken only over the region Ω, instead of $\bar{\Omega}$. Only in the case of the boundary control problem (8) must the discrete equation for the criterion functional be summed over all mesh points, since the function $\mathbf{f}$ is not known a priori on the boundary.

6-2-1 Consistency, Convergence, and Stability

Before going on to a discussion of optimization methods using multilevel methods, a brief discussion of the questions of consistency, convergence, and stability is warranted. Unfortunately we cannot make very general statements about these properties except for rather restricted classes of problems.

Consider the linear one-dimensional parabolic partial differential equation given by

$$L[u(x, t)] = u_t - a(x, t)\, u_{xx} - 2b(x, t)\, u_x + c(x, t)\, u = d(x, t) \tag{41}$$

where $a(x, t) > 0$ and having initial and boundary conditions

$$u(x, 0) = f(x) \qquad 0 \leqslant x \leqslant L$$

and

$$u(0, t) = g_0(t) \qquad t > 0$$

$$u(L, t) = g_1(t) \qquad t > 0$$

respectively. We now employ a semidiscrete approximation to Eq. (41) given by

$$L_j[v(x_j, t)] = \frac{dv_j}{dt} - h^{-2}a_j(t)[v_{j+1} - 2v_j + v_{j-1}]$$

$$- h^{-1}b_j(t)[v_{j+1} - v_{j-1}] + c_j(t)\, v_j = d_j(t) \tag{42}$$

$$v_j(0) = f(x_j) \qquad 0 \leqslant j \leqslant J + 1$$

$$v_0(t) = g_0(t) \qquad t > 0$$

$$v_{J+1}(t) = g_1(t) \qquad t > 0$$

where $v_j = v(x_j, t)$ is the solution to the ordinary differential equation (42). The semidiscrete approximation is said to be *consistent* with the original partial differential equation if

$$\lim_{h \to 0} \{L[u(x, t)] - L_j[u(x, t)]\} = 0 \tag{43}$$

This condition ensures that the semidiscrete equations actually do approximate the actual partial differential equation and implies that a solution to the partial differential equation is needed in order to verify consistency. However, by using an appropriate Taylor series expansion it is easy to verify that Eq. (42) is consistent if the third- and fourth-order space derivatives of u as well as the coefficients a and b are bounded.[13] The verification of consistency merely amounts to determining the truncation error of the difference scheme and ensuring that the truncation error goes to zero as the grid spacing goes to zero. Clearly similar results can be obtained for semidiscrete approximations of linear partial differential equations in more than one space variable and also for discrete approximations of parabolic- and elliptic-type partial differential equations.

The ordinary differential equations (42) are said to be *convergent* if their solution satisfies

$$\lim_{h \to 0} | u(x_j, t) - v_j(t)| = 0 \tag{44}$$

This condition simply requires that the approximate solution approach the exact solution as the grid is refined. By using the consistency result shown above, the solution to Eq. (42) can be shown to be convergent over any grid satisfying the inequalities

$$\begin{aligned} 2h^{-2}a(x, t) + c(x, t) &\geqslant 0 \\ a(x, t) - h \mid b(x, t)| &\geqslant 0 \end{aligned} \tag{45}$$

Thus Eq. (45) specifies a maximum grid spacing and a range of values for $c(x, t)$ over which the difference solution is convergent. This property too can usually be determined for other types of linear systems, e.g., (1) parabolic partial differential equations using other discrete or semi-discrete difference schemes and possibly having higher space dimensions, or (2) elliptic partial differential equations. For example, the elliptic equation corresponding to the steady-state solution of Eq. (42) is also convergent under the conditions (45).

The literature on stability theory of differential equations is very vast. If we consider $d_j(t)$ as a time-varying input, the following definition of stability is applicable: A differential equation of the form (42) is said to be *stable* if for all initial times t_0 and initial states $f(x_j)$ and for all bounded inputs $d_j(t)$, the solution $v_j(t)$ is bounded on the interval $[t_0, \infty)$. It is difficult to state conditions which guarantee stability for the general class of equations given by Eq. (42). However, by considering the zero-input case where $d_j(t) = 0$ for all j, a sufficient condition for stability is that

$$\lim_{t \to \infty} \int_{t_0}^{t} \Lambda(t')\, dt' < \infty \tag{46}$$

where $\Lambda(t)$ is the largest eigenvalue of the Jacobi matrix which arises when the equations (42) are put into vector-matrix form. The eigenvalue $\Lambda(t)$ depends upon h as well as $a_j(t)$, $b_j(t)$, and $c_j(t)$. Thus the stability of Eq. (42) depends upon the discretization interval h. If the coefficients a_j, b_j, and c_j are constants, the sufficient condition for stability becomes

$$\Lambda < 1 \tag{47}$$

where Λ is again the largest eigenvalue and in general is dependent upon h. It can be shown that for the case when $b_j = c_j = 0$, the magnitude of Λ is always less than unity regardless of the mesh spacing and the corresponding semidiscrete approximation (42) is always stable.[12]

In the case of a discrete approximation of parabolic partial differential equations, we must deal with the question of numerical stability of difference equations rather than the stability of ordinary differential equations. This question is answered by the following well-known theorem due to Lax:[7] Given a properly posed initial-value problem and a corresponding discrete approximation which satisfies the consistency condition, (numerical) stability is a necessary and sufficient condition for convergence. Thus if we were to solve Eq. (42) by discrete methods, the conditions (45) which ensure convergence would also guarantee numerical stability.

In this section we have discussed the practical details related to the formulation of several approximations to a class of partial differential equations. In the next section we concentrate on the aspects of these semidiscrete models which especially facilitate the multilevel approach. Various second-level controller configurations will be discussed.

6-3 SECOND-LEVEL CONTROLLERS

In the previous section we discussed discrete and semidiscrete models for elliptic and parabolic partial differential equations. In this section we will formulate the optimization problem for a general class of parabolic partial differential equations in terms of a semidiscrete approximation. We can then show that the discrete model formulation and the elliptic partial differential equation formulation can be regarded as special cases of this problem. The optimization problem is as described by Eqs. (16) and (34) to (38). When the criterion functional is of the form (34), we have the Bolza problem of the calculus of variations. Here we consider the slightly less general form known as the Lagrange problem which can be obtained by a suitable redefinition of the Bolza problem. We wish to minimize $J(\mathbf{m}_i)$ where

$$J(\mathbf{m}_i) = \sum_{i \in \Omega} \int_{t_0}^{t_1} P_{1i}(\mathbf{u}_i, \mathbf{m}_i, t)\, dt \tag{48}$$

such that

$$\dot{\mathbf{u}}_i = \mathbf{G}_i(\mathbf{u}_i, \mathbf{u}_{i\pm \mathbf{I}_k}, \mathbf{m}_i, t) \tag{49}$$

with initial conditions

$$\mathbf{u}_i(t_0) = \mathbf{u}_{0i} \tag{50}$$

and final conditions

$$\boldsymbol{\psi}_i(\mathbf{u}_i(t_1)) = \mathbf{0} \tag{51}$$

and subject to the inequality constraints

$$\mathbf{R}_i(\mathbf{u}_i, \mathbf{m}_i, t) \geqslant \mathbf{0} \tag{52}$$

Here $\mathbf{X}_i \in \Omega$ and we are dealing with the distributed control problem in which one and only one control vector $\mathbf{m}_i(t)$ is associated with each mesh point. We have assumed that the boundary conditions along with the inequality constraints completely determine the control on the boundaries as discussed previously and hence the boundary domain and the boundary conditions are excluded from the problem formulation. Furthermore, we assume that the final time t_1 is specified.

The optimization problem (48) to (52) is a straightforward problem in the calculus of variations or other relevant theory. The main potential difficulty comes from the problem size, which may be prohibitive for large space regions or small mesh spacings or both. The underlying computational problem involves the solution of a two-point boundary-value problem of a size at least twice as large as the number of internal mesh points. It is in the reduction of this computational problem to a manageable size that the methods of multilevel optimization can possibly benefit.

In Example 6-1 we partitioned a two-state distributed parameter system into two systems of vector ordinary differential equations in which the dynamics of each independent "subsystem" were clearly separated from the dynamics which coupled the systems together. In that example, each subsystem was taken over the entire space region. Alternatively, a further reduction in the dimensionality of each subsystem can be obtained by a partitioning of its spatial domain into smaller sections. The main requirement is that the convergence properties of the various levels be maintained. Clearly the partitioning problem is not affected in substance by the number of states of the distributed parameter system. By listing each state variable separately at each mesh point when forming the $\mathbf{U}$ vector in Example 6-1, we increased the dimensionality in proportion to the number of states; however, the concept of partitioning remains the same. In fact a single-state problem taken over two space dimensions may have a coefficient matrix similar to that of Example 6-1.

Consider now the above problem and assume that we have arranged all elements of $\mathbf{u}_i$ in some convenient fashion, e.g., the order described in Example 6-1. Now partition the resulting $\mathbf{U}$ vector consisting of all ordered elements into N subvectors. Assuming that the elements of the vectors $\mathbf{u}_i$ and $\mathbf{m}_i$ are separable between subsystems in the functions P_{1j}, ψ_j, and $\mathbf{R}_j$ appearing below, we can express the problem (48) to (52) as N independent problems of the form

$$\min_{\mathbf{M}_j} \sum_{j=1}^{N} \int_{t_0}^{t_1} P_{1j}(\mathbf{U}_j, \mathbf{M}_j, t)\, dt \tag{53}$$

such that

$$\dot{\mathbf{U}}_j = \mathbf{G}_j(\mathbf{U}_j, \mathbf{M}_j, \mathbf{S}_j, t) \tag{54}$$

with initial conditions

$$\mathbf{U}_j(t_0) = \mathbf{U}_{j0} \tag{55}$$

and final conditions

$$\boldsymbol{\psi}_j(\mathbf{U}_j(t_1)) = \mathbf{0} \tag{56}$$

and subject to the inequality constraints

$$\mathbf{R}_j(\mathbf{U}_j, \mathbf{M}_j, t) \geqslant \mathbf{0} \tag{57}$$

In Eq. (54) we have introduced a new vector $\mathbf{S}_j$ called a *pseudocontrol vector* in order to achieve the subsystem independence. For each element or function of elements $\mathbf{u}_{\mathbf{i}\pm\mathbf{I}_k}$ in Eq. (49) which is not contained in $\mathbf{U}_j$ but represents a coupling between $\mathbf{U}_j$ and some $\mathbf{U}_k$, a corresponding element s_j is introduced on a one-for-one basis. The variable s_j may represent a single variable u_k, a function of u_k, or a function of several variables. By collecting all such interactions into a vector variable $\mathbf{S}_j$, the interaction between subsystems can be separated out and the N lower-dimensional subsystems can be treated independently. Clearly the resultant solution will be a minimum for the overall problem if and only if the constraint

$$\mathbf{S}_j = \mathbf{F}_j(\mathbf{U}_1, \mathbf{U}_2, \ldots, \mathbf{U}_k, \ldots, \mathbf{U}_N)_{k \neq j} \tag{58}$$

is satisfied. In Eq. (58) $\mathbf{S}_j$ is an $\mathbf{m}_j$-dimensional vector and $\mathbf{F}_j$ is a vector-valued function of dimension $\mathbf{m}_j$. Obviously $\mathbf{F}_j$ cannot be a function of $\mathbf{U}_j$. Since by specific assumption we have excluded the possibility of any coupling between subsystems in $\mathbf{R}_j$ and $\boldsymbol{\psi}_j$, Eqs. (56) and (57) may complicate the subsystem optimization problem but do not affect the multilevel problem. For simplicity in the subsequent discussion, they will not be considered further.

Employing the canonical form of the calculus of variations, we define the Hamiltonian functional H as follows:

$$H = \sum_{j=1}^{N} \{P_{1j}(\mathbf{U}_j, \mathbf{M}_j, t) + \boldsymbol{\lambda}_j' \mathbf{G}_j(\mathbf{U}_j, \mathbf{M}_j, \mathbf{S}_j, t) + \boldsymbol{\rho}_j'[\mathbf{F}_j(\mathbf{U}_1, \mathbf{U}_2, ..., \mathbf{U}_k, ..., \mathbf{U}_N)_{k \neq j} - \mathbf{S}_j]\} \tag{59}$$

where the vectors $\boldsymbol{\lambda}_j$ are adjoint variables of dimension n_j and the vectors $\boldsymbol{\rho}_j$ are Lagrange multipliers of dimension m_j. The necessary conditions for a minimum can now be expressed in terms of H.

$$\dot{\mathbf{U}}_j = \frac{\gamma H}{\gamma \boldsymbol{\lambda}_j} = \mathbf{G}_j(\mathbf{U}_j, \mathbf{M}_j, \mathbf{S}_j, t) \tag{60}$$

$$\dot{\boldsymbol{\lambda}}_j = -\frac{\gamma H}{\gamma \mathbf{U}_j} = -\frac{\gamma P_{1j}}{\gamma \mathbf{U}_j} - \left(\frac{\gamma \mathbf{G}_j}{\gamma \mathbf{U}_j}\right)' \boldsymbol{\lambda}_j - \sum_{\substack{k=1 \\ k \neq j}}^{N} \left(\frac{\gamma \mathbf{F}_k}{\gamma \mathbf{U}_j}\right)' \boldsymbol{\rho}_k \tag{61}$$

$$\frac{\gamma H}{\gamma \mathbf{M}_j} = \frac{\gamma P_{1j}}{\gamma \mathbf{M}_j} + \left(\frac{\gamma \mathbf{G}_j}{\gamma \mathbf{M}_j}\right)' \boldsymbol{\lambda}_j = \mathbf{0} \tag{62}$$

$$\frac{\gamma H}{\gamma \mathbf{S}_j} = \left(\frac{\gamma \mathbf{G}_j}{\gamma \mathbf{S}_j}\right)' \boldsymbol{\lambda}_j - \boldsymbol{\rho}_j = \mathbf{0} \tag{63}$$

$$\frac{\gamma H}{\gamma \boldsymbol{\rho}_j} = \mathbf{F}_j - \mathbf{S}_j = \mathbf{0} \tag{64}$$

where the vector partial derivatives are defined by

$$\frac{\gamma \mathbf{G}}{\gamma \mathbf{U}} = \begin{bmatrix} \frac{\gamma g_1}{\gamma u_1} & \frac{\gamma g_1}{\gamma u_2} & \cdots & \frac{\gamma g_1}{\gamma u_n} \\ \frac{\gamma g_2}{\gamma u_1} & & \cdots & \vdots \\ \vdots & & & \\ \frac{\gamma g_n}{\gamma u_1} & & \cdots & \frac{\gamma g_n}{\gamma u_n} \end{bmatrix}$$

$$\frac{\gamma P}{\gamma \mathbf{U}} = \left[\frac{\gamma P}{\gamma u_1}, \frac{\gamma P}{\gamma u_2}, ..., \frac{\gamma P}{\gamma u_n}\right]'$$

where $\mathbf{G} = (g_1, g_2, ..., g_n)$ and $\mathbf{U} = (u_1, u_2, ..., u_n)$. The derivatives

$$\frac{\gamma \mathbf{G}_j}{\gamma \mathbf{M}_j} \quad \frac{\gamma \mathbf{G}_j}{\gamma \mathbf{S}_j} \quad \frac{\gamma \mathbf{F}_k}{\gamma \mathbf{U}_j} \quad \text{and} \quad \frac{\gamma P_{1j}}{\gamma \mathbf{M}_j}$$

are similarly defined.

If $\mathbf{F}_k$ is chosen so that its partial derivative with respect to $\mathbf{U}_j$ does not involve any state except $\mathbf{U}_j$, and if we consider $\boldsymbol{\rho}_k$ as a fixed parameter, the first four necessary conditions for the jth subsystem are completely independent of those for any other subsystem. In general there is no difficulty in choosing $\mathbf{F}_k$ in this way since for example $\mathbf{F}_k$ can always be taken as a matrix of zeros and ones where each u_k is designated as a single element s_j. In this case

$$\mathbf{S}_j = \sum_{\substack{k=1 \\ k \neq j}}^{N} \mathbf{C}_{jk} \mathbf{U}_k \tag{65}$$

where $\mathbf{C}_{jk}$ is an $(m_j \times u_k)$-dimensional matrix of ones and zeros where a one is used to indicate a coupling constraint and a zero otherwise. In order to provide a one-for-one substitution, each row of the composite matrix

$$\left[\mathbf{C}_{j1} \middle| \mathbf{C}_{j2} \middle| \cdots \middle| \mathbf{C}_{jk} \middle| \cdots \middle| \mathbf{C}_{jN}\right]_{j \neq k}$$

must have one and only one nonzero element. The ordering of the rows in $\mathbf{C}_{jk}$ (but not the elements within the rows) is arbitrary. In this case the coupling terms in the Hamiltonian can be rearranged using the adjoint relationship as

$$\sum_{j=1}^{N} \boldsymbol{\rho}_j' \sum_{\substack{k=1 \\ k \neq j}}^{N} \mathbf{C}_{jk} \mathbf{U}_k = \sum_{k=1}^{N} \mathbf{U}_k' \sum_{\substack{j=1 \\ k \neq j}}^{N} \mathbf{C}_{jk}' \boldsymbol{\rho}_j \tag{66}$$

Interchanging subscripts in Eq. (66), we can now write H in the form

$$H = \sum_{j=1}^{N} H_j \tag{67}$$

where

$$H_j = P_{j1}(\mathbf{U}_j, \mathbf{M}_j, t) + \boldsymbol{\lambda}_j' \mathbf{G}_j(\mathbf{U}_j, \mathbf{M}_j, \mathbf{S}_j, t) + \mathbf{U}_j' \sum_{\substack{k=1 \\ k \neq j}}^{N} \mathbf{C}_{kj}' \boldsymbol{\rho}_k - \boldsymbol{\rho}_j' \mathbf{S}_j$$

The functionals H_j are called the subsystem Hamiltonian functions. The last two terms may be regarded as part of the subsystem criterion function since they are no longer in the form of a constraint attached with Lagrange multipliers. By minimizing each H_j independently and then coordinating the solutions by choosing $\boldsymbol{\rho}_k$, the decomposed minimization problems taken together will solve the overall problem.

The question of an efficient convergent algorithm for solving the coordination or second-level problem remains.

The goal-coordination and model-coordination methods of Chapter 1 are not well suited to this problem formulation. In particular, their subsystem problems are awkward to solve and the second-level controller may be slow to converge. In fact for the goal-coordination method, the second level does not converge in general. The convergence condition requires that each subsystem possess a local *minimum* with respect to both $\mathbf{M}_j$ *and* $\mathbf{S}_j$ and this is generally not the case for our problem formulation. This can be easily demonstrated by examining a second necessary condition for subsystem minima. The Legendre necessary condition requires that the Hessian matrix of second partial derivatives be nonnegative definite; i.e.,

$$\boldsymbol{\pi}'[\nabla^2_{\boldsymbol{\xi}_j} H_j]\,\boldsymbol{\pi} \geqslant 0 \tag{68}$$

for every $\boldsymbol{\pi} \neq \mathbf{0}$ where

$$\boldsymbol{\xi}_j = [\mathbf{M}_j\,,\,\mathbf{S}_j]'$$

Clearly Eq. (68) is not satisfied in the simple case where $\mathbf{S}_j$ appears linearly in H_j.

We will discard these two methods in favor of a simple second-level controller which is especially well suited to this problem formulation. The method is suggested by the Gauss-Seidel method of obtaining a numerical solution for elliptic partial differential equations and we will, therefore, term this second-level technique the *Gauss-Seidel controller*.

In the Gauss-Seidel method, Eqs. (60) to (62) are first solved for each subsystem using some appropriate estimates of $\boldsymbol{\rho}_k$ and $\mathbf{S}_j$. The resulting values of $\boldsymbol{\lambda}_k$ and $\mathbf{U}_k$ are then used in Eqs. (63) and (64) to obtain new values for $\boldsymbol{\rho}_k$ and $\mathbf{S}_j$ respectively. The interconnection between subsystems is provided by the second-level controller [Eqs. (63) and (64)] since the calculation of inputs $\boldsymbol{\rho}_k$ and $\mathbf{S}_j$ to the jth subsystem depends in general upon the outputs of several other subsystems. Changing subscripts in Eq. (66) for notational clarity, the second-level equations are solved explicitly for the subsystem inputs as

$$\boldsymbol{\rho}_k = \left(\frac{\gamma \mathbf{G}_k}{\gamma \mathbf{S}_k}\right)' \boldsymbol{\lambda}_k \tag{69}$$

$$\mathbf{S}_j = \mathbf{F}_j(\mathbf{U}_1\,,\mathbf{U}_2\,,\ldots,\mathbf{U}_k\,,\ldots,\mathbf{U}_N)_{k\neq j}$$

The second-level algorithm is thus seen to consist of *explicit* relations which are quickly and easily solved on a digital computer. One method of solution is to solve all N subsystems using estimates for the vector inputs $\boldsymbol{\rho}_k$ and $\mathbf{S}_j$ to each subsystem. The outputs of these N subsystems are then combined by the second-level controller to form new input values. This approach is analogous to the Jacobi method for solving elliptic differential equations. An alternative approach, the Gauss-Seidel method, utilizes each new input value $\boldsymbol{\rho}_k$ or $\mathbf{S}_j$ for the jth subsystem as soon as it arises from the solution of some other subsystem. This method requires initial estimates of only half as many input values as the Jacobi method. In addition, for the solution of partial differential equations the Gauss-Seidel method is known to converge exactly twice as fast as the Jacobi method. A third method for solving partial differential equations, the method of overrelaxation, can be extended to this multilevel application. This technique requires the determination of a relaxation factor which, if properly chosen, speeds convergence over the Gauss-Seidel method but at the expense of additional computation. However, computational efficiency is of extreme importance in solving large problems, and the Gauss-Seidel method is considered preferable.

6-3-1 Convergence Conditions

Kolmogorov[6] proves sufficient convergence conditions using contraction mapping which apply to the Gauss-Seidel controller. Assuming that the control vector $\mathbf{M}_j$ can be eliminated from Eqs. (62) to (66), this system of equations can be represented by

$$\begin{aligned} \dot{\mathbf{Z}} &= \mathbf{F}(\mathbf{Z}, t) \\ \mathbf{Z}(t_0) &= \mathbf{Z}_0 \end{aligned} \tag{70}$$

where $\mathbf{Z} = (\mathbf{U}_1, \mathbf{U}_2, \ldots, \mathbf{U}_N, \boldsymbol{\lambda}_1, \boldsymbol{\lambda}_2, \ldots, \boldsymbol{\lambda}_N)$ and $\mathbf{F}$ is a continuous vector function on some region of the $2N + 1$ dimensional Euclidean space which contains the point $(t_0, \mathbf{Z}_0)$. In addition, $\mathbf{F}$ satisfies a Lipschitz condition with respect to $\mathbf{Z}$; namely,

$$|f_i(\mathbf{Z}^1, t) - f_i(\mathbf{Z}^2, t)| \leqslant L \max(|z_i^1 - z_i^2| : 1 \leqslant i \leqslant 2N) \tag{71}$$

where $f_i \in \mathbf{F}$, $z_i \in \mathbf{Z}$, and L is the Lipschitz constant. Then letting $\mathbf{T}$ be the integral operator arising from Eq. (70), the set of iterative equations

$$\mathbf{Z}^{k+1} = \mathbf{T}(\mathbf{Z}^k) \tag{72}$$

will converge to the solution of Eq. (70) on the interval $[t_0, t_1]$ if

$$L(t_1 - t_0) < 1 \tag{73}$$

and $\mathbf{Z}_0$ is suitably chosen. For a linear system

$$\dot{\mathbf{Z}} = \mathbf{AZ}$$
$$\mathbf{Z}(t_0) = \mathbf{Z}_0 \tag{74}$$

a constant L can be determined which satisfies Eq. (71) as follows

$$|f_i(\mathbf{Z}^1) - f_i(\mathbf{Z}^2)| = \sum_{j=1}^{2N} a_{ij}(z_j^2 - z_j^1)$$

$$\leqslant \sum_{j=1}^{2N} | a_{ij}(z_j^2 - z_j^1)|$$

$$\leqslant \sum_{j=1}^{2N} | a_{ij} | \, | z_j^2 - z_j^1 |$$

by the Hölder inequality and then

$$\leqslant \sum_{j=1}^{2N} | a_{ij} | \max(| z_j^2 - z_j^1 | : 1 \leqslant j \leqslant 2N)$$

To find a value of L which satisfies Eq. (71) for all i, we choose

$$L = \max_i \sum_{j=1}^{2N} | a_{ij} | \tag{75}$$

$$\text{subject to } i = 1, 2, \ldots, 2N$$

It is found in practice that Eq. (75) gives a very loose bound on L; that is, the Gauss-Seidel method converges under much more restrictive conditions than indicated by Eq. (75). One reason for this is the sparse form of the $\mathbf{A}$ matrix in Eq. (74) which arises in most semidiscrete models of partial differential equations. To illustrate, we consider an example of minimizing a quadratic cost functional subject to a scalar parabolic partial differential equation over a two-dimensional region.

Example 6-5

$$\min_m \int_0^1 \int_0^1 \int_{t_0}^{t_1} [q(\mathbf{X})(u_d(\mathbf{X}) - u(\mathbf{X}, t))^2 + r(\mathbf{X})\, m^2(\mathbf{X}, t)]\, dt\, d\mathbf{X} \tag{76}$$

$$\text{subject to } \nabla_t u = K_1 \nabla_{x_1}^2 u + K_2 \nabla_{x_2}^2 u - \sigma u + bm \tag{77}$$

where $K_1 > 0$, $K_2 > 0$, $\sigma \geqslant 0$, $\mathbf{X} = [x_1, x_2]'$

$$q > 0 \qquad r > 0$$

$$0 \leqslant x_2 \leqslant 1$$

and t_1 is specified.

The initial conditions are

$$u(\mathbf{X}, t_0) = u_0(\mathbf{X})$$

and the boundary conditions are

$$\begin{aligned} u(\mathbf{X}, t) &= f(\mathbf{X}, t) \\ \mathbf{X} \in \Omega_b \qquad & t \geqslant t_0 \end{aligned} \tag{79}$$

The space domain is taken as a square region in two space and a uniform grid of (arbitrarily) 16 mesh points is imposed on the region. The mesh points are ordered from left to right and bottom to top as shown in Fig. 6-1. Using this same ordering for **U** and **M**, the semidiscrete problem formulation is given by

$$\min_{\mathbf{M}} \int_{t_0}^{t_1} [(\mathbf{U}_d - \mathbf{U})' \mathbf{Q}(\mathbf{U}_d - \mathbf{U}) + \mathbf{M}'\mathbf{R}\mathbf{M}]\, dt \tag{80}$$

$$\text{subject to } \dot{\mathbf{U}} = \mathbf{AU} + \mathbf{BM} + \mathbf{GF} \tag{81}$$

$$\mathbf{U}(t_0) = \mathbf{U}_0$$

where **A** is a square matrix of the form shown in Fig. 6-1 with

$$\begin{aligned} a_i &= h^{-2}(2K_1 + 2K_2 + h^2\sigma) \\ e_i &= h^{-2}K_1 \\ d_i &= h^{-2}K_2 \\ h &= \Delta x_1 = \Delta x_2 \end{aligned} \tag{82}$$

The vector **F** consists of all boundary mesh points taken in the order shown by the letters a to p in Fig. 6-1. The values of **F** are specified by the boundary conditions (79) of the partial differential equation. The square matrix **G** is composed of elements of the form (82) corresponding to the boundary mesh points. The diagonal matrix **B** consists entirely of elements b from Eq. (77). The matrices **A**, **B**, and **G** may be functions of $\mathbf{X}_i$ and t.

We will arbitrarily decompose the square space region into four

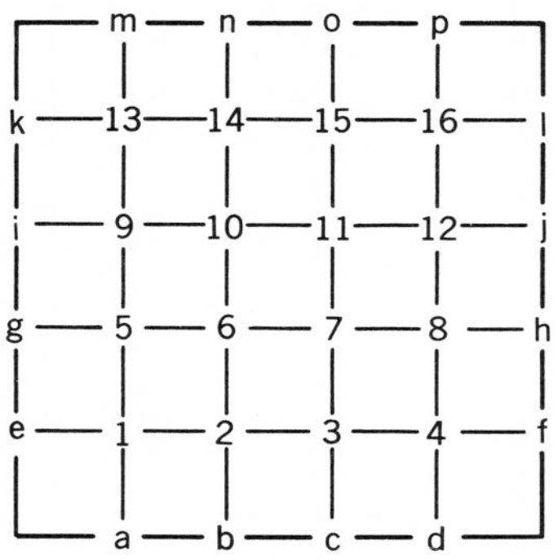

FIG. 6-1. The Ω region.

$$
\left[\begin{array}{cccc|cccc|cccc|cccc}
-a_1 & e_1 & & & d_1 \\
e_2 & -a_2 & e_2 & & & d_2 \\
& e_3 & -a_3 & e_3 & & & d_3 \\
& & e_4 & -a_4 & & & & d_4 \\
\hline
d_5 & & & & -a_5 & e_5 & & & d_5 \\
& d_6 & & & e_6 & -a_6 & e_6 & & & d_6 \\
& & d_7 & & & e_7 & -a_7 & e_7 & & & d_7 \\
& & & d_8 & & & e_8 & -a_8 & & & & d_8 \\
\hline
& & & & d_9 & & & & -a_9 & e_9 & & & d_9 \\
& & & & & d_{10} & & & e_{10} & -a_{10} & e_{10} & & & d_{10} \\
& & & & & & d_{11} & & & e_{11} & -a_{11} & e_{11} & & & d_{11} \\
& & & & & & & d_{12} & & & e_{12} & -a_{12} & & & & d_{12} \\
\hline
& & & & & & & & d_{13} & & & & -a_{13} & e_{13} \\
& & & & & & & & & d_{14} & & & e_{14} & -a_{14} & e_{14} \\
& & & & & & & & & & d_{15} & & & e_{15} & -a_{15} & e_{15} \\
& & & & & & & & & & & d_{16} & & & e_{16} & -a_{16}
\end{array}\right]
$$

The **A** matrix corresponding to Ω.

fourth-order subsystems corresponding to the four rows of mesh points in Ω. The partitioning of $\mathbf{U}$ is given by

$$\mathbf{U} = [\mathbf{U}_1, \mathbf{U}_2, \mathbf{U}_3, \mathbf{U}_4]'$$

and the corresponding partitioning of $\mathbf{A}$ is shown by the solid grid lines in Fig. 6-1. The nonzero elements d_i outside the block diagonal matrices represent coupling constraints in our multilevel formulation. Rewriting the problem in decomposed form yields

$$\min_{\substack{\mathbf{M}_j \\ j=1,2,3,4}} \sum_{j=1}^{4} J(\mathbf{M}_j) \tag{83}$$

where

$$J(\mathbf{M}_j) = \int_{t_0}^{t_1} [(\mathbf{U}_{dj} - \mathbf{U}_j)' \, \mathbf{Q}_j(\mathbf{U}_{dj} - \mathbf{U}_j) + \mathbf{M}_j'\mathbf{R}_j\mathbf{M}_j] \, dt$$

and subject to the constraints

$$\begin{aligned} \dot{\mathbf{U}}_j &= \mathbf{A}_{jj}\mathbf{U}_j + \mathbf{B}_j\mathbf{M}_j + \mathbf{S}_j + \mathbf{G}_j\mathbf{F}_j \\ \mathbf{U}_j(t_0) &= \mathbf{U}_{0j} \end{aligned} \tag{84}$$

In Eq. (84) $\mathbf{S}_j$ is an n_j-dimensional vector described by the coupling constraint.

$$\mathbf{S}_j = \sum_{\substack{k=1 \\ k \neq j}}^{4} \mathbf{A}_{jk}\mathbf{U}_k \tag{85}$$

From Fig. 6-1 we see that $\mathbf{A}_{jk}$ equals the null matrix $(j \neq k)$ except for $\mathbf{A}_{j,j-1}$ and/or $\mathbf{A}_{j,j+1}$ which are diagonal.

For this example, the Hamiltonian becomes

$$\begin{aligned} H = \sum_{j=1}^{4} \Big[&(\mathbf{U}_{dj} - \mathbf{U}_j)' \, \mathbf{Q}_j(\mathbf{U}_{dj} - \mathbf{U}_j) + \mathbf{M}_j'\mathbf{R}_j\mathbf{M}_j \\ &+ \boldsymbol{\lambda}_j'(\mathbf{A}_{jj}\mathbf{U}_j + \mathbf{B}_j\mathbf{M}_j + \mathbf{S}_j + \mathbf{G}_j\mathbf{F}_j) + \boldsymbol{\rho}_j' \Big(\sum_{\substack{k=1 \\ k \neq j}}^{4} \mathbf{A}_{jk}\mathbf{U}_k - \mathbf{S}_j \Big) \Big] \end{aligned} \tag{86}$$

and the necessary condition (63) yields

$$2\mathbf{R}_j\mathbf{M}_j + \mathbf{B}_j'\boldsymbol{\lambda}_j = \mathbf{0}$$

which can be used to eliminate $\mathbf{M}_j$ from the remaining conditions. Then

$$\dot{\mathbf{U}}_j = \mathbf{A}_{jj}\mathbf{U}_j - 2^{-1}\mathbf{B}_j\mathbf{R}_j^{-1}\mathbf{B}_j'\boldsymbol{\lambda}_j + \mathbf{S}_j + \mathbf{G}_j\mathbf{F}_j \tag{87}$$

$$\dot{\boldsymbol{\lambda}}_j = -2\mathbf{Q}_j(\mathbf{U}_{dj} - \mathbf{U}_j) - \mathbf{A}'_{jj}\boldsymbol{\lambda}_j - \sum_{\substack{k=1 \\ k \neq j}}^{4} \mathbf{A}'_{jk}\boldsymbol{\rho}_k \tag{88}$$

$$j = 1, 2, 3, 4$$

where $\mathbf{U}_j(t_0) = \mathbf{U}_{0j}$ and $\boldsymbol{\lambda}_j(t_1) = \mathbf{0}$. The parameters $\boldsymbol{\rho}_k$ and $\mathbf{S}_j$ are obtained from

$$\boldsymbol{\rho}_k = \boldsymbol{\lambda}_k \tag{89}$$

$$\mathbf{S}_j = \sum_{\substack{k=1 \\ k \neq j}}^{4} \mathbf{A}_{jk}\mathbf{U}_k \tag{90}$$

Utilizing the special form of $\mathbf{A}_{jk}$, Eqs. (88) and (90) can be expanded to

$$\dot{\boldsymbol{\lambda}}_j = -2\mathbf{Q}_j(\mathbf{U}_{dj} - \mathbf{U}_j) - \mathbf{A}'_{jj}\boldsymbol{\lambda}_j - \mathbf{A}'_{j,j-1}\boldsymbol{\rho}_{j-1} - \mathbf{A}'_{j,j+1}\boldsymbol{\rho}_{j+1} \tag{91}$$

$$\mathbf{S}_j = \mathbf{A}_{j,j-1}\mathbf{U}_{j-1} + \mathbf{A}_{j,j+1}\mathbf{U}_{j+1} \tag{92}$$

where we define $\mathbf{A}_{10} = \mathbf{A}_{45} = \boldsymbol{\phi}$, the null matrix. Thus we see that each $\boldsymbol{\lambda}_j$ is coupled only to $\boldsymbol{\lambda}_{j+1}$ and $\boldsymbol{\lambda}_{j-1}$ and similarly $\mathbf{U}_j$ is coupled to $\mathbf{U}_{j+1}$ and $\mathbf{U}_{j-1}$. This configuration arises for most scalar semidiscrete models in one or two space dimensions. For vector equations or more space dimensions, additional nonzero coupling matrices might arise but in general there will be fewer than the total number of subsystems. In particular, if the partial differential equation contains only one space dimension, the $\mathbf{A}$ matrix is a strictly tridiagonal matrix. On the other hand, if three space dimensions are present, the $\mathbf{A}$ matrix has four symmetrically placed diagonals in addition to the tridiagonal band. For any given problem, a particularly clever partitioning of $\mathbf{U}$ may yield additional simplifications.

The computational procedure for the Gauss-Seidel controller is illustrated in Fig. 6-2 for the case $N = 3$ and proceeds as follows:

1. Estimate $\mathbf{U}_{j+1}$, $\boldsymbol{\lambda}_{j+1}$ for $j = 1, 2, \ldots, N - 1$ and set $j = 1$.
2. Determine $\boldsymbol{\rho}_{j+1}$ and $\mathbf{S}_j$ from Eqs. (89) and (92) respectively.
3. Solve subsystem j for $\mathbf{U}_j$, $\boldsymbol{\lambda}_j$ from Eqs. (87) and (91), using the latest information.
4. Is $j = N$?
   No; Set $j = j + 1$ and go to (2).
   Yes; Are Eqs. (89) and (92) satisfied to desired accuracy for $j = 1, 2, \ldots, N$?
      No; go to (5).
      Yes; Stop.

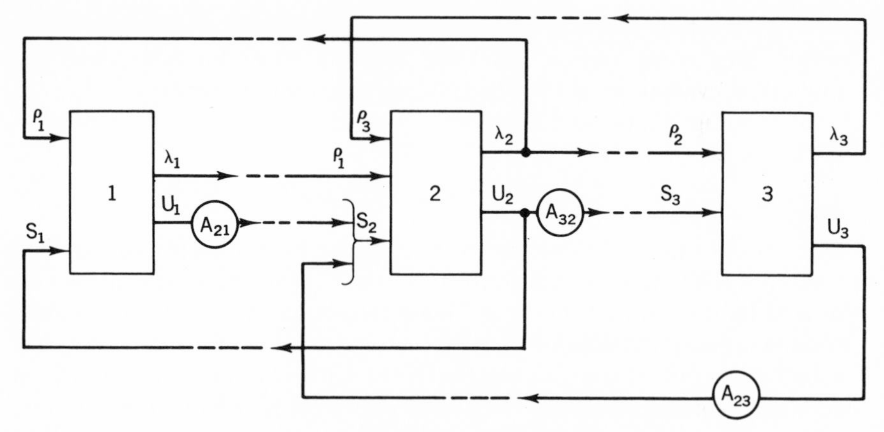

FIG. 6-2. Gauss-Seidel procedure for $N = 3$.

5. Determine $\boldsymbol{\rho}_{j-1}$ and $\mathbf{S}_j$ from Eqs. (89) and (92) respectively.
6. Solve subsystem j for $\mathbf{U}_j$, $\boldsymbol{\lambda}_j$ from Eqs. (87) and (91) using the latest information.
7. Is $j = 1$?
   No; Set $j = j - 1$ and go to (5).
   Yes; Go to (2).

This controller has been found to have excellent convergence properties for both linear and nonlinear problems. The good convergence is attributed to the strongly diagonal character of the $\mathbf{A}$ matrix. This same factor speeds convergence of the subsystems over a wide range of initial estimates for $\boldsymbol{\rho}_k$ and $\mathbf{S}_j$. It is because of these convergence properties that this method of solution is attractive.

In principle, the boundary control problem can be solved by the multilevel methods mentioned here. However, the system must be decomposed so that each subsystem contains at least as many boundary control variables as terminal conditions. Without this restriction, insufficient degrees of freedom are present in the state equations to satisfy both the terminal constraints (57) and the coupling constraints. Alternatively, if the state terminal conditions are not specified, the above restriction does not apply and the state equations can be integrated in the forward direction while the adjoint equations are integrated in the backward direction. This procedure considerably simplifies the solution of the two-point boundary-value problem.

One interesting application of multilevel methods to the optimization of distributed parameter systems occurs when inequality constraints involving only state variables are present. For such problems, it is well known that the adjoint variables possess discontinuities at points where they enter and/or exit from the constraint boundary and the numerical evaluation of this discontinuity is often difficult.[1]

In treating distributed parameter systems having a state inequality of the form

$$u_{\min}(\mathbf{X}, t) \leqslant u(\mathbf{X}, t) \leqslant u_{\max}(\mathbf{X}, t)$$

by decomposition, it is convenient to consider the inequality constrained variables $u(\mathbf{X}_i, t)$ as pseudocontrol variables. The problem can then be treated by the simpler theory applying to inequality constrained control variables. Since the adjoint variables are continuous when the inequality constraints contain controls explicitly, the treatment of state inequalities as pseudocontrol inequalities is not expected to yield exact results; in particular, it does not yield discontinuous adjoint variables. This application is similar to the penalty function approach discussed by Kelley.[5]

In this section, we have not emphasized the solution of the subsystem optimization problems. However, it should be clear that by a suitable decomposition, particularly simple subproblems may often be defined. For example, a linear subsystem possessing a closed-form solution might be extracted from an otherwise nonlinear system. In any case, different optimizing algorithms can be used in each subproblem. Thus dynamic programming, gradient, conjugate gradient, or other schemes could all be used in solving the independent subproblems if their use would prove advantageous. Clearly an efficient algorithm for solving the subproblems is extremely important since they must be solved many times in converging to the overall system optimum.

In case the discretized model of a distributed parameter system takes the form of pure difference equations, as is the case for elliptic or parabolic partial differential equations where both space and time are discrete, the application of multilevel control techniques requires only slight modification. Since the discrete system model takes the same form as the right-hand side of the differential equations discussed earlier, the characteristic form of the $\mathbf{A}$ matrix shown in Fig. 6-1 is maintained. The optimization problem is now static in nature and the mathematical programming theory discussed in Chapter 1 is applicable. The Hamiltonian function of this section is replaced by a Lagrange function which may include products of inequality constraints and Kuhn-Tucker multipliers in addition to the usual equality constraints. The subsystem optimization problems do not require the solution of two-point boundary-value problems but rather the satisfaction of certain simultaneous nonlinear algebraic or transcendental equations. The Gauss-Seidel controller requires almost no modification since it does not involve the solution of differential equations. Of course in this case the parameter values ρ_k and $\mathbf{S}_j$ are no longer time varying. Of importance is the fact that the system model is still characterized by strongly diagonal matrices and the good convergence properties of the Gauss-Seidel controller are maintained. A numerical example of this type of system will be given in the next section.

6-4 NUMERICAL EXAMPLES

In this section we will formulate and solve three numerical examples; a nonlinear distributed control problem, a state inequality constrained problem, and a system identification problem. In the first two examples, the subsystem optimization problems are dynamic and the resulting two-point boundary-value problems are solved by the method of *quasilinearization*.[4] This method is an iterative technique which satisfies the boundary conditions and the maximum principle exactly along a given trajectory and iterates until the subsystem state and adjoint differential equations are satisfied. Convergence depends upon the initial estimates of the state and adjoint solutions and, when obtained, the convergence is quadratic. For the examples discussed here convergence was rapid over a wide range of initial estimates. For simplicity, we consider a class of scalar partial differential equations in one space variable. This distributed parameter system is modeled by eight differential-difference equations divided between two subsystems consisting of four mesh points each. The resulting subsystem two-point boundary-value problem is of eighth order (four state and four adjoint variables) and represents a convenient size for the method of quasilinearization.

The third example consists of a diffusion equation over two space dimensions which models the pressure distribution in an underground reservoir. Using a transformation of variables, a closed-form expression is obtained in terms of unknown parameters. The resulting subsystem optimization problems are static and the necessary conditions for a minimum are provided by the Lagrange and Kuhn-Tucker theory discussed in Chapter 1. The subsystems are optimized by a direct search method resulting in a global solution. In all three examples, the Gauss-Seidel second-level controller is used.

Example 6-6

We consider the minimum-effort problem defined by

$$\min_{m} \int_0^1 \int_0^{t_1} m^2(x, t)\, dt\, dx \tag{93}$$

subject to the nonlinear diffusion equation

$$\frac{\gamma u}{\gamma t} = \alpha u \frac{\gamma^2 u}{\gamma x^2} + m \tag{94}$$

with boundary and initial conditions given by

$$u(0, x) = u(1, t) = 0$$

and

$$u(x, 0) = u_0(x)$$

respectively. It is desired to attain the terminal state given by

$$u(x, t_1) = u_1(x)$$

at a specified time t_1.

The semidiscrete approximation to this problem is stated as

$$\min_{\mathbf{M}} \int_0^{t_1} \mathbf{M}'\mathbf{M}\, dt$$

subject to the differential equation

$$\dot{\mathbf{U}} = \mathbf{A}(\mathbf{U})\mathbf{U} + \mathbf{M}$$

$$\mathbf{U}(t_0) = \mathbf{U}_0$$

$$\mathbf{U}(t_1) = \mathbf{U}_1$$

where

$$\mathbf{U} = [u_1, u_2, \ldots, u_n]'$$

$$\mathbf{M} = [m_1, m_2, \ldots, m_n]'$$

If we partition $\mathbf{U}$ into two fourth-order subsystems, the state-dependent matrix $\mathbf{A}(\mathbf{U})$ becomes

$$\mathbf{A}(\mathbf{U}) = k \left[\begin{array}{cccc|cccc} -2u_1 & u_1 & & & & & & \\ u_2 & -2u_2 & u_2 & & & 0 & & \\ & u_3 & -2u_3 & u_3 & & & & \\ & & u_4 & -2u_4 & u_4 & & & \\ \hline & & & u_5 & -2u_5 & u_5 & & \\ & & & & u_6 & -2u_6 & u_6 & \\ & 0 & & & & u_7 & -2u_7 & u_7 \\ & & & & & & u_8 & -2u_8 \end{array}\right]$$

where

$$k = \alpha h^{-2}$$

The Hamiltonian is given by

$$H = \sum_{j=1}^{2} \Big\{ \mathbf{M}_j'\mathbf{M}_j + \boldsymbol{\lambda}_j'[\mathbf{A}_{jj}(\mathbf{U}_j)\,\mathbf{U}_j + \mathbf{M}_j + \mathbf{D}_j(\mathbf{U}_j)\,\mathbf{S}_j]$$

$$+ \boldsymbol{\rho}_j' \Big(\sum_{\substack{k=1 \\ k \neq j}}^{2} \mathbf{C}_{jk}\mathbf{U}_k - \mathbf{S}_j \Big) \Big\} \tag{95}$$

where

$$\mathbf{D}_1 = \mathbf{A}_{12} \qquad \mathbf{D}_2 = \mathbf{A}_{21}$$

and $\mathbf{C}_{12}$, $\mathbf{C}_{21}$ are 4×4 matrices of zeros except for a single one in the upper-left and lower-right corners respectively. The one-for-one decomposition is required in this case because otherwise the multiplicative form of the nonlinearity couples the subsystems through the adjoint equation (61). The pseudocontrol variables and Lagrange multipliers are defined by

$$\mathbf{S}_1 = [s_1, s_2, s_3, s_4]' \qquad \mathbf{S}_2 = [s_5, s_6, s_7, s_8]'$$

$$\boldsymbol{\rho}_1 = [\rho_1, \rho_2, \rho_3, \rho_4]' \qquad \boldsymbol{\rho}_2 = [\rho_5, \rho_6, \rho_7, \rho_8]'$$

By applying the necessary conditions, the subsystem two-point boundary-value problems are readily seen to be

SUBSYSTEM 1:

$$\begin{aligned}
\dot{u}_1 &= k[\qquad - 2u_1^2 + u_1u_2] - 2^{-1}\lambda_1 \\
\dot{u}_2 &= k[u_1u_2 - 2u_2^2 + u_2u_3] - 2^{-1}\lambda_2 \\
\dot{u}_3 &= k[u_2u_3 - 2u_3^2 + u_3u_4] - 2^{-1}\lambda_3 \\
\dot{u}_4 &= k[u_3u_4 - 2u_4^2 + u_4s_1] - 2^{-1}\lambda_4 \\
\dot{\lambda}_1 &= -k[\qquad (-4u_1 + u_2)\,\lambda_1 + u_2\lambda_2] \\
\dot{\lambda}_2 &= -k[u_1\lambda_1 + (u_1 - 4u_2 + u_3)\,\lambda_2 + u_3\lambda_3] \\
\dot{\lambda}_3 &= -k[u_2\lambda_2 + (u_2 - 4u_3 + u_4)\,\lambda_3 + u_4\lambda_4] \\
\dot{\lambda}_4 &= -k[u_3\lambda_3 + (u_3 - 4u_4 + s_1)\,\lambda_4] - \rho_8
\end{aligned}$$

with boundary conditions

$$u_i(0) = u_{i0} \qquad u_i(t_1) = u_{i1}$$

$$i = 1, 2, 3, 4$$

SUBSYSTEM 2:

$$\dot{u}_5 = k[-2u_5^2 + u_5u_6 + u_5s_8] - 2^{-1}\lambda_5$$

$$\dot{u}_6 = k[\ u_5u_6 - 2u_6^2 + u_6u_7] - 2^{-1}\lambda_6$$

$$\dot{u}_7 = k[\ u_6u_7 - 2u_7^2 + u_7u_8] - 2^{-1}\lambda_7$$

$$\dot{u}_8 = k[\ u_7u_8 - 2u_8^2\ ] - 2^{-1}\lambda_8$$

$$\dot{\lambda}_5 = -k[(s_8 - 4u_5 + u_6)\,\lambda_5 + u_6\lambda_6] - \rho_1$$

$$\dot{\lambda}_6 = -k[u_5\lambda_5 + (u_5 - 4u_6 + u_7)\,\lambda_6 + u_7\lambda_7]$$

$$\dot{\lambda}_7 = -k[u_6\lambda_6 + (u_6 - 4u_7 + u_8)\,\lambda_7 + u_8\lambda_8]$$

$$\dot{\lambda}_8 = -k[u_7\lambda_7 + (u_7 - 4u_8)\,\lambda_8]$$

with boundary conditions

$$u_i(0) = u_{i0} \qquad u_i(t_1) = u_{i1}$$

$$i = 5, 6, 7, 8$$

In solving these subproblems, s_i and ρ_k are considered parameters which are determined by the second-level controller. Using the Gauss-Seidel controller, these values are determined by Eq. (69) as

$$\begin{aligned} s_1 &= u_5 \qquad & s_8 &= u_4 \\ \rho_8 &= ku_5\lambda_5 \qquad & \rho_1 &= ku_4\lambda_4 \end{aligned} \tag{96}$$

The computational scheme requires the initial estimation of s_1 and ρ_8 only. The cyclical procedure described for Example 6-5 can then be applied to these equations directly.

Numerical results have been obtained for this problem using the method of quasilinearization to solve the two-point boundary-value problems. The problem can be regarded as one of taking a thin rod from some initial temperature profile to some final temperature profile in a given length of time. The temperature is controlled by a distributed source or sink which must be controlled using minimum energy. The thermal diffusivity of the rod is proportional to temperature. Taking k as 0.27, and $[0, t_1]$ as $[0, 5]$, the controls $\mathbf{M}_j$ and states $\mathbf{U}_j$ are given in Fig. 6-3 for the initial and final conditions shown. By symmetry, both subsystems have the same response. For initial estimates of

$$s_1 = 50$$

$$\rho_8 = 0$$

the convergence of the Gauss-Seidel controller was monitored by the norms

$$\| e_1 \|_i = \int_0^{.5} \{| u_4 - s_8 |_i + | ku_4\lambda_4 - \rho_1 |_i\} \, dt$$
$$\| e_2 \|_i = \int_0^{.5} \{| u_5 - s_1 |_i + | ku_5\lambda_5 - \rho_8 |_i\} \, dt \tag{97}$$

where i represents the iteration number. The values obtained for this example are shown in Fig. 6-4.

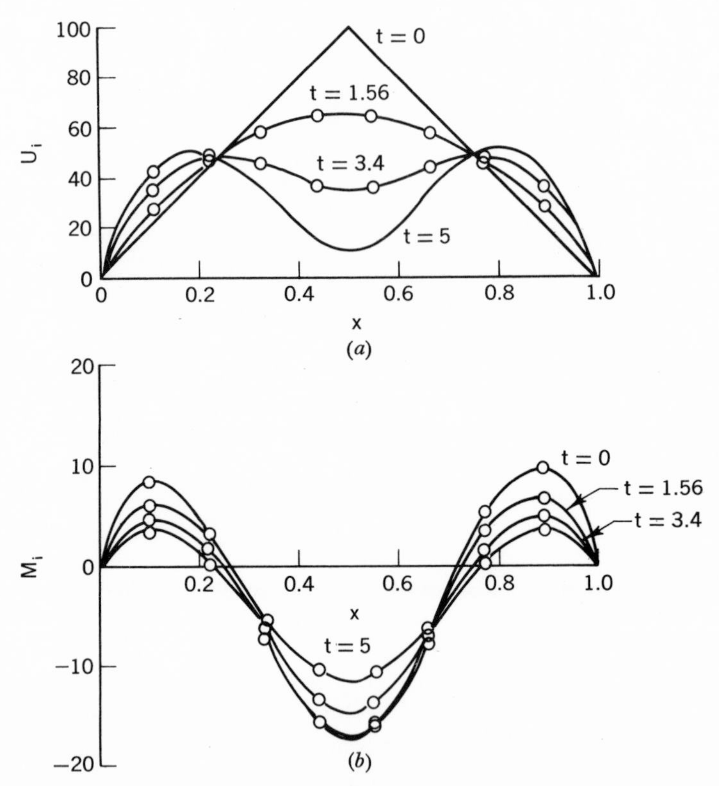

FIG. 6-3. Optimal states and controls for Example 6-6. (*a*) Optimal state response. (*b*) Optimal distributed controls.

It is sometimes of interest to monitor the Hamiltonian function since for autonomous systems with prescribed terminal conditions H is constant at the optimum. Checking H for this condition serves as a convenient check of the numerical techniques. Of course the subsystem Hamiltonians H_j of Eq. (67) are not necessarily constant since the optimum controls for the overall problem are not in general optimum for the subsystems taken separately.

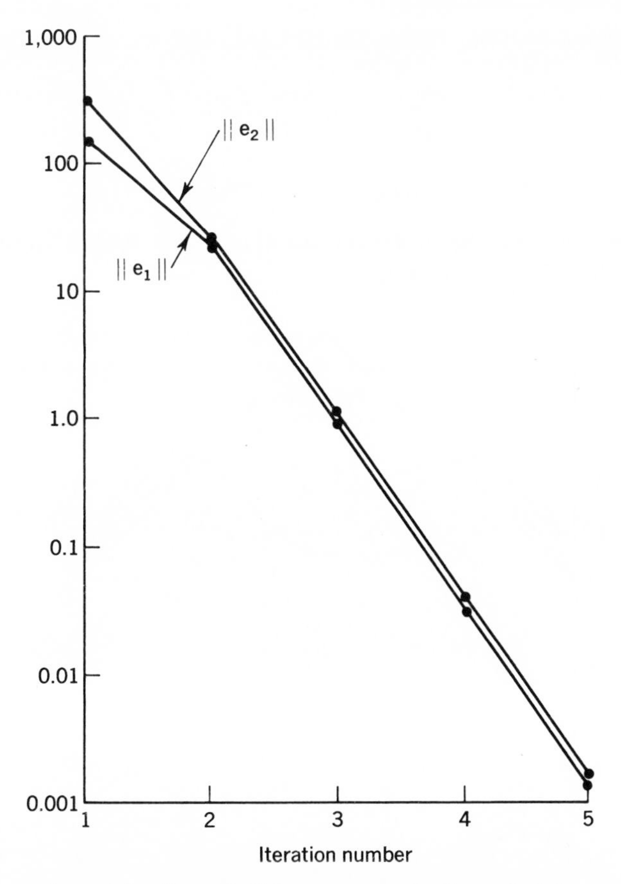

FIG. 6-4. Gauss-Seidel convergence for Example 6-6.

Example 6-7

We consider now a problem similar to Example 6-6 except that the system equation is linear

$$\frac{\gamma u}{\gamma t} = \alpha \frac{\gamma^2 u}{\gamma x^2} + m \tag{98}$$

and a certain region $\mathscr{I}$ of x is subject to the inequality constraint

$$u(x_i, t) \geqslant c(t) \qquad i \in \mathscr{I}$$

As discussed previously, we wish to avoid the difficulty of determining the discontinuities in the adjoint variables by satisfying the inequality with a pseudocontrol variable. Arbitrarily taking $u(x_i, t)$ as u_4 in

Example 6-6 and employing Valentine's[10] method of slack variables, we can write the Hamiltonian as

$$H = \sum_{j=1}^{2} \Big[\mathbf{M}_j'\mathbf{M}_j + \boldsymbol{\lambda}_j' \Big(\mathbf{A}_{jj}\mathbf{U}_j + \mathbf{M}_j + \sum_{\substack{k=1 \\ k \neq j}}^{2} \mathbf{A}_{jk}\mathbf{S}_j\Big)$$

$$+ \boldsymbol{\rho}_j' \Big(\sum_{\substack{k=1 \\ k \neq j}}^{2} \mathbf{C}_{jk}\mathbf{U}_k - \mathbf{S}_j\Big)\Big] + \nu(\xi^2 - \boldsymbol{\beta}'\mathbf{S}_2 + c) \qquad (99)$$

where $\boldsymbol{\beta}' = [0, 0, 0, 1]$, ν is a scalar Lagrange multiplier, and ξ is a real scalar slack variable. The **C** matrices are defined in Example 6-6. The subsystem necessary conditions are given by

SUBSYSTEM 1:

$$\dot{\mathbf{U}}_1 = \mathbf{A}_{11}\mathbf{U}_1 + \mathbf{A}_{12}\mathbf{S}_1 - 2^{-1}\boldsymbol{\lambda}_1$$

$$\dot{\boldsymbol{\lambda}}_1 = \mathbf{A}_{11}'\boldsymbol{\lambda}_1 - \mathbf{C}_{21}'\boldsymbol{\rho}_2$$

$$\mathbf{U}_1(t_0) = \mathbf{U}_{10} \qquad \mathbf{U}_1(t_1) = \mathbf{U}_{11}$$

SUBSYSTEM 2:

$$\dot{\mathbf{U}}_2 = \mathbf{A}_{22}\mathbf{U}_2 + \mathbf{A}_{21}\mathbf{S}_2 - 2^{-1}\boldsymbol{\lambda}_2$$

$$\dot{\boldsymbol{\lambda}}_2 = -\mathbf{A}_{22}'\boldsymbol{\lambda}_2 - \mathbf{C}_{12}'\boldsymbol{\rho}_1$$

$$\mathbf{U}_2(t_0) = \mathbf{U}_{20} \qquad \mathbf{U}_2(t_1) = \mathbf{U}_{21}$$

The remaining necessary conditions needed for the second-level control are

$$\begin{aligned} s_1 &= u_5 & s_8 &= u_4 \\ \rho_8 &= k\lambda_5 - \nu & \rho_1 &= k\lambda_4 \\ & & \xi^2 &= s_8 - c \\ & & \xi\nu &= 0 \end{aligned} \qquad (100)$$

The Legendre necessary condition requires that $\nu \geqslant 0$. If $\nu = 0$ the strict inequality $u_4 > c$ is satisfied and the effect of the constraint can be ignored. In this case the Gauss-Seidel equations as given above are similar to those for Example 6-6. Alternatively, if $\nu > 0$, Eq. (100) requires that $\xi = 0$, and $u_4 = c$. It remains to control subsystem 1 by determining ν in order to satisfy this relationship on the boundary. Taking the first variation of H with respect to ν and substituting Eq. (100) yields

$$\delta \mathbf{H} = (c - u_4)\, \delta\nu$$

By the saddle-value properties shown in Chapter 4, $\delta\nu$ should be chosen to minimize H or

$$\delta\nu = L(c - u_4) \qquad L > 0 \tag{101}$$

Equation (101) forms the basis for a gradient-type algorithm which must be combined with the Gauss-Seidel controller. Since the Gauss-Seidel equations are solved in an iterative fashion, ρ_8 can be modified to include calculation of ν as follows:

$$\rho_8^{(i)} = k\lambda_5^{(i-1)} + L(c - u_4)^{(i-1)} + \nu^{(i-1)} \tag{102}$$

where i is the iteration number. In Eq. (102), L, the gradient step size, must be specified along with the initial estimate $\nu^{(0)}$ on the boundary. In general, Eq. (102) will bring u_4 arbitrarily close to the boundary but will not attain it exactly.

This problem was solved numerically taking $c(t)$ as the ellipse shown in Fig. 6-5. The Gauss-Seidel controller was used with the addition of the gradient terms along the boundary. The step size L was chosen as 0.95 and the initial estimates were as in Example 6-6. This procedure converged in eight iterations to the values shown in Fig. 6-5.

The remaining example in this chapter also deals with the diffusion equation; however, the multilevel optimization is formulated as a static problem involving the identification of unknown system parameters.

Example 6-8

The pressure distribution in an underground reservoir is modeled by a linear parabolic partial differential equation of the form

$$\frac{\gamma}{\gamma x}\left(\tau \frac{\gamma P}{\gamma x}\right) + \frac{\gamma}{\gamma y}\left(\tau \frac{\gamma P}{\gamma y}\right) = \sigma \frac{\gamma P}{\gamma t} + q \tag{103}$$

Pressure is denoted by P and q represents a source strength (production rate per unit area) at some producing well. The coefficients in the equation characterize the porous medium. Transmissibility $\tau(x, y)$ is a measure of the ease with which fluid moves through the system and storage $\sigma(x, y)$ is a measure of system capacity. Vertical fluid flow is assumed to be negligible.

We wish to determine the distributed parameters $\tau(x, y)$ and $\sigma(x, y)$ by fitting the model in a least-squares sense to observed pressure data obtained at a number of wells in the reservoir. When each well is producing, boundaries with zero potential gradient (no flow) must exist between wells. Fluid on opposite sides of these "boundaries" flows to opposite wells just as if the line of separation were impermeable. A convenient approximation in modeling the system is to separate

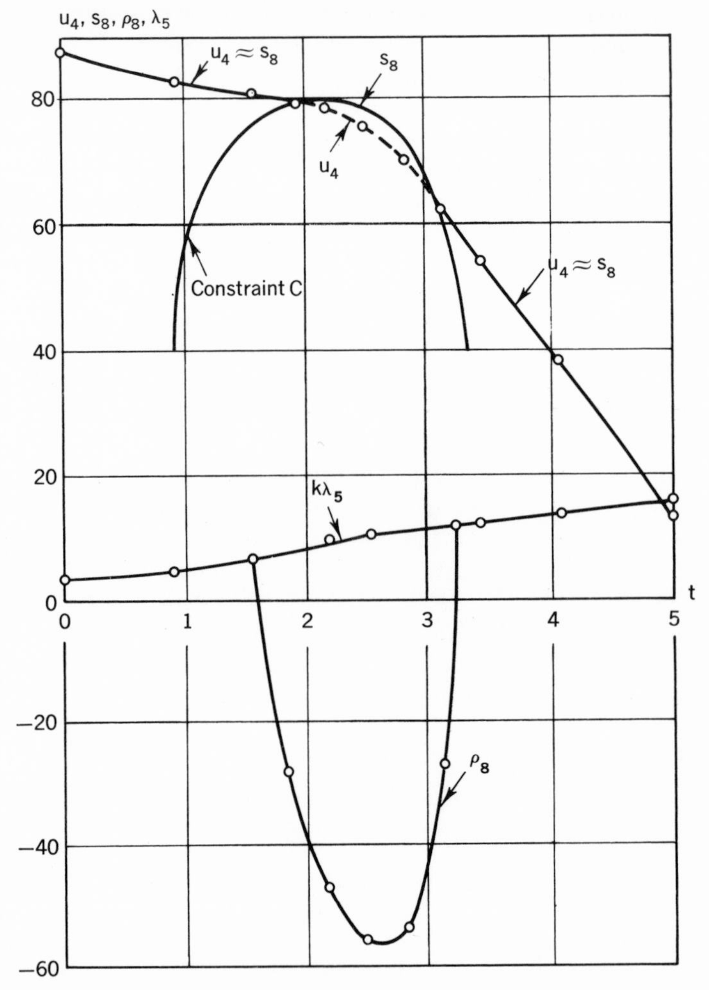

FIG. 6-5. Final values of second-level variables for Example 6-7.

regions containing a single well by straight-line boundaries. A pair of boundary lines on opposite sides of a well will intersect to form a wedge. As a result the model for N wells in a large reservoir is obtained by dividing the system into N wedge-shaped homogeneous regions, each radiating from a single arbitrarily selected origin and each enclosing a single well. We denote the ith well location by the angle θ_i. It is bounded by two boundaries at locations given by angles α_i and α_{i-1}, where

$$\theta_i < \alpha_i < \theta_{i+1}$$

The distributed parameters $\tau(x, y)$ and $\sigma(x, y)$ are approximated by constant "average" values in each wedge-shaped region. The identification scheme must then determine values for the N-dimensional vectors $\boldsymbol{\tau}$ and $\boldsymbol{\sigma}$ composed of elements from each region along with the location of the N well boundaries $\boldsymbol{\alpha}$. Thus the geometry of the model is actually provided by the system behavior.

Indicating the model pressure by u to distinguish it from observed pressure values, we can write Eq. (103) in a more convenient coordinate system as follows

$$\tau\left[\frac{1}{r}\frac{\gamma}{\gamma r}\left(r\frac{\gamma u}{\gamma r}\right)\right] = \sigma\frac{\gamma u}{\gamma t} \tag{104}$$

where the initial and boundary conditions are given by[9]

$$\lim_{t\to 0} u(r, t) = \lim_{r\to\infty} u(r, t) = P_1 = \text{const}$$

$$\lim_{r\to 0} r\left(\frac{\gamma u}{\gamma r}\right) = \frac{q}{2\pi\tau} = \text{const}$$

Equation (104) can be transformed into an ordinary differential equation by using the Boltzmann transformation:

$$\lambda = \frac{\sigma r^2}{4\tau t}$$

The system is now represented by

$$\lambda\frac{d^2u}{d\lambda^2} + (1 + \lambda)\frac{du}{d\lambda} = 0 \tag{105}$$

with boundary conditions

$$\lim_{\lambda\to\infty} u(\lambda) = P_1 \qquad \lim_{\lambda\to 0} 2\lambda\frac{du}{d\lambda} = \frac{q}{2\pi\tau}$$

The solution of Eq. (105) is readily obtained in terms of the exponential integral $E(\xi)$:

$$u = P_1 - \frac{q}{4\pi\tau}\int_{\xi=\lambda}^{\infty} \xi^{-1}\exp(-\xi)\,d\xi = P_1 + \frac{q}{4\pi\tau}E\left(-\frac{\sigma r^2}{4\tau t}\right)$$

Since $E(-\xi)$ is negative, the pressure decreases as production occurs.

The method of images can be used to extend the solution to problems in a bounded, wedge-shaped, homogeneous region.[2] Suppose the ith actual image includes an angle of $2\pi/m_i$ radians between azimuths α_i and α_{i+1}, where m_i is an even integer. The image system then contains m_i wedges filling the entire plane. Each wedge has the same origin and contains a well which is the mirror image of the actual well and/or image wells across each of its adjacent boundaries. The angle between the ith well with azimuth θ_i and its kth image is given by $2\zeta_{ik}$ where

$$\zeta_{ik} = \left[\frac{k}{2}\right](\alpha_i - \theta_i) + \left[\frac{k-1}{2}\right](\theta_i - \alpha_{i-1} + 2\pi\delta_{i1}) \tag{106}$$

$$i = 1, 2, \ldots, N \qquad k = 1, 2, \ldots, m_i$$

and δ_{ik} is the Kronecker delta. The notation $[k/2]$ and $[(k-1)/2]$ denotes that these quantities are truncated to integer values. The pressure computed at the ith well and the jth time corresponds to production q_i and is denoted by u_{ij}. Summing the pressure effects caused by m_i image wells for each of N actual wells yields

$$u_{ij} = P_1 + \frac{q_i}{4\pi\tau} \sum_{k=1}^{m_i} E_i\left(-\frac{\sigma_i r_i^2 \sin^2 \zeta_{ik}}{4\tau_i t_j}\right) \tag{107}$$

where r_i is the radial distance between the origin and the ith well. The trigonometric term arises directly from the model geometry.

The static optimization problem can now be posed as

$$\min_{\tau_i, \sigma_i, \alpha_i} \sum_{i=1}^{N} \sum_{j=1}^{K} (u_{ij} - P_{ij})^2 \tag{108}$$

subject to the constraints given by Eqs. (106) and (107) and

$$\left.\begin{aligned} m_i(\alpha_i - \alpha_{i-1}) &= 2\pi \\ \theta_i < \alpha_i &< \theta_{i+1} \\ \tau_i &> 0 \\ \sigma_i &> 0 \end{aligned}\right\} \quad i = 1, 2, \ldots, N \tag{109}$$

where

$$\begin{aligned} \alpha_0 &\triangleq \alpha_N - 2\pi \\ \theta_{N+1} &\triangleq \theta_1 \\ m_i &\triangleq \text{even integer} \end{aligned}$$

It is convenient to rewrite Eq. (109) as

$$\begin{aligned} &\alpha_i \leqslant \theta_{i+1} - \epsilon_i \\ &\alpha_i \geqslant \theta_i + \epsilon_i \\ &\tau_i \geqslant \beta_i \\ &\sigma_i \geqslant \gamma_i \qquad i = 1, 2, \ldots, N \end{aligned}$$

where ϵ_i, β_i, and γ_i are arbitrarily small constants to be specified.

We can now formulate this problem as a two-level optimization problem where each wedge-shaped region is considered as a subsystem. Since individual subsystems are coupled only by ζ_{ik}, we can regard these subsystems as independent after making the simple change of variables

$$\alpha_{i-1} = s_i \qquad i = 1, 2, \ldots, N \tag{110}$$

where the s_i are the pseudocontrol variables. The decomposed static optimization problem can now be written as follows:

$$\min_{\tau_i, \sigma_i, \alpha_i, s_i} \left\{ f = \sum_{i=1}^{N} \sum_{j=1}^{K} \left[P_1 + \frac{q_i}{4\pi\tau_i} \sum_{k=1}^{m_i} E_i \left(\frac{-\sigma_i r_i^2 \sin^2 \zeta_{ik}}{4\tau_i t_j} \right) - P_{ij} \right]^2 \right\} \tag{111}$$

where

$$\zeta_{ik} = \left[\frac{k}{2}\right] (\alpha_i - \theta_i) + \left[\frac{k-1}{2}\right] (\theta_i - s_i + 2\pi\delta_{i1})$$

$$m_i = \frac{2\pi}{\alpha_i - s_i} = \text{even integer}$$

and such that

$$\mathbf{G}_i(\alpha_i, \tau_i, \sigma_i) = \begin{bmatrix} \theta_{i+1} & -\epsilon_i - \alpha_i \\ \alpha_i & -\epsilon_i - \theta_i \\ \tau_i & -\beta_i \\ \sigma_i & -\gamma_i \end{bmatrix} \geqslant 0 \qquad i = 1, 2, \ldots, N \tag{112}$$

Equation (111) is separable and can be written

$$f(\mathbf{X}; s) = \sum_{i=1}^{N} f_i(\mathbf{X}_i; s_i)$$

$$\mathbf{X}_i \triangleq (\tau_i, \sigma_i, \alpha_i)$$

$$\mathbf{X} = [\mathbf{X}_1, \mathbf{X}_2, \ldots, \mathbf{X}_N]'$$

Regarding s_i as a known parameter in the ith subsystem, we see that the subsystems are uncoupled. Hence each subsystem optimization is performed by

$$\min_{\mathbf{X}_i} f_i(\mathbf{X}_i; s_i)$$

subject to the constraints (112).

In order to ensure the minimization of Eq. (111) for N wells, we define the Lagrangian

$$L(\mathbf{X}, \boldsymbol{\lambda}, \mathbf{s}, \mu) = \sum_{i=1}^{N} L_i(\mathbf{X}_i, \boldsymbol{\lambda}_i ; s_{i+1}, s_i, \mu_i)$$

where

$$L_i = f_i(\mathbf{X}_i ; s_i) + \boldsymbol{\lambda}_i'\mathbf{G}_i(\mathbf{X}_i) + \mu_i(\alpha_i - s_{i+1})$$

$$\mathbf{s} = [s_1, s_2, \ldots, s_N]'$$

$$\boldsymbol{\lambda} = [\lambda_1, \lambda_2, \ldots, \lambda_N]'$$

$$\boldsymbol{\mu} = N\text{-dimensional vector Lagrange multiplier}$$

$$s_{N+1} \triangleq s_1$$

Assuming that the Kuhn-Tucker constraint qualification holds, the necessary conditions for a minimum of each subsystem are

$$\begin{aligned} \nabla_{\mathbf{x}_i} L_i(\mathbf{X}_i^*, \boldsymbol{\lambda}_i^*; s_i, s_{i+1}, \mu_i) &= \mathbf{0} \\ \boldsymbol{\lambda}_i' \nabla_{\boldsymbol{\lambda}_i} L_i(\mathbf{X}_i^*, \boldsymbol{\lambda}_i^*; s_i, s_{i+1}, \mu_i) &= 0 \\ \lambda_{ij}^* \leqslant 0 \qquad i &= 1, 2, \ldots, N \\ j &= 1, 2, 3, 4 \end{aligned} \tag{113}$$

The solution of Eq. (113) proceeds for given values of the parameters s_i, s_{i+1}, and μ_i by any of the standard nonlinear programming algorithms. It remains to determine these parameter values by satisfying the remaining necessary conditions for a minimum; namely,

$$\begin{aligned} \nabla_{s_{i+1}} L = \nabla_{s_{i+1}} f_{i+1} - \mu_i^* = 0 \qquad i = 1, 2, \ldots, N \\ \nabla_{\mu_i} L = \alpha_i - s_{i+1}^* = 0 \qquad i = 1, 2, \ldots, N \end{aligned} \tag{114}$$

Equations (114) are just the Gauss-Seidel second-level controller equations previously discussed. The solution proceeds iteratively until

$$\| \mathbf{s}^{(k+1)} - \mathbf{s}^{(k)} \| \leqslant \delta_1$$

$$\| \boldsymbol{\mu}^{(k+1)} - \boldsymbol{\mu}^{(k)} \| \leqslant \delta_2$$

where k is the iteration number and δ_1 and δ_2 are specified scalar values.

A numerical example was solved using four wells and seven pressure observations for each well. A direct-search technique was found to be efficient for solving the subsystems and produced global subsystem minima. The Gauss-Seidel algorithm was used for coordinating the subsystems and converged in six iterations. These results are given in detail by Haimes, Perrine, and Wismer.[3]

REFERENCES:

1. Berkovitz, L. D.: On Control Problems with Bounded State Variables, *J. Math. Anal. Appl.*, vol. 5, pp. 488–498, 1962.
2. DeWiest, R. J. M.: "Geohydrology," Wiley, 1965.
3. Haimes, Y. Y., R. L. Perrine, and D. A. Wismer: Identification of Aquifer Parameters by Decomposition and Multilevel Optimization, *Univ. Calif., Los Angeles, Dept. Eng. Rept.* 67-63, March, 1968.
4. Kalaba, R.: On Nonlinear Differential Equations: The Maximum Operation and Monotone Convergence, *J. Math. Mech.*, vol. 8, no. 4, pp. 519–574, July, 1959.
5. Kelley, H. J.: Method of Gradients, in George Leitman (ed.), "Optimization Techniques," vol. 5 of "Mathematics in Science and Engineering," Academic, 1962.
6. Kolmogorov, A. N., and S. V. Fomin: "Elements of the Theory of Functions and Functional Analysis," vol. 1, Graylock, 1957 (translated from 1954 Russian edition).
7. Lax, P. D., and R. D. Richtmyer: Survey of Stability of Linear Finite Difference Equations, *Commun. Pure Appl. Math.*, vol. 9, pp. 267–293, 1956.
8. Ogata, K.: "State Space Analysis of Control Systems," Prentice-Hall, 1967.
9. Polubarinova-Kochina, P. Ya.: "Theory of Ground Water Movement," p. 549, translated from the Russian by R. J. M. DeWiest, Princeton University Press, 1962.
10. Valentine, F. A.: The Problem of Lagrange with Differential Inequalities as Added Side Constraints, in "Contributions to the Calculus of Variations," University of Chicago Press, 1937.
11. Varga, R. S.: "Matrix Iterative Analysis," Prentice-Hall, 1962.
12. Wang, P. K. C., and F. Tung: Optimum Control of Distributed Parameter Systems, *Preprints Joint Autom. Control Conf.*, 1963.
13. Wismer, D. A.: Optimal Control of Distributed Parameter Systems Using Multilevel Techniques, *Univ. Calif., Los Angeles, Dept. Eng. Rept.* 66–55, 1966.
14. Young, D. M.: The Numerical Solution of Elliptic and Parabolic Partial Differential Equations, in John Todd (ed.), "Survey of Numerical Analysis," chap. 11, McGraw-Hill, 1962.

7

TRAJECTORY DECOMPOSITION

EDWARD J. BAUMAN

7-1 INTRODUCTION

This chapter deals with the decomposition and optimization of the system equations along subintervals of the independent variable (assumed to be time) rather than decomposing the system equations into subsets of equations over the entire time interval. Thus, subsystems previously discussed correspond to subarcs (subintervals) in this chapter. A somewhat simpler second-level controller problem results since the subarcs are matched only at boundaries corresponding to discrete times rather than subsystems that are matched over the entire time interval of interest.

The motivation to decompose equations in time arises from the fact that optimization techniques which relate to the calculus of variations are based on certain continuity assumptions concerning the state variables and their derivatives. If these assumptions are violated then extensions of these techniques must be developed. The extension described here is to decompose the equations into subarcs which are determined by the occurrence of the discontinuities and then to optimize each subarc independently. By using decomposition each subarc has the necessary continuity requirements and thus standard optimization techniques can be used for each subarc. A second-level controller is then used to force these subarc solutions to result in an optimal solution for the entire trajectory.

The objective of this chapter is to present the mathematical formulation of this technique and then to apply it to a rocket trajectory problem.

7-2 DECOMPOSITION INTO SUBARCS AND SUBARC OPTIMIZATION

Consider the following optimal control problem formulation

$$\min_{\mathbf{u}} J(\mathbf{u}) = \min_{\mathbf{u}} \int_{t_0}^{t_f} F(\mathbf{x}, \mathbf{u}, t)\, dt \tag{1}$$

$$\text{subject to } \dot{\mathbf{x}} = \mathbf{f}(\mathbf{x}, \mathbf{u}, t) \qquad \mathbf{x}(t_0) = \mathbf{x}_0 \tag{2}$$

Except at N known sets of boundary conditions of the form

$$\boldsymbol{\psi}^i(\mathbf{x}^i, t_f^{\,i}) = 0 \qquad i = 1,\ldots, N \tag{3}$$

(where each component $\psi_j^{\,i}$ of $\boldsymbol{\psi}^i$ is of class C^2 for $j = 1, 2,\ldots, q \leqslant n + 1$), the terms are defined as follows:

$\mathbf{f}$ = vector function with components f_j of class C^2, $j = 1,\ldots, n$

$\mathbf{u}$ = vector function with components u_j , $j = 1,\ldots, m$

F = scalar function of class C^2

t = the independent variable time

$\mathbf{x}^i$ = a vector function with components $x_j^{\,i}$, $j = 1,\ldots, n$, and $i = 1,\ldots, N$

$t_f^{\,i}$ = the time point at which the ith discontinuity occurs and the final time of the ith subarc

At the boundary conditions, given by Eq. (3), discontinuities may occur in $\mathbf{x}, \mathbf{f}, F$, or any combination thereof. The trajectory can be broken into N subarcs (it is assumed that the order of occurrence of the boundary conditions is known). Thus Eqs. (1) and (2) become

$$\min_{\mathbf{u}} J(\mathbf{u}) = \sum_{i=1}^{N} \min_{\mathbf{u}^i} \int_{t_0^{\,i}}^{t_f^{\,i}} F^i(\mathbf{x}^i, \mathbf{u}^i, t)\, dt \tag{4}$$

$$\text{subject to } \dot{\mathbf{x}}^i = \mathbf{f}^i(\mathbf{x}^i, \mathbf{u}^i, t^i) \qquad \mathbf{x}(t_0^{\,i}) = \mathbf{x}_0^{\,i} \qquad i = 1,\ldots, N \tag{5}$$

where $t_0^{\,i}$ is the initial time of the ith subarc and $\mathbf{x}^i$, $\mathbf{u}^i$, $\mathbf{f}^i$, and F^i are the variables and functions of the ith subarc. The interface constraints for the state variables and time are assumed known and given by

$$t_f^i - t_0^{i+1} = 0 \tag{6}$$

$$\mathbf{h}^i(\mathbf{x}^i, t_f^i) - \mathbf{x}_0^{i+1} = \mathbf{0} \tag{7}$$

where each component h_j^i of $\mathbf{h}^i$ is of class C^2, $j = 1,\ldots, n$ and $i = 1,\ldots, N - 1$.

If inequality constraints are present, they may also be decomposed as follows:

$$\mathbf{R}^i(\mathbf{x}^i, \mathbf{u}^i, t^i) \geqslant \mathbf{0} \qquad i = 1,\ldots, N \tag{8}$$

where each component R_j^i of $\mathbf{R}^i$ is of class C^2, $j = 1,\ldots, r$. In fact there may be different inequality constraints prescribed along each subarc. These inequality constraints may now be included as differential equation equality constraints using Jazwinski's[4] technique; i.e., let

$$g_j^i = \begin{cases} (R_j^i)^4 & \text{if } R_j^i < 0 \\ 0 & \text{if } R_j^i \geqslant 0 \end{cases} \tag{9}$$

and replace the inequality constraints by differential equality constraints

$$\dot{\mathbf{y}}^i = \mathbf{g}^i \qquad \mathbf{y}^i(t_0^i) = \mathbf{0} \qquad \mathbf{y}^i(t_f^i) = \mathbf{0} \tag{10}$$

Because of the end conditions of Eq. (10), the differential constraints are clearly equivalent to those of Eq. (8) since g_j^i is nonnegative. However, the dimension of the state vector $\mathbf{x}^i$ has been increased by the dimension of $\mathbf{y}^i$. The second-level feasible controller is not changed at all, as will be seen later, since the state of $\mathbf{y}^i$ at any boundary, Eq. (10), is always known to be zero.

Necessary conditions for a minimum of J are that the first variation of the following functional $\tilde{J}$ vanish

$$\tilde{J} = \sum_{i=1}^{N} \int_{t_0^i}^{t_f^i} \{F^i + (\boldsymbol{\lambda}^i)'(\mathbf{f}^i - \dot{\mathbf{x}}^i)\}\, dt$$

$$+ \sum_{i=1}^{N-1} (\boldsymbol{\rho}^i)'(\mathbf{h}^i - \mathbf{x}^{i+1}) + \sum_{i=1}^{N-1} l^i(t_f^i - t_0^{i+1}) + \sum_{i=1}^{N} (\boldsymbol{\gamma}^i)'\, \boldsymbol{\psi}^i \tag{11}$$

where $\boldsymbol{\lambda}^i$, $\boldsymbol{\rho}^i$, l^i, and $\boldsymbol{\gamma}^i$ are Lagrange multipliers used to attach the constraints to J. By applying the feasible method of decomposition[1,3] the initial conditions on each subarc, $\mathbf{x}_0^{i+1}$ and t_0^{i+1}, are assumed to be known quantities and then $\tilde{J}$ easily decomposes into N separate minimization problems

$$\tilde{J}^i = \int_{t_0{}^i}^{t_f{}^i} \{F^i + (\boldsymbol{\lambda}^i)'(\mathbf{f}^i - \dot{\mathbf{x}}^i)\}\, dt + (\boldsymbol{\rho}^i)'(\mathbf{h}^i - \mathbf{x}^{i+1})$$

$$+ \, l^i(t_f{}^i - t_0^{i+1}) + (\boldsymbol{\gamma}^i)' \, \boldsymbol{\psi}^i \qquad i = 1,\ldots, N-1 \tag{12}$$

$$\tilde{J}^N = \int_{t_0{}^N}^{t_f{}^N} \{F^N + (\boldsymbol{\lambda}^N)'(\mathbf{f}^N - \dot{\mathbf{x}}^N)\}\, dt + (\boldsymbol{\gamma}^N)' \, \boldsymbol{\psi}^N \tag{13}$$

In order that the first variation of Eqs. (12) and (13) vanish, the following equations must be satisfied for $i = 1,\ldots, N-1$.

$$\dot{\mathbf{x}}^i = \mathbf{f}^i \tag{14}$$

$$\dot{\boldsymbol{\lambda}}^i = -\left(\frac{\partial F^i}{\partial \mathbf{x}^i}\right)' - \left(\frac{\partial \mathbf{f}^i}{\partial \mathbf{x}^i}\right)' \boldsymbol{\lambda}^i \tag{15}$$

$$\frac{\partial F^i}{\partial \mathbf{u}^i} + (\boldsymbol{\lambda}^i)' \frac{\partial \mathbf{f}^i}{\partial \mathbf{u}^i} = \mathbf{0}' \tag{16}$$

$$\left[l^i + F^i + (\boldsymbol{\lambda}^i)' \, \mathbf{f}^i + (\boldsymbol{\gamma}^i)' \frac{\partial \boldsymbol{\psi}^i}{\partial t^i} + (\boldsymbol{\rho}^i)' \frac{\partial \mathbf{h}^i}{\partial t^i}\right]_{t^i = t_f{}^i} = 0 \tag{17}$$

$$\left[-(\boldsymbol{\lambda}^i)' + (\boldsymbol{\rho}^i)' \frac{\partial \mathbf{h}}{\partial \mathbf{x}^i} + (\boldsymbol{\gamma}^i)' \frac{\partial \boldsymbol{\psi}^i}{\partial \mathbf{x}^i}\right]_{t^i = t_f{}^i} = \mathbf{0}' \tag{18}$$

$$\mathbf{h}^i \,|_{t^i = t_f{}^i} = \mathbf{x}^{i+1} \,|_{t^{i+1} = t_0^{i+1}} \qquad t_f{}^i = t_0^{i+1} \tag{19}$$

$$\boldsymbol{\psi}^i \,|_{t^i = t_f{}^i} = \mathbf{0} \tag{20}$$

Since $\mathbf{x}_0^{i+1}$ and t_0^{i+1} are chosen by the second-level controller, there are only $n + 1$ independent end variables $\mathbf{x}_f{}^i$, $t_f{}^i$, and thus only $n + 1 - q$ equations of Eq. (19) are active in the ith subsystem. These active equations with Eq. (20) (q equations, which must be satisfied to determine the boundary) give a total of $n + 1$ equations which must be functionally independent (i.e., the Jacobian must not vanish at $t_f{}^i$) if a solution is to exist. The remaining q equations of Eq. (19) are used in the second-level controller to determine q values of $\mathbf{x}_0^{i+1}$ and t_0^{i+1}. Thus these q equations, their partial derivatives, and associated Lagrange multipliers are removed from the subsystem conditions Eqs. (17) to (19).

The Nth subsystem has the same form as Eqs. (14) to (20), except that Eq. (19) is missing along with the partial derivatives and Lagrange multipliers associated with Eq. (19) which appear in Eqs. (17) and (18).

7-3 SECOND-LEVEL CONTROLLERS FOR FEASIBLE DECOMPOSITION

Bauman[1] shows that by satisfying these necessary conditions for a minimum on each of two subarcs, all the terms of the first variation vanish with the possible exception of those multiplying the variations of t_0^1 and $\mathbf{x}_0^1$. These results are generalized to N subarcs as shown below

$$\delta \tilde{J} = \sum_{i=1}^{N-1} \{[(\boldsymbol{\lambda}^{i+1})' - (\boldsymbol{\rho}^i)']\, \delta \mathbf{x}_0^{i+1} + [-F^{i+1} - (\boldsymbol{\lambda}^{i+1})'\, \mathbf{f}^{i+1} - l^i]\, dt_0^{i+1}\} \tag{21}$$

where $i = 1,\ldots, N - 1$ and δ represents the first variation. Now the q equations of Eq. (19) used in the second level determine q of the values of $\mathbf{x}_0^{i+1}$ and t_0^{i+1} after the ith subarc has been optimized. Therefore, those $\delta \mathbf{x}_0^{i+1}$ and dt_0^{i+1} are zero and only the remaining $n + 1 - q$ values of $\mathbf{x}_0^{i+1}$ and t_0^{i+1} need be changed. Since all the constraints have been satisfied by subarc minimization,

$$J = \tilde{J} \tag{22}$$

Also, since J is being minimized, δJ should be negative. Therefore, by Eq. (22) $\delta \tilde{J}$ should be negative. A sufficient condition for $\delta \tilde{J} < 0$ follows from Eq. (21); i.e.,

$$\delta \mathbf{x}_0^{i+1} = -k(\boldsymbol{\lambda}^{i+1} - \boldsymbol{\rho}^i) \tag{23}$$

$$dt_0^{i+1} = -k[-F^{i+1} - (\boldsymbol{\lambda}^{i+1})'\, \mathbf{f}^{i+1} - l^i] \tag{24}$$

where $k > 0$. The other terms on the right-hand side of Eqs. (23) and (24) come from the appropriate subarc after it has been minimized. Using Eqs. (23) and (24), the new $n + 1 - q$ values of $\mathbf{x}_0^{i+1}$ and t_0^{i+1} are determined for the next iteration as

$$(\mathbf{x}_0^{i+1})_{\text{new}} = (\mathbf{x}_0^{i+1})_{\text{old}} + \delta \mathbf{x}_0^{i+1} \tag{25}$$

$$(t_0^{i+1})_{\text{new}} = (t_0^{i+1})_{\text{old}} + dt_0^{i+1} \tag{26}$$

Equations (25) and (26) provide a simple second-level gradient controller that will, under the stated assumptions, drive the solution toward the one which satisfies necessary conditions for a minimizing trajectory. However, in the vicinity of the minimum (if one exists), k must decrease or this iterative solution becomes oscillatory about the minimizing trajectory, as in the case of simple gradient methods.

A second-level controller that gives a linear estimate of the step size to be taken (i.e., the value of k in the gradient method) and also has excellent convergence properties near the minimizing trajectory is provided by the Newton-Raphson method for solving simultaneous nonlinear algebraic equations.[5] A formal (not rigorous) derivation of this controller is given in the next section for a two-subarc minimization problem (that is, $N = 2$).

7-4 DERIVATION OF A NEWTON-RAPHSON SECOND-LEVEL CONTROLLER FOR FEASIBLE DECOMPOSITION

For this controller, normally $n + 1 - q$ values of $\mathbf{x}_0^{i+1}$ and t_0^{i+1} will be changed so that the coefficients of $\delta\mathbf{x}_0^{i+1}$ and dt_0^{i+1} in Eq. (21) are driven to zero. Thus the second-level controller strives to satisfy the corresponding $n + 1 - q$ equations at t_0^{i+1}.

$$\boldsymbol{\lambda}^{i+1} - \boldsymbol{\rho}^{i} = \mathbf{0} \tag{27}$$

$$-F^{i+1} - (\boldsymbol{\lambda}^{i+1})' \, \mathbf{f}^{i+1} - l^{i} = 0 \tag{28}$$

where $i = 1,..., N - 1$. To simplify notation and the complexity of the derivation let $N = 2$ and define:

$$\mathbf{g}(\mathbf{z}) = \begin{pmatrix} \boldsymbol{\lambda}^2 - \boldsymbol{\rho}^1 \\ -H^2 - l^1 \end{pmatrix} \tag{29}$$

where $\mathbf{z} = \binom{\mathbf{x}_0^2}{t_0^2}$, $\mathbf{z}$ and $\mathbf{g}$ are of dimension $n + 1 - q$, $\mathbf{g}$ is *implicitly* a function of $\mathbf{z}$, and $H^2 = F^2 + (\boldsymbol{\lambda}^2)' \, \mathbf{f}^2$. Note that $t_0{}^2$ does not appear in $\mathbf{z}$ and $-H^2 - l^1$ does not appear in Eq. (29) if $t_f{}^1$ is fixed. In general $\mathbf{g}(\mathbf{z})$ will not equal zero as desired. However, the Newton-Raphson technique can be used to find the zeros of Eq. (29). This technique is based upon the classical algorithm for finding the roots of a polynomial. A first-order Taylor series expansion of $\mathbf{g}$ about a particular value of $\mathbf{z}$ is

$$\mathbf{g}(\mathbf{z} + \mathbf{dz}) = \mathbf{g}(\mathbf{z}) + \frac{\partial \mathbf{g}}{\partial \mathbf{z}} \, \mathbf{dz} \tag{30}$$

If $\mathbf{z} + \mathbf{dz}$ is a root of $\mathbf{g}$ then the left side of Eq. (30) is zero. Solving the resulting equation for $\mathbf{dz}$ gives the increment that must be added to the present value of $\mathbf{z}$ such that $\mathbf{z} + \mathbf{dz}$ is a first-order approximation to a root of $\mathbf{g}$. Then let

$$(\mathbf{z})_{\text{new}} = (\mathbf{z})_{\text{old}} - c \left[\frac{\partial \mathbf{g}}{\partial \mathbf{z}} \right]^{-1} \mathbf{g}(\mathbf{z}) \qquad 0 < c \leqslant 1 \tag{31}$$

A solution of Eq. (31) depends on the existence and nonsingularity of $\partial \mathbf{g}/\partial \mathbf{z}$.

Equation (29) shows that $\mathbf{g}$ is *explicitly* a function of $\boldsymbol{\lambda}^2$, $\boldsymbol{\rho}^1$, l^1, $F^2(\mathbf{x}^2, \mathbf{u}^2, t^2)$, and $\mathbf{f}^2(\mathbf{x}^2, \mathbf{u}^2, t^2)$. By the implicit differentiation rule the required derivatives are

$$\frac{\partial \mathbf{g}}{\partial z_j} = \frac{\partial \mathbf{g}}{\partial \boldsymbol{\lambda}^2}\frac{\partial \boldsymbol{\lambda}^2}{\partial z_j} + \frac{\partial \mathbf{g}}{\partial \boldsymbol{\rho}^1}\frac{\partial \boldsymbol{\rho}^1}{\partial z_j} + \frac{\partial \mathbf{g}}{\partial l^1}\frac{\partial l^1}{\partial z_j} + \frac{\partial \mathbf{g}}{\partial \mathbf{x}^2}\frac{\partial \mathbf{x}^2}{\partial z_j} + \frac{\partial \mathbf{g}}{\partial t^2}\frac{\partial t^2}{\partial z_j} \tag{32}$$

$j = 1, 2, \ldots, n + 1 - q$ where all derivatives are evaluated at the boundary between subarcs 1 and 2 and the subarc optimality condition $\partial \mathbf{g}/\partial \mathbf{u}^2 = \mathbf{0}$ has been used. The partial derivatives with respect to z_j can be calculated from the appropriate subarcs by forming the partial derivatives of the boundary and transversality conditions with respect to z_j. Thus for *subarc* 1 using Eq. (17) and noting that certain terms cancel we get

SUBARC 1:

$$\left\{\mathbf{f}'\frac{\partial \boldsymbol{\lambda}}{\partial z_j} + \left(\frac{\partial H}{\partial \mathbf{x}} + \frac{\partial^2 \Phi}{\partial \mathbf{x}\,\partial t}\right)\frac{\partial \mathbf{x}}{\partial z_j} + \left(\frac{\partial \boldsymbol{\Omega}}{\partial t}\right)'\frac{\partial \boldsymbol{\nu}}{\partial z_j} + \left[\frac{\partial H}{\partial t} + \frac{D}{Dt}\left(\frac{\partial \Phi}{\partial t}\right)\right]\frac{\partial t}{\partial z_j} + \frac{\partial \Phi}{\partial z_j}\right\}_{t=t_f} = 0 \tag{33}$$

where the subarc superscript has been dropped for notational clarity. Also, the following definitions have been made:

$$H = F + \boldsymbol{\lambda}'\mathbf{f}$$

$$\frac{D}{Dt}(\ ) = \frac{\partial}{\partial t}(\ ) + \frac{\partial}{\partial \mathbf{x}}(\ )\,\mathbf{f}$$

$$\Phi = \boldsymbol{\nu}'\boldsymbol{\Omega}$$

$$\boldsymbol{\Omega} = \begin{pmatrix} \boldsymbol{\psi} \\ \mathbf{h} \\ t \end{pmatrix}_{t=t_f} \qquad \boldsymbol{\nu} = \begin{pmatrix} \mathbf{g} \\ \boldsymbol{\rho} \\ l \end{pmatrix}$$

where only active equations of h and t [see discussion following Eq. (20)] and their associated multipliers appear in $\boldsymbol{\Omega}$ and $\boldsymbol{\nu}$ respectively. Similarly, by taking the partial derivatives of Eqs. (18) to (20) and using the above definitions,

$$\left\{-\frac{\partial \boldsymbol{\lambda}}{\partial z_j} + \frac{\partial^2 \Phi}{\partial \mathbf{x}^2}\frac{\partial \mathbf{x}}{\partial z_j} + \left(\frac{\partial \boldsymbol{\Omega}}{\partial \mathbf{x}}\right)'\frac{\partial \boldsymbol{\nu}}{\partial z_j} + \left[\frac{\partial H}{\partial \mathbf{x}} + \frac{D}{Dt}\left(\frac{\partial \Phi}{\partial \mathbf{x}}\right)\right]'\frac{\partial t}{\partial z_j} + \frac{\partial^2 \Phi}{\partial \mathbf{x}\,\partial z_j}\right\}_{t=t_f} = 0 \tag{34}$$

$$\left\{\frac{\partial \boldsymbol{\Omega}}{\partial \mathbf{x}}\frac{\partial \mathbf{x}}{\partial z_j} + \frac{D\boldsymbol{\Omega}}{Dt}\frac{\partial t}{\partial z_j} + \frac{\partial \boldsymbol{\Omega}}{\partial z_j}\right\}_{t=t_f} = 0 \tag{35}$$

Equations (33) to (35) represent $2n + 2$ equations in $3n + 2$ unknowns, namely, $\partial t/\partial z_j$, $\partial \boldsymbol{\lambda}/\partial z_j$, $\partial \mathbf{x}/\partial z_j$, and $\partial \boldsymbol{\nu}/\partial z_j$. However, by using the second variational scheme discussed by Breakwell, Speyer, and Bryson,[2] differential equations can be formed to solve for $\partial \boldsymbol{\lambda}/\partial z_j$ in terms of $\partial \mathbf{x}/\partial z_j$ as follows:

$$\frac{d}{dt}\begin{bmatrix} \dfrac{\partial \mathbf{x}}{\partial z_j} \\ \dfrac{\partial \boldsymbol{\lambda}}{\partial z_j} \end{bmatrix} = \begin{bmatrix} \mathbf{C}_1(t) & \mathbf{C}_2(t) \\ \mathbf{C}_3(t) & -\mathbf{C}_1'(t) \end{bmatrix} \begin{bmatrix} \dfrac{\partial \mathbf{x}}{\partial z_j} \\ \dfrac{\partial \boldsymbol{\lambda}}{\partial z_j} \end{bmatrix} \tag{36}$$

where

$$\mathbf{C}_1(t) = \frac{\partial^2 H}{\partial \boldsymbol{\lambda}\, \partial \mathbf{x}} - \frac{\partial^2 H}{\partial \boldsymbol{\lambda}\, \partial \mathbf{u}} \left(\frac{\partial^2 H}{\partial \mathbf{u}^2} \right)^{-1} \frac{\partial^2 H}{\partial \mathbf{u}\, \partial \mathbf{x}}$$

$$\mathbf{C}_2(t) = -\frac{\partial_2 H}{\partial \boldsymbol{\lambda}\, \partial \mathbf{u}} \left(\frac{\partial^2 H}{\partial \mathbf{u}^2} \right)^{-1} \frac{\partial^2 H}{\partial \mathbf{u}\, \partial \boldsymbol{\lambda}}$$

$$\mathbf{C}_3(t) = -\frac{\partial^2 H}{\partial \mathbf{x}^2} + \frac{\partial^2 H}{\partial \mathbf{x}\, \partial \mathbf{u}} \left(\frac{\partial^2 H}{\partial \mathbf{u}^2} \right)^{-1} \frac{\partial^2 H}{\partial \mathbf{u}\, \partial \mathbf{x}}$$

Since the $\mathbf{C}$ terms are calculated along the optimized subarc 1 trajectory, Eq. (36) is a linear differential matrix equation and the solution can be written in terms of the transition matrix $\mathbf{Y}$

$$\begin{bmatrix} \dfrac{\partial \mathbf{x}}{\partial z_j} \\ \dfrac{\partial \boldsymbol{\lambda}}{\partial z_j} \end{bmatrix}_{t_f} = \begin{bmatrix} \mathbf{Y}_1 & \mathbf{Y}_2 \\ \mathbf{Y}_3 & \mathbf{Y}_4 \end{bmatrix} \begin{bmatrix} \dfrac{\partial \mathbf{x}}{\partial z_j} \\ \dfrac{\partial \boldsymbol{\lambda}}{\partial z_j} \end{bmatrix}_{t_0} \tag{37}$$

Since the initial states are fixed, Eq. (2), it follows that

$$\left(\frac{\partial \mathbf{x}}{\partial z_j} \right)_{t_0} = \mathbf{0}$$

and substituting this in Eq. (37), we get

$$\left(\frac{\partial \boldsymbol{\lambda}}{\partial z_j} \right)_{t_0} = \mathbf{Y}_2^{-1} \left(\frac{\partial \mathbf{x}}{\partial z_j} \right)_{t_f} \tag{38}$$

or

$$\left(\frac{\partial \boldsymbol{\lambda}}{\partial z_j} \right)_{t_f} = \mathbf{Y}_4 \mathbf{Y}_2^{-1} \left(\frac{\partial \mathbf{x}}{\partial z_j} \right)_{t_f} \tag{39}$$

Now Eq. (39) is substituted into Eqs. (33) to (35) to yield for subarc 1

$$\begin{bmatrix} \dfrac{\partial \boldsymbol{\Omega}}{\partial \mathbf{x}} & \mathbf{0} & \dfrac{D\boldsymbol{\Omega}}{Dt} \\ -\mathbf{Y}_4\mathbf{Y}_2^{-1} + \dfrac{\partial^2 \Phi}{\partial \mathbf{x}^2} & \left(\dfrac{\partial \boldsymbol{\Omega}}{\partial \mathbf{x}}\right)' & \left[\dfrac{\partial H}{\partial \mathbf{x}} + \dfrac{D}{Dt}\left(\dfrac{\partial \Phi}{\partial \mathbf{x}}\right)\right]' \\ \mathbf{f}'\mathbf{Y}_4\mathbf{Y}_2^{-1} + \dfrac{\partial H}{\partial \mathbf{x}} + \dfrac{\partial^2 \Phi}{\partial \mathbf{x}\,\partial t} & \left(\dfrac{\partial \boldsymbol{\Omega}}{\partial t}\right)' & \dfrac{\partial H}{\partial t} + \dfrac{D}{Dt}\left(\dfrac{\partial \Phi}{\partial t}\right) \end{bmatrix} \begin{bmatrix} \dfrac{\partial \mathbf{x}}{\partial z_j} \\ \dfrac{\partial \boldsymbol{\nu}}{\partial z_j} \\ \dfrac{\partial t}{\partial z_j} \end{bmatrix} = \begin{bmatrix} -\dfrac{\partial \boldsymbol{\Omega}}{\partial z_j} \\ -\dfrac{\partial^2 \Phi}{\partial \mathbf{x}\,\partial z_j} \\ -\dfrac{\partial \Phi}{\partial z_j} \end{bmatrix} \tag{40}$$

We note that the matrix in Eq. (40) is written only once for each second-level iteration since only the right-hand side of the equation varies as z_j varies. The last row and column of the matrix in Eq. (40) are not needed if t_f is fixed. By solving Eq. (40) for the appropriate components of $\partial \boldsymbol{\nu}/\partial z_j$, the $\partial \boldsymbol{\rho}/\partial z_j$ and $\partial l/\partial z_j$ terms are calculated for use in Eq. (32). The other terms of Eq. (32) are calculated from subarc 2 after it has been optimized.

SUBARC 2:

The values of $\mathbf{x}_0^2$ and t_0^2 for subarc 2 are fixed by the second-level controller. Therefore, this subarc optimization is almost identical with that of subarc 1, except that no coupling equations appear at the end of the second subarc; i.e., Eqs. (33) to (35) are again solved, but now $\boldsymbol{\Omega} = \boldsymbol{\psi}$ and $\boldsymbol{\nu} = \boldsymbol{\gamma}$.

After the second subarc has been optimized, the terms for the second-level controller, Eq. (32), are calculated. The *total* variations of $(\partial \boldsymbol{\lambda}/\partial z_j)_{\text{total}}$, $(\partial \mathbf{x}/\partial z_j)_{\text{total}}$, and $\partial t/\partial z_j$ must be calculated for Eq. (32) by allowing the appropriate $n + 1 - q$ terms z_j to vary. The total variations result when time variations as well as function variations are considered. Thus

$$\left(\frac{\partial \boldsymbol{\lambda}}{\partial z_j}\right)_{\text{total}} = \left(\frac{\partial \boldsymbol{\lambda}}{\partial z_j} - \dot{\boldsymbol{\lambda}}\frac{\partial t}{\partial z_j}\right)_{t=t_0} \tag{41}$$

$$\left(\frac{\partial \mathbf{x}}{\partial z_j}\right)_{\text{total}} = \left(\frac{\partial \mathbf{x}}{\partial z_j} - \mathbf{f}\frac{\partial t}{\partial z_j}\right)_{t=t_0} \tag{42}$$

where the partial derivatives on the right side of Eqs. (41) and (42) are partial derivatives in the ordinary sense. For example, if $z_j = x_j$, then

$$\frac{\partial \mathbf{x}}{\partial z_j} = \left(\frac{\partial \mathbf{x}}{\partial x_j}\right)_{t_0} = [0,\ldots, 1,\ldots, 0]'$$

where the one appears as the jth component. The term $\partial \boldsymbol{\lambda}/\partial z_j$ is calculated using the second variational differential equations (36) applied to this subarc, which results in the same form of solution as Eq. (37). The final conditions at t_f are now fixed; i.e.,

$$\left(\frac{\partial \mathbf{x}}{\partial z_j}\right)_{t_f} = \mathbf{0}$$

Then

$$\left(\frac{\partial \boldsymbol{\lambda}}{\partial z_j}\right)_{t_0} = -\mathbf{Y}_2^{-1}\mathbf{Y}_1 \left(\frac{\partial \mathbf{x}}{\partial z_j}\right)_{t_0} \tag{43}$$

Substituting Eqs. (41) to (43) and

$$\left(\frac{\partial t}{\partial z_j}\right)_{t_0} = \begin{cases} 0 & \text{if } z_j \neq t_0 \\ 1 & \text{if } z_j = t_0 \end{cases} \tag{44}$$

into Eq. (32) allows us to compute the matrix of functions $\partial \mathbf{g}/\partial \mathbf{z}$ and thus the terms for the controller equation (31). Although these equations appear formidable, very little additional calculation is needed if the second variational method is used to optimize the subarcs. In the next section we apply this second variational (Newton-Raphson) controller to maximizing the range of a rocket.

7-5 A TRAJECTORY OPTIMIZATION PROBLEM USING THE NEWTON-RAPHSON CONTROLLER

We will consider the problem of maximizing the range of a rocket vehicle during powered flight outside of the earth's atmosphere. Specifically we maximize the range angle between two given altitudes for a fixed time of flight. The equations of motion are given by

$$\dot{r} = v \sin\phi \qquad r(t_0) = r_0 \tag{45}$$

$$\dot{v} = \frac{-g_0 R_e^2}{r^2} \sin\phi + \frac{T}{m}\cos\alpha \qquad v(t_0) = v_0 \tag{46}$$

$$\dot{\phi} = \left(\frac{-g_0 R_e^2}{v r^2} + \frac{v}{r}\right)\cos\phi + \frac{T}{vm}\sin\alpha \qquad \phi(t_0) = \phi_0 \tag{47}$$

where

r = distance between the vehicle and the earth's center

v = vehicle velocity

ϕ = angle between the local horizontal and the velocity vector

R_e = radius of the earth

g_0 = acceleration of gravity at the earth's surface

α = angle between the velocity vector and the thrust vector

T = thrust

m = mass of the vehicle

The problem is to

$$\max_{\alpha} \int_{t_0}^{t_f} \frac{v}{r} \cos\phi \, dt \tag{48}$$

where the integrand is the range angle rate.

A discontinuity will be considered to occur in thrust at $t_f^1 = t_A$ where $t_0 < t_A < t_f$ and t_A is fixed. Thus the first subarc has terminal condition

$$\psi_1^1 = t_f^1 - t_A = 0 \tag{49}$$

and the thrust at this point changes from T^1 to T^2. The second subarc has terminal conditions

$$\psi_1^2 = t_B - t_f^2 = 0 \tag{50}$$

$$\psi_2^2 = r_B - r_f^2 = 0 \tag{51}$$

where t_B is the fixed final time and r_B is the fixed final radius. Since the state variables are continuous, the coupling equations between subarcs are

$$t_f^1 = t_0^2 \tag{52}$$

$$r_f^1 = r_0^2 \tag{53}$$

$$v_f^1 = v_0^2 \tag{54}$$

$$\phi_f^1 = \phi_0^2 \tag{55}$$

SUBARC 1:

The functional to be maximized is

$$\max_{\alpha^1} \int_{t_0^1}^{t_A} \frac{v^1}{r^1} \cos\phi^1 \, dt \tag{56}$$

and the state differential equations are given by Eqs. (45) to (47) with all state variables having the superscript 1. The feasible method of coordination will be used. Since the final time is fixed, Eq. (52) is not functionally independent of Eq. (49). The terminal constraints of subarc 1 are thus given by Eqs. (49) and (53) to (55) and

$$\Omega = \begin{bmatrix} t_f^1 - t_A \\ r_f^1 \\ v_f^1 \\ \phi_f^1 \end{bmatrix} \tag{57}$$

Initially the second-level controller must estimate r_0^2, v_0^2, and ϕ_0^2 (t_A is already known) before subarc 1 is optimized.

7-5-1 Newton-Raphson Second-level Controller

Since t_A is fixed, Eq. (29) becomes

$$\mathbf{g}(\mathbf{z}) = \begin{pmatrix} \lambda_1^2 - \rho_1^1 \\ \lambda_2^2 - \rho_2^1 \\ \lambda_3^2 - \rho_3^1 \end{pmatrix} \tag{58}$$

where

$$\mathbf{z} = \begin{pmatrix} x_1^2 \\ x_2^2 \\ x_3^2 \end{pmatrix}_{t_0^2}$$

Therefore Eq. (32) becomes simply

$$\frac{\partial \mathbf{g}}{\partial z_j} = \frac{\partial \boldsymbol{\lambda}^2}{\partial z_j} - \frac{\partial \boldsymbol{\rho}^1}{\partial z_j} \tag{59}$$

The second-level controller (31) evaluated at t_0^2 is

$$(\mathbf{x}_0^2)_{\text{new}} = (\mathbf{x}_0^2)_{\text{old}} - c\left(\frac{\partial \boldsymbol{\lambda}^2}{\partial \mathbf{z}} - \frac{\partial \boldsymbol{\rho}^1}{\partial \mathbf{z}}\right)^{-1} (\boldsymbol{\lambda}^2 - \boldsymbol{\rho}^1) \tag{60}$$

where $0 < c \leqslant 1$ and where all terms of Eq. (60) are evaluated along the minimized subarcs. The terms $\partial \boldsymbol{\rho}^1/\partial \mathbf{z}$ and $\boldsymbol{\rho}^1$ must be calculated from subarc 1 using Eq. (40) as follows:

SUBARC 1:

Due to the simplicity of the constraint equations and since the final time is fixed, Eq. (40) yields

$$\begin{bmatrix} \mathbf{I} & \mathbf{0} \\ -\mathbf{Y}_4\mathbf{Y}_2^{-1} & \mathbf{I} \end{bmatrix} \begin{pmatrix} \dfrac{\partial \mathbf{x}^1}{\partial \mathbf{z}} \\ \dfrac{\partial \boldsymbol{\rho}^1}{\partial \mathbf{z}} \end{pmatrix} = \begin{pmatrix} \mathbf{I} \\ \mathbf{0} \end{pmatrix} \tag{61}$$

where $\mathbf{z} = \begin{pmatrix} r \\ v \\ \phi \end{pmatrix}_{t_0^2}$ and $\mathbf{I}$ and $\mathbf{0}$ are each of dimension 3×3. The solution of Eq. (61) is

$$\frac{\partial \boldsymbol{\rho}^1}{\partial \mathbf{z}} = \mathbf{Y}_4 \mathbf{Y}_2^{-1} \tag{62}$$

where the transition matrices $\mathbf{Y}_2$ and $\mathbf{Y}_4$ are calculated for subarc 1. From Eq. (18) we see that

$$\boldsymbol{\rho}^1 = \boldsymbol{\lambda}^1 \tag{63}$$

SUBARC 2:

The functional to be maximized is

$$\max_{\alpha^2} \int_{t_0^2}^{t_B} \frac{v^2}{r^2} \cos \phi^2 \, dt \tag{64}$$

and the state differential equations are Eqs. (45) to (47) with initial conditions $\mathbf{z}$ and $t_0^2 = t_A$. The altitude at $t_f^2 = t_B$ is also fixed. Subarc 2 is optimized using these conditions.

Since t_0 is fixed, we can combine Eqs. (41) and (43) to get

$$\left(\frac{\partial \boldsymbol{\lambda}^2}{\partial \mathbf{z}}\right)_{\text{total}} = \frac{\partial \boldsymbol{\lambda}^2}{\partial \mathbf{z}} = -\mathbf{Y}_2^{-1} \mathbf{Y}_1 \tag{65}$$

where the partitioned transition matrices $\mathbf{Y}_1$ and $\mathbf{Y}_2$ are calculated for subarc 2. The feasible multilevel optimization method for the example problem is illustrated in Fig. 7-1. The second-level controller determines

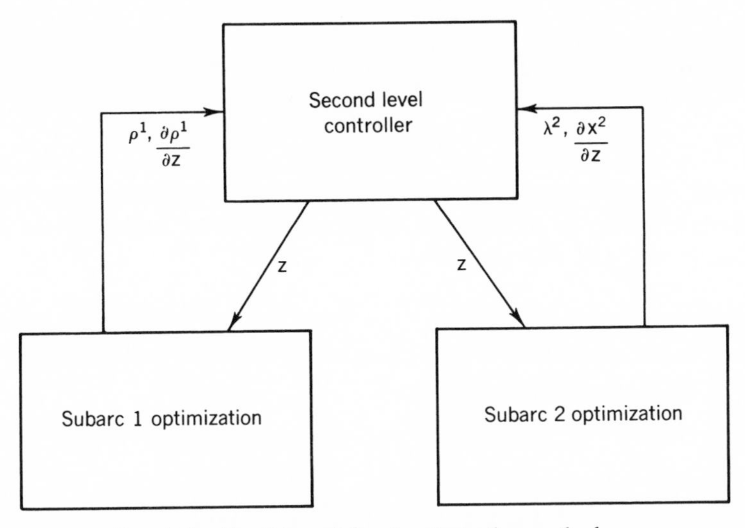

FIG. 7-1. Feasible multilevel optimization method.

values of $\mathbf{z}$ which are the final conditions for subarc 1. These values are used in the optimization of subarc 1. Then subarc 1 transmits $\boldsymbol{\rho}^1$ and $\partial\boldsymbol{\rho}^1/\partial\mathbf{z}$ to the second-level controller. These $\mathbf{z}$ values are also transmitted to subarc 2 and it optimizes its subarc problem using $\mathbf{z}$ as the initial conditions. After optimization, subarc 2 transmits $\boldsymbol{\lambda}^2$ and $\partial\mathbf{x}^2/\partial\mathbf{z}$ to the second-level controller. The controller, Eq. (60), now has the necessary information to calculate the new $\mathbf{z}$ and the process is continued until no appreciable improvement is obtained in the range angle. The Newton-Raphson controller was used to improve upon the results obtained by a gradient controller given in the numerical example by Bauman.[1] In that example a near-optimum solution was given by

SUBARC 1:

$$t_0 = 66 \text{ sec}$$

$$h(t_0) = r(t_0) - R_e = 100{,}000 \text{ ft}$$

$$v(t_0) = 5{,}000 \text{ fps}$$

$$\phi(t_0) = 0.7 \text{ rad}$$

$$T = 34{,}500 \text{ lb}$$

$$m(t_0) = 340 \text{ slugs}$$

$$\dot{m} = 3.5 \text{ slugs per second}$$

$$t_f = 120 \text{ sec}$$

$$R_e = 20{,}902{,}400 \text{ ft}$$

$$\lambda_1(t_0) = 0.295 \times 10^{-7}$$

$$\lambda_2(t_0) = 0.311 \times 10^{-5}$$

$$\lambda_3(t_0) = -0.234 \times 10^{-2}$$

The final conditions on subarc 1 are the same as the initial conditions on subarc 2 as given below.

SUBARC 2:

$$t_0 = 120 \text{ sec}$$

$$h(t_0) = 328{,}035 \text{ ft}$$

$$v(t_0) = 12{,}059.5 \text{ ft}$$

$$\phi(t_0) = 0.5031 \text{ rad}$$

$$T = 17{,}250 \text{ lb}$$

$$\lambda_1(t_0) = 0.294 \times 10^{-7}$$

$$\lambda_2(t_0) = 0.222 \times 10^{-5}$$
$$\lambda_3(t_0) = 0.143 \times 10^{-2}$$
$$m(t_0) = 151 \text{ slugs}$$
$$\dot{m} = 1.75 \text{ slugs per second}$$
$$t_f = 160 \text{ sec}$$
$$h(t_f) = 600{,}000 \text{ ft}$$

The values above were used as the initial conditions for the Newton-Raphson controller. Table 7-1 shows the state and adjoint variables

Table 7-1. State and Adjoint Variables at Discontinuity versus the Value of the Constant c [See Eq. (60)]

Value of c	$\lambda_1 \times 10^7$ Arc 1	$\lambda_1 \times 10^7$ Arc 2	$\lambda_2 \times 10^5$ Arc 1	$\lambda_2 \times 10^5$ Arc 2	$\lambda_3 \times 10^2$ Arc 1	$\lambda_3 \times 10^2$ Arc 2	h ft	v fps	ϕ rad	Range Angle rad
...	0.302	0.294	0.012	0.222	−0.001	0.143	328,035	12059.5	0.5031	0.0412158
0.08	0.301	0.293	0.035	0.222	0.014	0.138	327,869	12059.5	0.5038	0.0412168
0.08	0.301	0.294	0.072	0.222	0.037	0.141	327,735	12059.6	0.5038	0.0412171
0.08	0.303	0.295	0.152	0.222	0.086	0.145	327,654	12059.6	0.5038	0.0412172
0.10	0.303	0.296	0.194	0.222	0.112	0.147	327,596	12059.7	0.5037	0.0412172
0.10	0.300	0.296	0.114	0.222	0.065	0.147	327,571	12059.7	0.5038	0.0412173
0.10	0.301	0.295	0.174	0.222	0.104	0.147	327,547	12059.6	0.5039	0.0412173
0.10	0.301	0.296	0.218	0.222	0.133	0.147	327,523	12059.6	0.5039	0.0412173
0.10	0.301	0.296	0.230	0.222	0.141	0.147	327,507	12059.6	0.5039	0.0412173
0.10	0.300	0.296	0.241	0.222	0.149	0.147	327,494	12059.6	0.5039	0.0412173
0.10	0.300	0.296	0.249	0.222	0.156	0.147	327,483	12059.6	0.5039	0.0412173

at the time of discontinuity ($t = 120$ sec) and the total range angle achieved by both subarcs. Considerable difficulty was experienced in obtaining convergence of the first subarc to the four specified end constraints (h, v, ϕ, and t_A). Table 7-1 lists the variables versus cumulative iterations with values of c which total 0.08 and 0.1. These cumulative values were initially the results of several iterations with c equal to 0.02 or 0.05. In fact, only for latter iterations could a single iteration be taken with c equal to 0.1. Larger values of c caused the second variational optimization method for subarc 1 to diverge. This divergence forced the second-level controller to reduce its step size c, and many subarc iterations were required.

These results show that multilevel optimization techniques can be applied in a straightforward way to solve optimization problems that have discontinuities at points along the trajectory. However, the convergence difficulties experienced with the second variational optimization method on subarc 1 cause the computation to be very lengthy. This difficulty occurred only when $n + 1$ terminal conditions were specified. In cases where n or fewer terminal conditions were present, no convergence problems were encountered.

REFERENCES:

1. Bauman, E. J.: Multilevel Optimization Techniques with Application to Trajectory Decomposition, in C. T. Leondes (ed.), "Advances in Control Systems," vol. 6, Academic, 1968.
2. Breakwell, J. V., J. L. Speyer, and A. E. Bryson: Optimization and Control of Nonlinear Systems Using the Second Variation, *SIAM J. Control*, ser. A, vol. 1, no. 2, pp. 193–223, 1963.
3. Brosilow, C. B., L. S. Lasdon, and J. D. Pearson: Feasible Optimization Methods for Interconnected Systems, *Proc. Joint Autom. Control Conf.*, 1965.
4. Jazwinski, A. J.: Optimal Trajectories and Linear Control of Nonlinear Systems, presented at Joint AIAA-IMS-SIAM-ONR Symposium on Control and System Optimization, U.S. Naval Postgraduate School, Monterey, Calif., Jan. 27, 1967.
5. Nielsen, K. L.: "Methods in Numerical Analysis," Macmillan, 1956.

8

ON-LINE MULTILEVEL SYSTEMS

JAMES D. SCHOEFFLER

8-1 INTRODUCTION

On-line multilevel systems arise in a natural way in response to a serious problem in the design and operation of complex systems. An on-line or real-time system may be defined as one which responds to input disturbances and produces an output in a time interval short enough to permit some compensation for the disturbance.[38] Thus if the definition of disturbance is taken in a very wide sense, real-time systems can be considered to be control systems, operating so as to optimize some implicit or explicit criterion in an environment including unpredictable disturbances.[37] These disturbances include variations in physical variables (pressure, temperature, composition, etc.), variations in operating conditions (load in a power system or volume in a production system), variations in the parameters of a system (wages, costs, reflectivity, etc.), changing economic conditions and markets, laws affecting manufacture or distribution, patents, orders, and any other input which a complex system may receive.[31]

The original design, continued updating, and daily operation of such a system are complex. In this discussion, emphasis is placed on the *design* of such systems since the design must take into account the eventual method of operation and expansion and consequently essentially determines the operation of the system. Moreover, the emphasis is further placed on systems which depend critically on real-time, on-line computers for their operation, for it is in this situation that the benefits of on-line multilevel systems are greatest.

Five problems inherent in such a system make the design difficult. These are summarized below.

1. The class of disturbances is very broad, implying that the class of methods or controls used to compensate for these disturbances

must be equally broad. That is, the various disturbances which are listed above differ completely in the methods available for their control. For example, ordinary feedback control is effective for variations in process variables but is less so for variations in economic conditions or markets due to the time delays involved. Rather, restructuring of the system or its operation using mathematical programming techniques (linear programs for example) is often used to respond to these disturbances. Similarly, the other disturbances listed above are handled by other quite different methodologies. Naturally, this also leads to complications in the operation of the system, since the people or computers involved must respond to widely different disturbances by widely different strategies.

2. The structure or configuration of the system is never self-evident except in simple cases. That is, design of the system includes specification of the components of a system and the way they are interconnected. This includes location of plants, people, computers, and communication links. In a complex system, there are many possible configurations and the problem is to choose the one most appropriate to the given control problem. Conversely, selection of a poor structure complicates the design and inhibits smooth growth and operation of the system.

3. Real systems are not static but rather evolve in time. Any design must take into consideration the fact that future disturbances may arise which are not present in the existing system, and the control system must itself evolve in order to respond to them. Consequently, the system cannot be designed once and for all as in a textbook example. Moreover, the system changes even during the design and implementation, and this must be considered in selection of structure and control schemes.

4. The data about the disturbances and state of the system needed in order to adequately perform the control task are formidable. That is, each portion of the control system responds to a different set of disturbances which require different data in order to perform the control strategy. For example, response to a rush order requires information about the inventory and production schedules as well as the state of the manufacturing plants, whereas response to an upset in the flow rate of a process stream requires only dynamic data about that stream and its local environment. Poor choice of structure for the system and control scheme may result in large amounts of communication of data among subsystems, decreasing reliability, and increasing costs.

5. Realization of the control strategy in the complex hardware configuration poses major programming or software problems, ranging from complexity of operating systems to debugging in an on-line, real-time situation. Since the system evolves in time, changes to this solution may become necessary, resulting in inefficient hardware and software utilization.

The solution to these problems is the on-line, real-time, multilevel system but with the *decomposition* and *coordination* taking place *not only in the system structure* (hardware and communication) *but also in the design of the control strategy.*

First consider the decomposition of the control strategy into a multilevel structure with each subsystem responsible for compensation for a subset of the disturbance, including those introduced through decomposition (i.e., compensation for neglected interaction).

Decomposition of the control system design problem into a multilevel system produces a set of subsystems or tasks each of which is capable of complete specification. That is, its four essential characteristics may be determined explicitly. These are

1. *The method of solution:* Since each task is designed to handle a specific disturbance or set of disturbances, the control methodology involved is well defined.
2. *Input/output data requirements:* Specification of the task and the method of carrying it out implies that the input data needed are known and the outputs produced are specified. In particular, data not needed by the task are then implicitly specified.
3. *The computer load:* From the specified task and methodology, the amount of computer time and capacity to carry out the task once can be estimated.
4. *Frequency of the task:* From the frequency of the disturbances associated with this task, the frequency at which the control program must be executed is known.

Decomposition of the structure of the system permits delineation of the individual components and communication links in order to permit evaluation of a configuration. The design problem may then proceed as shown in Fig. 8-1. Here the control design problem is defined and decomposed or partitioned using any of the techniques to be described here. The four characteristics of each task listed above are defined carefully. A structure for the system is assumed and the tasks assigned arbitrarily to the subsystems of the assumed structure. Using the

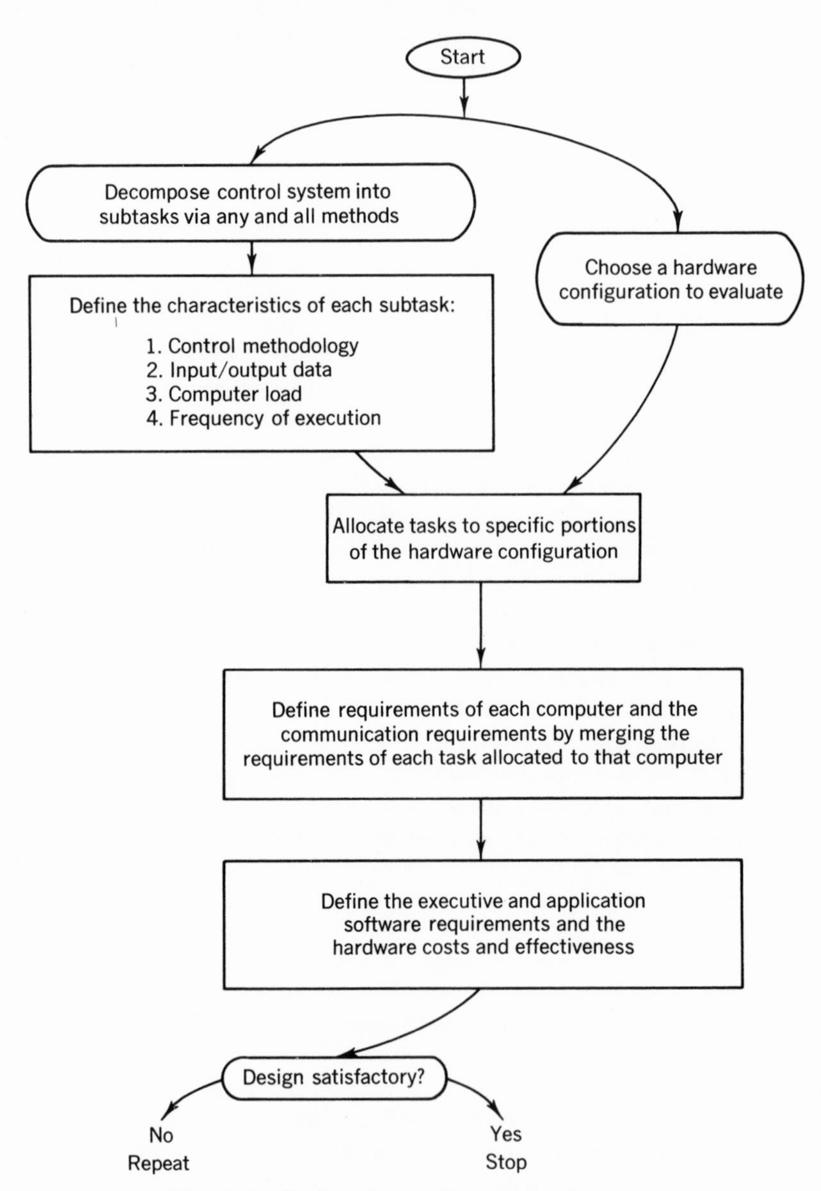

FIG. 8-1. Design of an on-line multilevel system.

characteristics determined from each subtask, the requirements for that subsystem are determined by merging the requirements of each task assigned to it. In addition, the communication requirements between subsystems can be evaluated from the input and output data requirements of the task assigned to each structured subsystem. In the case in which the subsystem is a computer, the merging of the characteristics of each task assigned to that computer defines the total computing load, executive requirements, input/output rates and volume, bulk storage, reliability, operator communication, and intercomputer communication. This in turn defines the size of computer, its cost, and ultimately the cost of producing software for that subsystem. This then permits a cost-benefit evaluation of the given decomposition of the problem, the assumed structure, and the allocation of tasks to the structure.

The decomposition of on-line control problems may be based on analytical techniques, as, for example, when the problem can be formulated as a static or dynamic optimization problem.[35,36] In this case, points of weak interaction may be selected and the interconnection variables either "cut" or "fixed," permitting solution of that optimization problem by a multilevel structure with either goal coordination or model coordination. However, in general the on-line control problem is not realistically a straightforward optimization problem but rather a set of complex interacting problems, and the decomposition into a multilevel on-line system is based less strictly on analytical problems in which convergence and existence can be demonstrated. Even in these cases, the concepts developed for coordination of partitioned optimization problems are extremely useful. In particular, most decomposition methods use a model-coordination scheme since this allows one level to treat a variable as known and the upper or coordinating levels to either compute, predict, or estimate these variables. This method is particularly effective when uncertainties plague a system, for the task of estimating or predicting becomes localized to a specific subsystem.

From the above analysis, any weak links (excessive size, excessive communication, etc.) can be modified and the design process repeated until a satisfactory system results. Future expansion or development of the system involves addition or deletion of coordinatable subsystems rather than complete redesign of the existing system. This approach to the design of a multilevel, on-line, real-time system has the advantage that those facets of the problem which are inherently real time are carefully considered—especially the communication requirements between subsystems.

Of major importance are the methods for reducing the complex real-time control problems to a multilevel form which permits coordination. In this chapter, the various methods for decomposing the problem into subproblems and assigning these subproblems to hardware are discussed and illustrated.

8-2 DECOMPOSITION AND COORDINATION OF ON-LINE SYSTEMS

In order to achieve the advantages of a multilevel multigoal system, it is necessary to decompose the system and/or its control into a number of smaller parts whose individual design or implementation is straightforward (or at least feasible), and in addition to provide a mechanism for coordination of the parts in order to achieve some overall system goal (efficiency, reliability, etc.). The decomposition or partitioning of the system into subparts is very easy, and in fact there is no limit to the number of different ways in which this can be done. However, the problem of coordinating these parts is not always straightforward and places a practical limitation on the decomposition. In other words, like Humpty Dumpty, a system is easy to take apart, but hard to put together again. This is especially true in on-line systems since the constraint of response in real time may preclude many theoretically possible coordination algorithms (e.g., unfeasible methods).

Three major classifications of decomposition methods for on-line systems include all the practical approaches which lend themselves to efficient coordination and realization.[22,23] They are

1. Decomposition on the basis of *structure*
2. Decomposition on the basis of *levels of influence*
3. Decomposition on the basis of *levels of control*

Decomposition on the basis of structure implies a conceptual partitioning of a system or process into separate subsystems each with its own (perhaps conflicting) goals and with interaction among the subsystems. The decomposition may be in space (different physical parts of a system) or in time (different operational phases of a system).

Control systems are designed for each subsystem but in such a way that the individual first-level control systems may be coordinated, i.e., by including in each controller some coordinating variables. Individual second-level controllers are then assigned the task of coordinating groups of first-level controllers through manipulation of the first-level coordinating variables. Similarly, coordinating variables are designed into the second-level controllers so that third-level controllers may in turn control the second-level units, etc., until a hierarchy results.

In general, this approach to decomposition leads to coordination through *manipulation of goals* of the individual subsystem controllers or through *estimation or prediction of the interaction variables* when these are the coordinating variables.

Decomposition on the basis of *levels of influence* refers to a decomposition of the control system and implies a division of the overall decision making or control effort among several levels with a priority associated with each level, such that higher levels have the higher priority. Management control systems are often designed this way. Such a decomposition leads to coordination through *direct intervention*, *goal intervention* (manipulation of goals of lower-level units or parameters in those goals), and *constraint intervention* (restriction or modification of the domain of the control actions of the lower-level units).

The third approach, decomposition on the basis of *levels of control*, implies a division of the control objective into separate parts, each of which is solved by a different methodology. For example, division of the frequency range of the disturbances in a system into distinct bands can lead to quite different levels of control for the fast physical disturbances than for the slower economic disturbances. Coordination in this case is usually through *information intervention* or *goal intervention*.

Since in each approach there results a decomposition into *levels*, it is convenient to distinguish between them. Consequently, each level in a decomposition on the basis of *levels of influence* will be termed a *stratum*, whereas a *level of control* will be termed a *layer*. All three approaches to decomposition are often present in one system. For example, a control problem may first be decomposed into several strata or levels of influence and then the task or objective of each stratum realized in the form of a multilevel, multigoal control system by decomposing the problem associated with that stratum on the basis of structure. Finally, each control system in that multilevel structural realization of that stratum can itself be decomposed into layers or levels of control in order to simplify the final design and implementation.

It is important to determine techniques for coordination in each type of multilevel structure and to consider the methods by which each structure handles physical disturbances in the system together with disturbances introduced by the decentralized or decomposed nature of the system.

8-3 DECOMPOSITION ON THE BASIS OF STRUCTURE

The most common approach to decomposition in order to both minimize complexity of individual units in a system and minimize the effects of disturbances on the overall system operation is to decompose the system on the basis of its structure. This is essentially a *feature extraction* process, recognizing the portions of the system which form natural partitions. Four features commonly used to partition a system into subsystems are geographic location, legal or political boundaries, functional behavior or objective, or operational phase.[5]

The first two, geographic location and legal or political boundaries, are directly related to the physical structure of systems. For example, consider Fig. 8-2*a*, which shows an electric power system greatly simplified.[5,8,13] This system is a "pool," that is, a set of interconnected

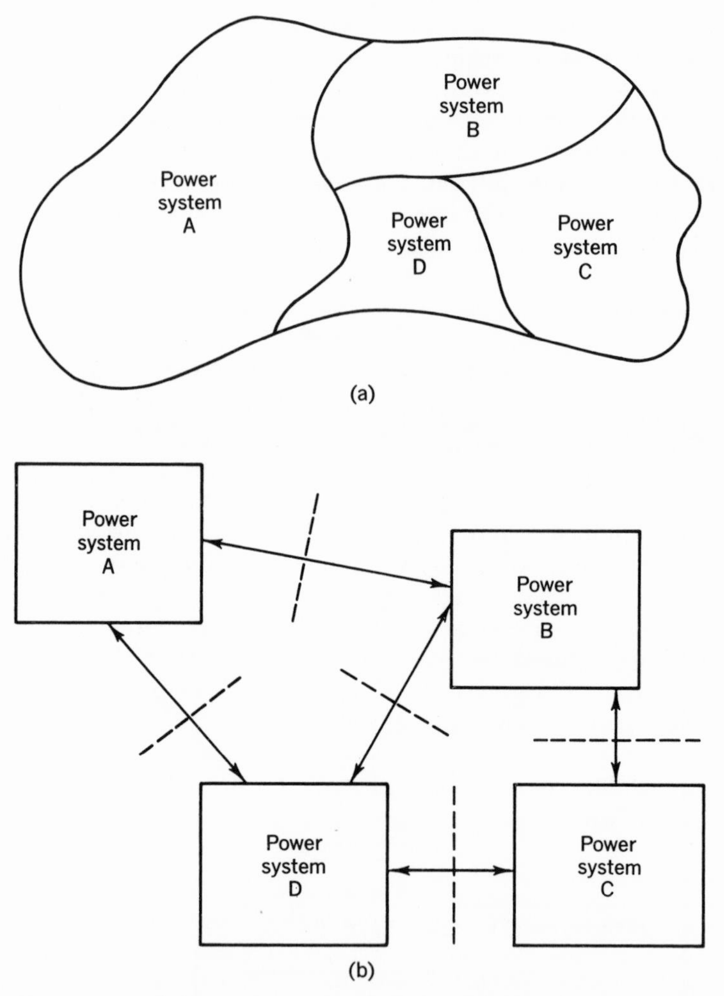

FIG. 8-2. (*a*) A pool of interconnected power systems; (*b*) decomposition of the power pool on the basis of structure (interconnections are tie lines).

power systems each of which is responsible for supplying power in its own distinct, nonoverlapping region but which cooperate to buy and sell power to one another through intersystem tie lines in order to make the overall pool as reliable and efficient as possible. Thus one company can supply its excess power to another if the load on the one system is more than the company can handle. In addition, nonnormal distributions of load can often be more economically supplied by purchasing part of the power rather than generating it, and this is possible if the companies are interconnected into power pools. But control of such a system leads naturally to a multilevel decomposition, for it is a fact that the individual companies are separately managed and owned and therefore have distinct objectives. Thus the system or pool is naturally partitioned along legal boundaries as shown in Fig. 8-2*b*. But individual systems are complex in their own right and it might be desirable to further partition the individual systems geographically by generating station, for example, and even further by principal loads within each major center as shown in Fig. 8-2*c*. Note that this block diagram form

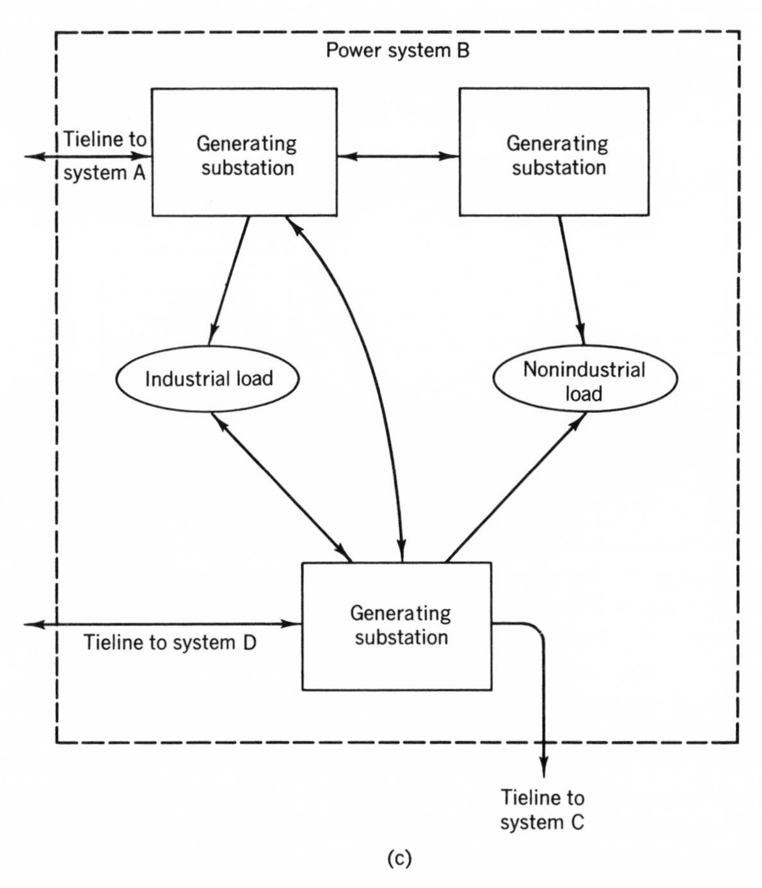

(c)

FIG. 8-2. (*c*) Further decomposition on the basis of structure within one power system.

is a partition of the system but not decomposition (that is, the blocks are not separated). However, based on this partitioning of the system, a decomposition of the management or control system may be made.

Other systems are of a similar nature with distinct parts of a process often being physically separated from others, leading to a natural partitioning of the system and decomposition of the control problem. Definition of the decomposed control problems and their coordination leads to multilevel, multigoal or hierarchical form for the control system and is discussed below.

Often features are readily extracted on the basis of different *functions* of a system. Consider for example Fig. 8-3, which shows a possible

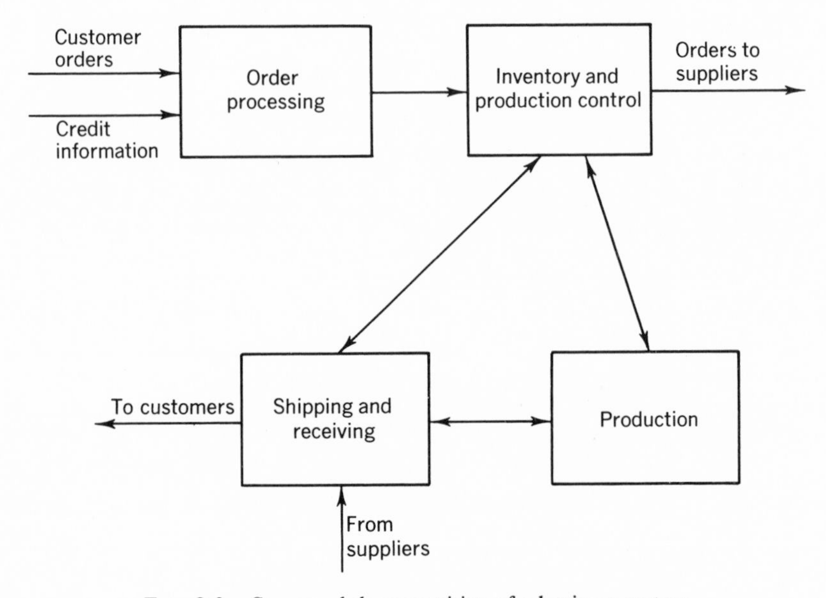

FIG. 8-3. Structural decomposition of a business system.

partitioning of a business system into distinct functional parts.[14] Notice that these parts are not physical like the generating plants of a power system but rather have different objectives and tasks in the overall business. It is clear that there is advantage to this partitioning in that the individual parts of the system can be carefully designed to carry out their own tasks as optimally as possible and in particular to respond to their own disturbances without upsetting other parts of the system. Of course this again does not correspond to a physical decomposition since the parts of the system interact considerably. Design of a hierarchical control system, however, can be performed by decomposing the system i.e., making the partitioned subsystems separate and distinct or independent as far as the first-level control or management systems are concerned, and designing control systems which can be coordinated by higher-level units in the hierarchy.

The fourth type of feature which can be used as a basis for structural decomposition is that of *operational phase*.[5] Chestnut points out that there is distinct advantage to structuring a system (while designing it) so that its operation may be divided into distinctive time periods with significant characteristics associated with each such time period.[5] For example, consider Fig. 8-4, which shows a division of the manufacturing

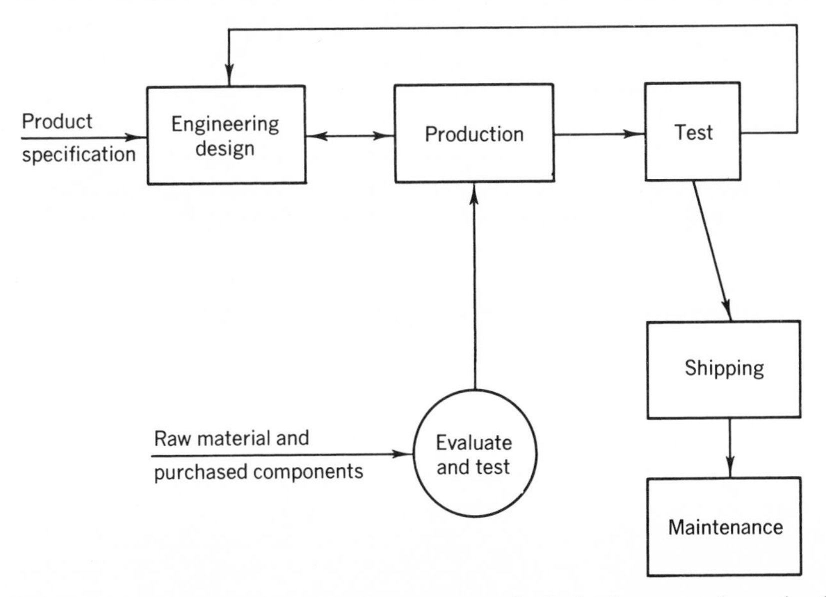

FIG. 8-4. Decomposition of an on-line system on the basis of structure (operational phase).

process into operational phases ranging from product design through manufacturing, testing, shipping, installation, and finally maintenance. This partitioning of the system leads to a natural multilevel, multigoal control system with individual first-level controls for each operational phase together with coordination in order to achieve overall system objectives. Notice that coordination may be critical in a decomposition such as this since first-level subsystems often have conflicting goals. For example, the objective of production is to ship units and if no coordination is present, units may be shipped which have not been sufficiently checked out in the plant, with the result that the maintenance division of the organization must spend an excessive amount of money and time to repair the product in the field. Coordination in this case must occur through control of accounting procedures of the individual subsystems, charging some of the maintenance costs against production, for example, so that production is heavily penalized for excessively poor quality control. This corresponds to coordination through *goal intervention.*

It is typical that the partitioning of a system might involve all four of these features and perhaps others peculiar to a particular class of system. The common denominator or thread is that this feature extraction process results in a structural partitioning of the system, or block diagram, which is in the form of a *set of interconnected subsystems.*[15,16]

The design of a multilevel, multigoal control system now proceeds in three steps. First the system is abstracted or simplified so that it can be handled. This includes specification of inputs and outputs of individual subsystems which are to be considered or neglected in the design of the control system. In addition, certain disturbances in the system are either neglected as inconsequential or else handled as special cases. For example, catastrophes such as the death of a key executive might be controlled through a life insurance program, etc. Finally, objectives are assigned to the overall system and subsystems. These objectives might be artificial (quota for the sales division for example) or in conflict with the objectives of other subsystems (manufacturing quality control versus maintenance for example) but serve to define the individual subsystem control problems.

The second step of the design consists in the selection of a decomposition and coordination scheme which then defines the first-level control problems for each subsystem completely. In the case of decomposition on the basis of structure, the methodology developed for the solution of static optimization problems for interconnected subsystems is the most common technique applied. This approach takes advantage of the fact that in most systems, the partitioning is such that the dynamics of the individual subsystems are not the cause of the difficult interaction. That is, efficient operation of individual subsystems is usually such that variables within the subsystem are controlled and all dynamics are accounted for by the individual noninteracting control systems. Consider the partitioned system in Fig. 8-5. Here a local control system called a direct control system is shown for each subsystem whose objective it is to control the dynamics of the subsystem, that is, to hold the subsystem variables at particular desirable values (which may change with time, but slowly compared to the frequency of control computation). The combination of a subsystem and the direct control system then represents a subsystem which is in the pseudosteady state, that is, one whose variables are changing slowly with time. For example, a natural structural partitioning of the paper-making process results

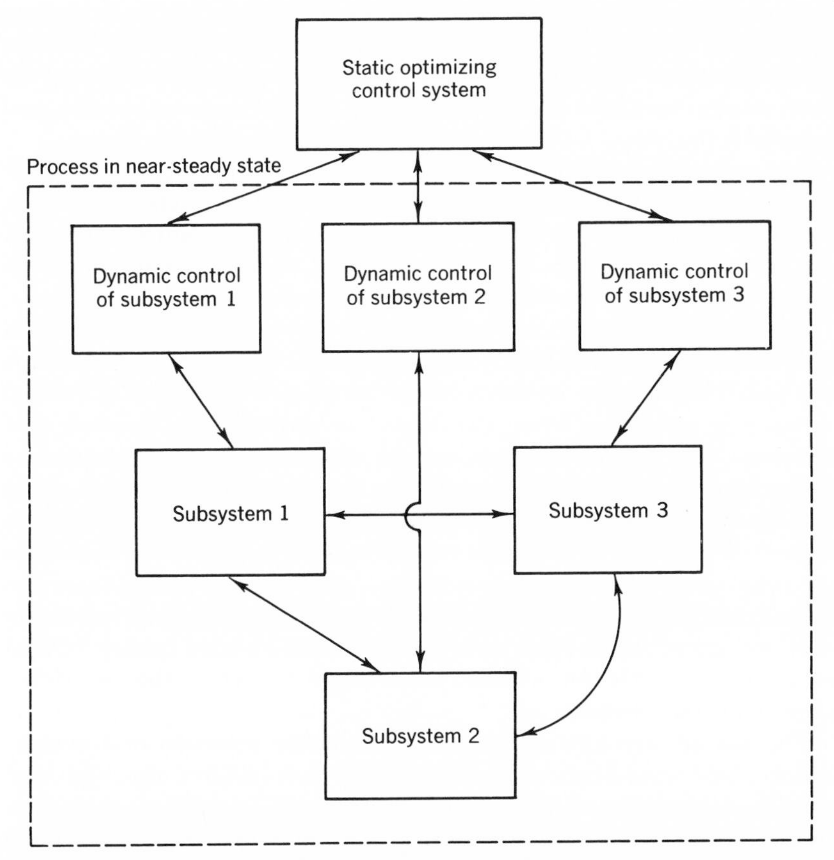

FIG. 8-5. Converting a dynamic system to a static system to use decomposition based on mathematical programming.

in the dryers as one subsystem and the Fourdrinier and white water system as a second subsystem.[40] The interaction between them is due to basis weight and moisture content of the sheet as it leaves the Fourdrinier and enters the presses and dryer sections. Within the Fourdrinier section, however, there are many dynamics which are locally controlled, including all the flows and pressures. Similarly, steam pressure and flow within the dryer section are carefully controlled. If each individual control system is operating properly, the integrated system operates in a steady state or near steady-state condition so that interaction is nondynamic, at least on a short-term basis. Thus basis weight and moisture content are relatively constant and are considered constant during the optimization cycle of the individual subsystems.

Assuming that the individual first-level control systems control the dynamics of the individual subsystems, overall system optimization can be considered to be a static optimization problem and decomposition and coordination handled by the methods developed in Chapter 1. Even when this is not exactly the case, most systems are designed so that their operation is nearly steady state and the approximation is a good one. Moreover, in the on-line situation, disturbances enter the system periodically which change the steady-state operation, and it is the function of the multilevel system to "track" those disturbances, that is, to change the steady-state operation of the system so that the overall system performance remains nearly optimal.

Decomposition under these conditions can take place if the overall system performance can be specified in a separable form or, equivalently, if individual goals can be assigned to the subsystems such that overall system performance is the sum of the performances of the individual subsystems. Then following the coupled subsystem procedure of Chapter 1, a set of coordinating variables is chosen corresponding to the interaction variables being considered. Then the subsystems are decomposed or decoupled by either fixing the coordinating variables (the *feasible method*) or "cutting" the coordinating variables (the *dual-feasible method*). In the former case, values of the coordinating or interacting variables are predicted or estimated by the second-level control systems and transmitted to the first-level systems, each of which optimizes its performance subject to the constraint on the values of the interconnecting or coordinating variables. This proceeds iteratively with the second-level units updating their estimates of the optimal values of the interacting variables until an overall system optimum is achieved. Of course in an on-line system, disturbances entering the system change this overall optimum and consequently the control units repeat their computations as fast and as often as necessary.

Unlike the off-line system, models used for the optimization are not perfect, and consequently the predicted and actually realized interacting variables are not the same. Thus the use of a multilevel control system has the effect of introducing a disturbance into the system, namely, the difference between the predicted and realized interacting or coordinating variables. Part of the function of the control systems at both the first and second levels is to compensate for these internal disturbances by (for example) using past information to update or adapt the models used for the optimization. Such a control system is best handled by a further decomposition into *layers* discussed below.

In the case of the dual-feasible method, a Lagrange multiplier or *price* is set on the coordinating or interacting variables as discussed in Chapter 1, and the cost or return associated with input and output intermediate process variables for each subsystem is added to the objective function of that unit. The second-level coordinating units then coordinate through goal intervention by adjusting the prices or values of the intermediate products until the amount produced by the one subsystem (the supply) is equal to the amount desired by the using subsystems (the demand). In the off-line solution of the static optimization problem by the dual-feasible method, these prices are adjusted iteratively but the subsystems are never connected until the optimum is reached. In the on-line situation, the outputs of the subsystems are actually inputs to the user subsystems. That is, the real system is not actually cut into noninteracting subsystems and consequently, nonoptimum values for the prices result in nonoptimum operation of the individual subsystems. Of course the system is designed so that the rate of computation of the control systems is such that if no disturbances entered the system, it would reach its optimum quickly, and if disturbances enter the system, the controllers are fast enough to "track" the optimum. That is, the prices can be adjusted quickly enough that supply and demand of intermediate or coordinating variables stay within satisfactory limits.

Not all systems correspond to the above example of an interacting static optimization problem being solved iteratively in time. For example, the business system shown in Fig. 8-3 is decoupled in general by prediction of the interaction variables (the feasible method). Specifically, control of manufacturing is according to a schedule which is based on orders on hand as well as a forecast of the demand and a model of the current inventory. Disturbances correspond either to "hot" orders which were unanticipated and must be met, or to internal disturbances such as an error in the forecast which causes the inventory to become depleted and penalties for late delivery to be applied. In any case, the prediction, forecasting, and scheduling are done at such a rate that these disturbances can be "tracked" as best possible or, in other words, so that the system operates at the optimum corresponding to those disturbances. Such an approach is successful unless the dynamics of the system are comparable to the frequency of the disturbances. In this case the system lags behind the changing disturbance and does not operate in a pseudosteady state. An example of this is the power generating system in an emergency condition. Here a severe fault or sequence of faults might cause subsystem dynamics to interact in such a way that the system actually becomes unstable. Successful decomposition of such a system then requires a special control system for this situation to ensure that it cannot occur or is quickly corrected when it does occur (by, for example, shedding load or physically decoupling the subsystems). Such a control system is called a *security control system* and is an example of a *stratum* of a multilevel system to be discussed later.

Design of an on-line multilevel, multigoal system then depends on the frequency of disturbance of the actual system. Thus implementation is greatly concerned with efficiency of on-line computers which do the on-line decision making or control and with the efficiency of the data acquisition or information gathering which is necessary for the decision making. This is the reason for management information systems, for without sufficiently rapid access to information, the higher levels of control cannot coordinate efficiently enough to track a moving system optimum.

An important problem in electric power generation provides a good example of an on-line multilevel system based on a structural decomposition, namely, the calculation of load flows in an interconnected network.[8,26] Consider the decomposition of an electric power system into areas as shown in Fig. 8-6. At any interconnection point (a bus

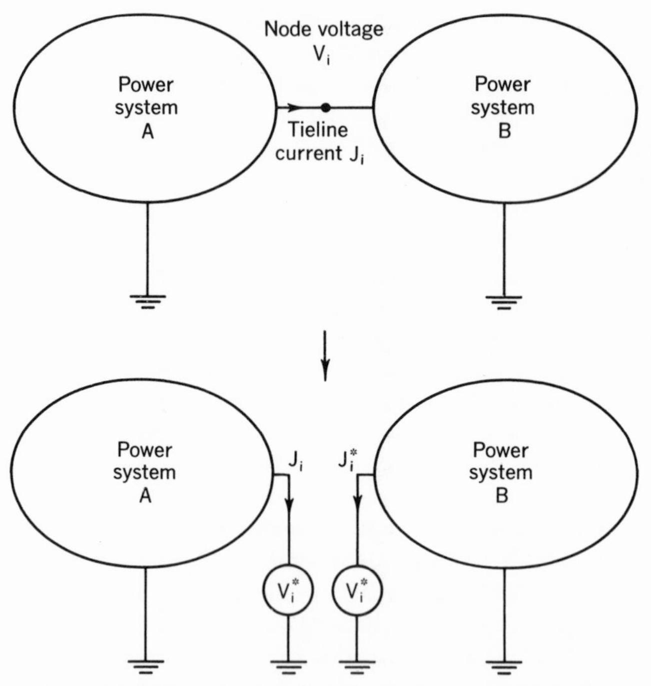

FIG. 8-6. Decomposition of a power system load flow problem. On-line computer chooses $V_1{}^*$ to force $J_k + J_1{}^* = 0$.

between two power systems or any transmission line within one system but between two areas or partitions) there may be considered to be a node of the network shared by the adjacent areas (two or more). Now it is well known in electrical network theory that any two-terminal subnetwork embedded in a larger network can be replaced by a voltage generator whose voltage is the same as the voltage across the replaced network without changing the voltages and currents anywhere in the external network (the compensation theorem).[25] This is exactly equivalent to the result in decomposition that fixing certain variables at their optimal values does not change the optimal solution of the system.

The network is decomposed according to the partitioned areas by *fixing the voltages at the interconnection nodes* (the feasible or model-coordination method) by replacing them by equivalent generators of fixed voltage. This of course separates the areas as shown in Fig. 8-6 so that the distribution of currents and voltages in each area may be independently determined. As in the model-coordination method, coordination is through estimation or prediction of the interconnecting variables (the voltages of the interconnection nodes). This is done on the basis that in the connected system, no current flows from an interconnection node to ground, as shown in Fig. 8-6. That is, according to the compensation theorem, the correct interconnection node voltages should produce the actual interconnected network currents and voltages and there is no current flow to ground at the interconnection nodes.

Thus for the example in Fig. 8-6, it is required that at each interconnection node,

$$j_i + j_i^* = 0$$

which means physically that the current flowing from one area to the other is actually j_i and none flows to ground at that point. This guarantees that the voltages and currents calculated from the individual area networks are identical to those which would exist in the integrated or connected network. In general there may be more than two areas sharing an interconnection node and in that case, the appropriate algebraic sum of the currents through the generators introduced must be zero.

The load flow problem for each subnetwork is now a straightforward network problem (although nonlinear) including generators, transmission lines, loads, and, in addition, voltage generators at each interconnection node. The writing of network equations even for nonlinear networks is relatively straightforward and, moreover, the computation of gradients of the dependent variables with respect to the independent variables is readily determined.[8] Consequently, the gradients of the currents flowing in the interconnection generators with respect to the voltages assumed at the interconnections are readily determined, so that any steepest descent or accelerated steepest descent algorithm (such as the Fletcher-Powell algorithm) may be used for the coordination algorithm on the second level.[2,12]

Such a system can be operated in an on-line fashion since it is possible to adjust bus voltages at predetermined points in a network to any desired value.[8] Thus if a load flow calculation is used in an on-line control system, it is possible to do it in a decentralized manner and even in a multilevel manner. That is, it is common to do this separately for each individual power company, but it is also possible to decompose the computation for a given power company into arbitrary subcomputations which are easily coordinated. The advantage of doing so lies in the reliability of the resulting decentralized system.

A second example of the use of multilevel on-line control in power systems is the problem of economic dispatch.[8,9,20] Consider again the partition of the system into a number of subsystems as shown in Fig. 8-6. Define

P_j^i = complex power produced by generator j in area i

$f_j^i(P_j^i)$ = cost of generating P_j^i

L_j^i = jth complex load or loss in area i

T_j^{ik} = complex power received by area i on the jth tie line between area i and k

The optimization problem associated with operation of the system is then

$$\min_{\mathbf{P},\mathbf{T}} \sum_i \sum_j f_j^i(P_j^i)$$

subject to the constraint in each area

$$\sum_j P_j^i + \sum_{k,j} T_j^{ik} - \sum_j L_j^i = 0 \qquad \text{for each } i$$

and the constraints

$$\text{real } P_j^i \geqslant 0 \qquad \text{for all } i \text{ and } j$$

The *feasible* method leads immediately to a decomposition for it is self-evident that the only interaction between partitioned areas is the interarea power T_j^{ik}, and fixing these variables immediately decouples the optimization problem into independent problems, one for each area. The function of the second level is then to determine the optimal values for the interconnection or tie-line power. Happ[13] points out that it is possible to create an on-line computer system with as many levels as desired in order to decompose as large a power system or pool of power systems as desired.

If we take the vector $\mathbf{P}$ to consist of all the powers P_j^i and the vector $\mathbf{T}$ to consist of all the interconnection powers T_j^{ik} and the overall system objective to be minimized $\phi(\mathbf{P}, \mathbf{T})$, then the optimization problem may be stated as

$$\min_{\mathbf{P},\mathbf{T}} \phi(\mathbf{P}, \mathbf{T})$$

subject to the constraints

$$\mathbf{P} \in S_1 = \{\text{set of permissible generator complex power vectors}\}$$

$$\mathbf{T} \in S_2 = \{\text{set of permissible tie-line or interconnection complex power vectors}\}$$

and the area constraints

$$g_i(\mathbf{P}, \mathbf{T}) = 0$$

For each area i this constraint corresponds to conservation of power.

The multilevel form can be written as

$$H(\mathbf{T}) = \min_{\mathbf{P}\in S_1} \phi(\mathbf{P}, \mathbf{T})$$

Let the domain of H be S_3 :

$$S_3 = \{\mathbf{T} \mid H(\mathbf{T}) \text{ exists for all } \mathbf{T} \in S_2\}$$

Then the second level or coordinating level solves the problem

$$\min_{\mathbf{T}\in S_3} H(\mathbf{T})$$

and this is a well-formed problem which is amenable to any hill-climbing procedure, provided that the gradients of the function $H(\mathbf{T})$ exist and can be calculated. This problem was considered in detail in the discussion of the differentiation of the dual function in the dual-feasible decomposition method in Ref. 35, and the results obtained are directly applicable here.

Forming the constrained optimization problem in terms of Lagrange multipliers yields

$$L(\mathbf{P}, \mathbf{T}, \lambda) = \phi(\mathbf{P}, \mathbf{T}) + \sum_i \lambda_i g_i(\mathbf{P}, \mathbf{T})$$

If the first-level system minimizes the objective function $\phi(\mathbf{P}, \mathbf{T})$ subject to the constraints $g_i(\mathbf{P}, \mathbf{T}) = 0$, then it is clear that the Lagrangian L is equal to the objective function ϕ, and furthermore, $H(\mathbf{T}) = L(\mathbf{P}^*, \mathbf{T}, \lambda^*)$ where $\mathbf{P}^*$ and λ^* are the results of the first-level optimizations.

In Sec. 1-2-5, it is shown that provided

1. The Lagrangian is continuous with continuous partial derivatives.
2. S_1 is compact.
3. The $\mathbf{P}^*$ which minimizes $\phi(\mathbf{P}, \mathbf{T})$ subject to the constraints is unique for a given $\mathbf{T}$.

then the partial derivatives of $H(\mathbf{T})$ exist and are given by

$$\nabla_{\mathbf{T}} H(\mathbf{T}) = \nabla_{\mathbf{T}} L(\mathbf{P}^*, \mathbf{T}, \lambda^*)$$

and thus for the problem at hand

$$\nabla_{\mathbf{T}} H(\mathbf{T}) = \sum \lambda_i^* g_i(\mathbf{P}^*, \mathbf{T})$$

Thus each first-level control system solves its constrained optimization problem, minimizing its generating cost subject to its own area constraint and transmitting to the second-level coordinating system both the optimizing $\mathbf{P} = \mathbf{P}^*$ and also the values of the Lagrange multipliers λ^* which result from the constrained optimization. First-level subsystems are decoupled since $\mathbf{T}$ is assumed known. The second level can then form the gradient of $H(\mathbf{T})$ in order to determine the direction to change $\mathbf{T}$ in order to decrease the integrated system performance function. With the gradient readily available, efficient convergence of the multilevel system is possible.

Once having determined the optimal power generation and tie-line schedules for a pool and a company, it is necessary to translate this into individual generator voltages and phase angles in order to produce this desired distribution.[8] The proper formulation of the load flow problem already discussed solves this problem and is a further illustration of the interaction in an on-line system of two multilevel systems.[8]

The on-line system operates in real time, tracking disturbances and changing loads, losses, and configurations and enforcing the current $\mathbf{T}$ at the interconnection nodes of the network. If an interconnection node corresponds to the connection between power companies, this is straightforward, but at internal nodes, it may be easier not to control $\mathbf{T}$ at the interconnection nodes but rather to track fast enough that the optimal value of $\mathbf{T}$ is realized by adjusting the remaining $\mathbf{P}$ to the optimal values in the generators. In the former case, the penalty for not computing fast enough to be at the optimum is simply suboptimal performance of the system, but in the latter case, the operation may become unsatisfactory in the sense that some loads may not be satisfied and the system is projected into the emergency state.

Other formulations of the multilevel multigoal solution to these problems are possible, and, in fact, since they amount to constrained static optimization problems (repeated as often as disturbances change), any of the methods or combinations of methods applicable to such problems can be used to derive a great variety of solutions. It is important in the on-line problem, however, to investigate as many of these alternative solutions as possible in order to take advantage of any simplification or efficiency peculiar to one formulation over another. That is, although all such solutions are conceptually or theoretically equivalent, they are quite different from a practical point of view with respect to computing requirements, information requirements, communication requirements, operator intervention, etc. Perhaps of most importance, but not yet discussed, is the inclusion of constraints of *secure operation* in the optimization which may affect greatly the choice of solution method. Other approaches to this problem are discussed in the power system literature and especially in Refs. 9 and 13.

8-4 DECOMPOSITION ON THE BASIS OF LEVELS OF INFLUENCE

Decomposition on the basis of structure leads to a multilevel control or management system because individual subsystems have their own control systems and these must be coordinated by upper levels in order to optimize overall system objectives. In contrast to this approach is decomposition on the basis of levels of influence which, instead of partitioning the physical system into parts, directly partitions the control or management system itself into multiple levels. Thus a control for a physical process might be decomposed into several levels of influence, called *strata* to differentiate them from the other two uses of the term *level*, but with or without partitioning or decomposing the system itself.[21,23] The partitioning into strata again has the objective of simplifying an overall systems engineering problem by separating the problem into a number of smaller, better defined subproblems each of which is solved separately.

Two characteristics or features of the decomposition on the basis of levels of influence are the following: First, individual strata have quite different objectives or tasks, so that there is little direct interaction as characterizes the interaction between subsystems partitioned on the basis of structure; and second, the division into strata usually results in a priority associated with each stratum, with the higher strata having priority over the lower ones. The result is that the decision making or control carried out in a higher stratum uses information transmitted to it from the system itself, the outside world, and the lower levels. The result of its decision making is an almost direct intervention into the decision-making process taking place on the lower strata.

Thus decomposition on the basis of levels of influence results in coordination through

1. Direct intervention
2. Constraint intervention
3. Goal intervention

The first means, *direct intervention*, is the overriding of a decision made in a lower-level stratum. This might happen, for example, by an upper-stratum decision to revise a schedule picked by a lower-level stratum (on the basis of an optimization procedure, for example) in order to handle some emergency order or condition.

Constraint intervention is a change in the domain of allowable control actions of a lower-level stratum. For example, an upper-stratum control system concerned with security of a process might restrict the mode of operation of the normal control system being carried out in a lower stratum in order to ensure that the process will continue to operate successfully despite some anticipated contingency. This is exactly the mode of intervention by the security strata in an electric power system in order to ensure that the power generating system will remain stable despite the occurrence of anticipated faults.

Goal intervention takes place between strata usually by change of the goal itself. For example, the decision to change the goal of an on-line computer control system from maximum production subject to the constraint of adequate quality to a fixed and less than maximum production but with maximum quality may be made by a management stratum in response to changing market conditions for a product. If a catastrophe occurs in an electric power system, the goal changes from minimum cost production of power to maximum load capacity in order that the system be capable of delivering the desired load without being shut down by the disturbance.

Unlike decomposition on the basis of structure, decomposition on the basis of stratum is not based on the decomposition of a well-formulated mathematical programming problem and consequently does not make use of a rigorous theory to justify its performance. Rather a control problem is partitioned or structured in such a way that it can be solved sequentially in levels or strata with the result or output of one stratum (a higher-numbered one) serving as partial input to another lower-numbered stratum. Thus from the top level down, the decision-making process is similar to a staged process rather than a completely interacting one. Of course all strata are acting in parallel, but in general the higher the level, the longer the horizon or time over which a decision is made and, consequently, the less often the control action takes place. The result is a more-or-less equal distribution of the total effort in decision making or control among the various strata with the lower levels performing a simpler task more often and the higher levels performing a more complex task less often. The decomposition is best illustrated by examples.

Consider first the overall control of the steel-making process, which is typical of a large-scale industrial process.[24,27] Figure 8-7 illustrates the process schematically together with a possible decomposition of the control problem into four strata or levels. The four strata are discussed in detail below.

The first stratum is concerned with process control, that is, the actual operation of the process given the commands from the upper strata as to what to produce and with what objectives or goals. This stratum is concerned with the actual process variables (speeds, temperatures, etc.) in order that the process units, blast furnaces through rolling mills, operate correctly as well as optimally. This level must be concerned with operators and management interactions and must respond to the fast-acting disturbances (breakdowns of equipment, changes in characteristics of raw material, etc.). This stratum is itself highly complex and is usually implemented by further decomposition of the process control problem in both space and level. That is, the process itself is usually partitioned on the basis of its structure to create an on-line, multilevel, multigoal process control system. Furthermore, each control system on each level is usually further decomposed on the basis of levels of control in order to make feasible the implementation in an on-line computer system. This stratum is itself so important that it is discussed separately in Sec. 8-5.

The second stratum, as defined by Miller, is the *information collection stratum*.[24] The problem of concern here is the tracking of the progress of material as it passes through the process as well as its characteristics. For example, the amount and type of iron arriving at the steel plant from the iron works significantly affect the operation of the steel plant, just as the output of the steel plant affects the operation of the soaking pits, etc., throughout the process. The output of each portion of the process (imagined decomposed on the basis of structure) represents an input to another portion of the plant, and the function of this stratum is to collect and transmit this information. Thus much of the information collected by the second stratum is used by process control systems on the first stratum. However, this information also constitutes the current

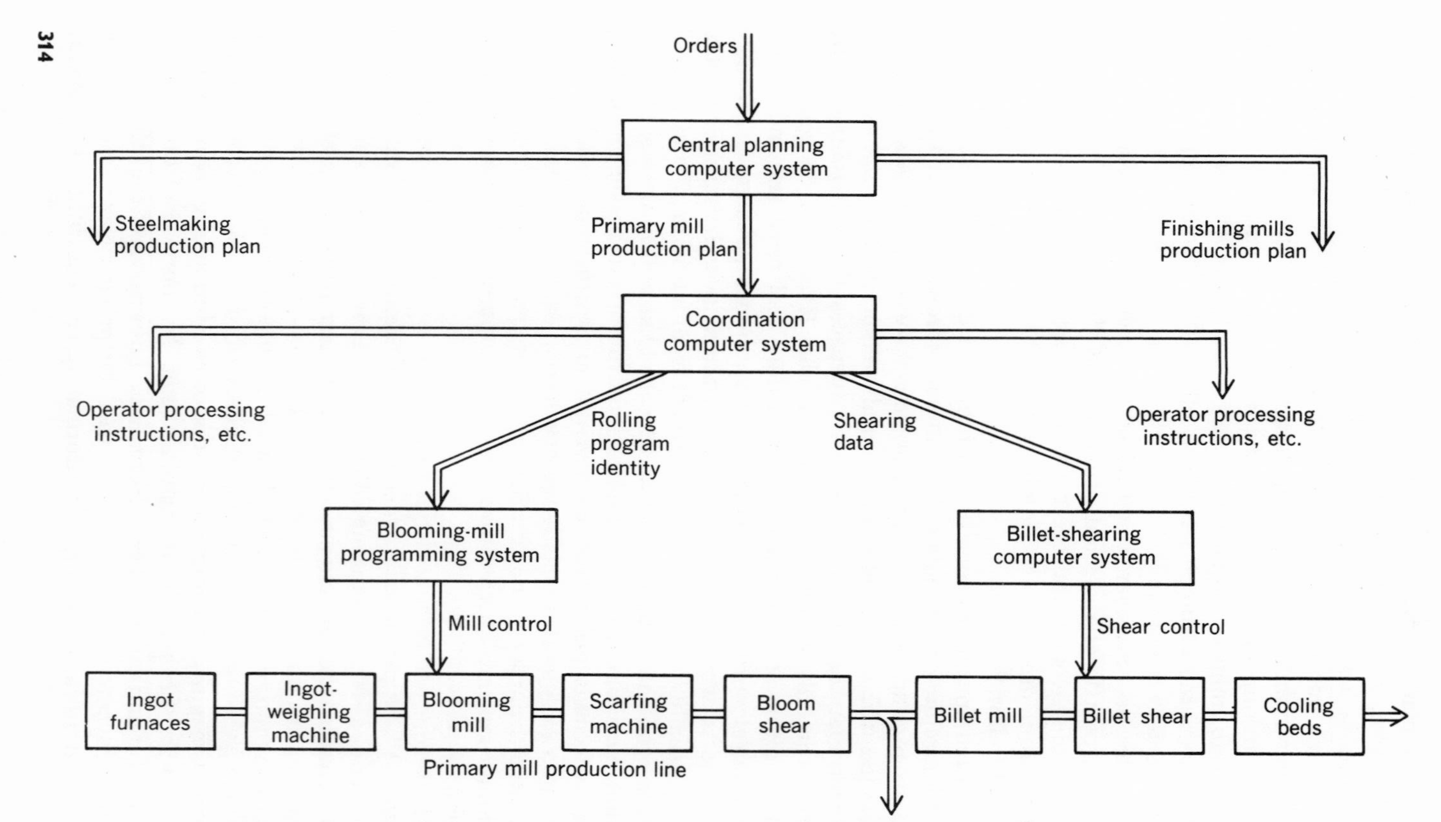

FIG. 8-7. Decomposition into strata of an on-line control system for the steel industry. (*From Ref.* 24.)

state of the plant, indicating the status of orders, raw material, equipment, workers, etc., and consequently much of the information collected by this strata is processed and transmitted to higher-level strata to aid in achieving overall plant optimal performance.

The next stratum, the third, is termed the *scheduling stratum* by Miller.[24] Information as to the state of the plant is obtained from the information collection strata and the orders to be produced from a higher stratum, and the orders are scheduled and the resulting schedule transmitted to the second and first strata for implementation. Changes in the process—as during emergencies—necessitate rescheduling using the current state of the plant. It is third stratum which is responsible for achieving the overall economic optimum during normal operation.

Miller's fourth stratum is concerned with digesting the information transmitted upward from the lower three strata and producing accounting and management information as well as overall scheduling and planning.[24] The scheduling and planning is over a longer time horizon than the scheduling stratum and provides the goals or inputs to the third-stratum scheduling operation. Notice that the coordination is complex. The fourth stratum controls the third stratum by changing its goals (the results of overall scheduling and planning) and by setting constraints (e.g., orders must be delivered). The third stratum supplies detailed schedules to the process control strata which is a direct intervention. In all cases, intervention is in a downward direction with little interaction between strata, a characteristic of decomposition on the basis of levels of influence. Moreover, the decision making at the higher levels takes precedence over that of the lower levels (higher priority) and is usually done over a longer time period or horizon.

The mapping of this decomposition of the overall steel control problem onto a hardware system is also usually hierarchical. Figure 8-8 shows a section of the steel process described by Miller. The fourth stratum is off-line with no direct connection to the on-line three-stratum system. The first stratum, process control, is decomposed on the basis of structure into analog control sections, operator controlled sections, and a computer controlled section. The second stratum, information collection, is done also by decomposing on the basis of structure, using computers for a portion of the collection and hand data collection for the rest. The third stratum, scheduling, is not decomposed but is assigned to a single computer. This decomposition and assignment of tasks to the hardware is partially dictated by communication requirements. In particular, where data rates and volumes are high, direct connection between computers is used, whereas in other cases, paper tape or operators using typewriters are used. It is clear that the hardware

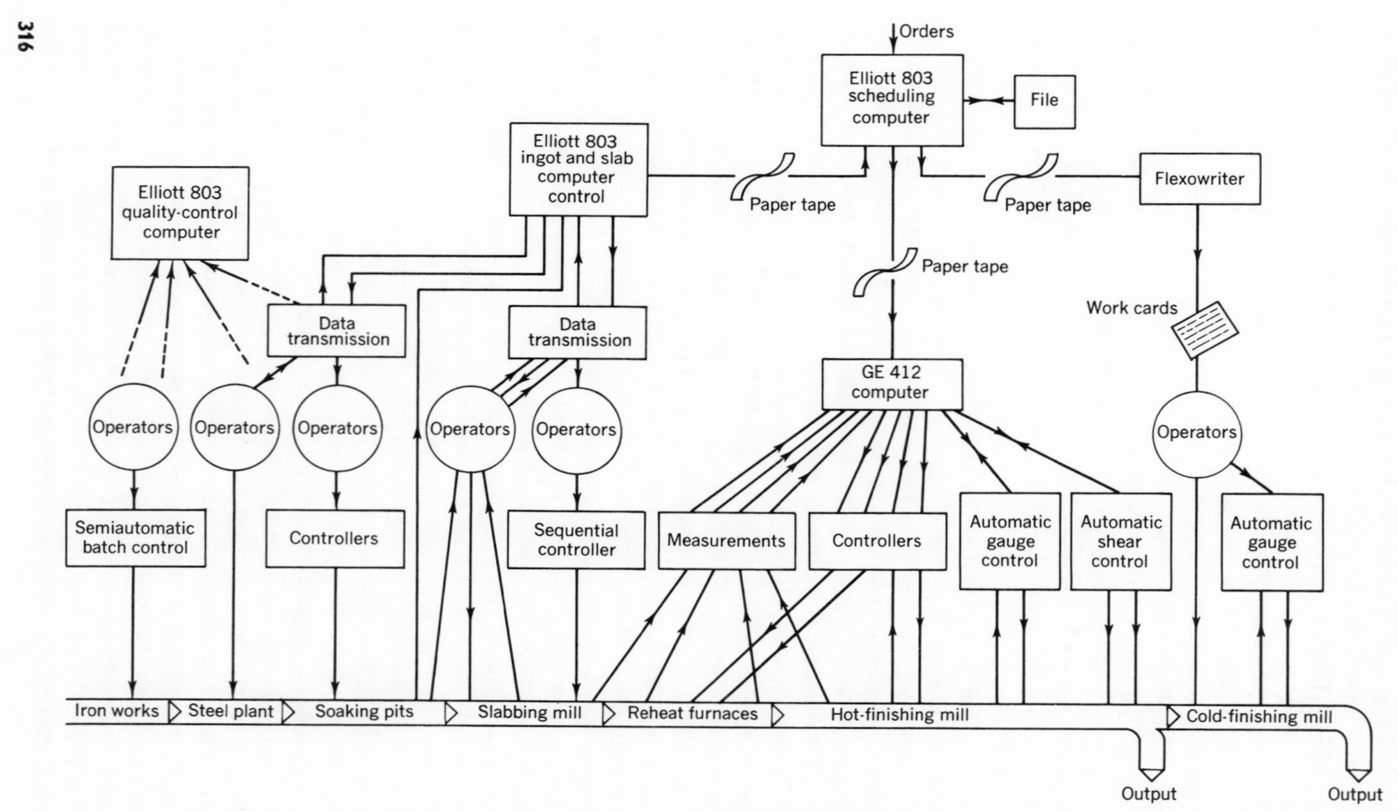

FIG. 8-8. Further decomposition of the process control stratum on the basis of structure. (*From Ref.* 24.)

decomposition has been done to minimize the amount of information transmitted and stored in each computer.

One of the largest and most important systems in the world is the electric power generating system, which is fundamental to a technological society.[8,19,13,20] The objective of such a system is to supply the power demanded by the loads on the system at any instant of time without exceeding any operating constraints (frequency variation, voltage variation, current levels, etc.) and with maximum economy. The disturbances to the system are highly varied, including unpredictable load changes, a multitude of faults, demands for power through tie lines to neighboring power companies, equipment failures, etc. Yet emergencies do occur when the system cannot supply the desired power, and in this case the system must protect itself so as to supply as much power as possible without self-damage and finally must restore service as quickly as possible. The design of a reliable, economic control for such a complex system is greatly aided through the use of on-line multilevel decomposition of the control system itself as well as the hardware realization, and provides an excellent example of decomposition using all the methods discussed here.

The task of a control system varies widely depending on whether the system is normal or not. Define normal operation of the system as any operation which simultaneously satisfies all load constraints, i.e., all load desired is being supplied, and all operating constraints, i.e., no operating conditions are being violated and all system variables are within tolerance.[8,20]

During normal operation, the system has time to anticipate possible troubles, called contingencies, which may force it to leave this desirable operating condition. If any such contingency is discovered and if the system operation can be modified to lessen the expected trouble should such a contingency occur, but without violating any load or operating constraint, then the system operation is modified. In general, subject to the load and operating constraints and taking into account contingencies, system operation is economically optimized (optimal dispatching). A system in normal operation anticipating contingencies is said to be in the *preventive state*.[8,20]

If an emergency occurs which causes either a load or an operating constraint to be violated, then the system passes out of the preventive state and into the *emergency state*.[8,20] The objective here is not economic but rather to eliminate as quickly as possible any operating constraint violation so that the power system can continue to operate even though all load is not being supplied, the philosophy being that the system must survive the emergency if it is to recover at all. Thus load is either redistributed or else shed until the emergency can be cleared.

A system operation which violates some load constraint, i.e., not all demanded load being supplied, but which satisfies all operating constraints is said to be in the *restorative state*.[8,20] The objective here of course is to modify operation of the system or its configuration if necessary in order to satisfy the load constraints, that is, supply the load which is desired but not yet supplied.

The tasks of the control system are different in each of these three states and lead to a natural decomposition on the basis of levels of influence. The control systems for the preventive, emergency, and restorative states can each be assigned to one stratum and independently designed provided coordination between the strata is considered.

Figure 8-9 shows the three strata resulting from this decomposition.

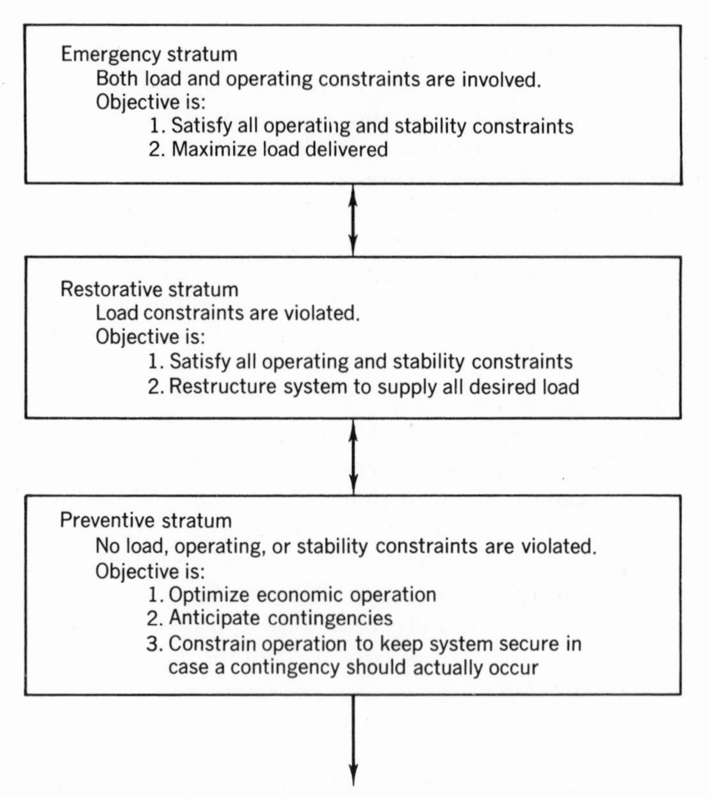

FIG. 8-9. Multistratum on-line control for an electric power system.

The coordination between these strata is direct intervention in the case of the emergency stratum (fault clearing, load shedding, generator switching, etc.) as well as goal coordination (changing from maximum economy to maximum load servicing when a fault occurs). The restorative stratum coordinates mainly through goal intervention and constraint intervention, ensuring that the shed load is restored when feasible but also ensuring through constraints that the emergency will not reoccur.

The tasks of each of these strata are different and involve different methodology for their solution and moreover have the usual priority ordering associated with levels of influence. Figures 8-10, 8-11, and 8-12

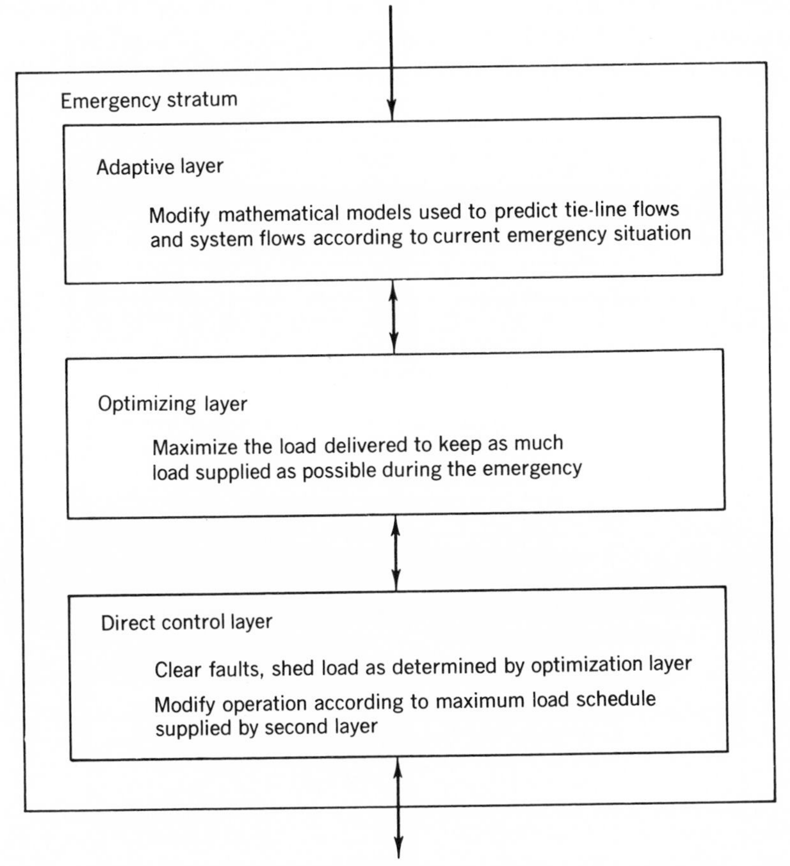

FIG. 8-10. Decomposition of the emergency stratum control into three on-line control layers.

show a further decomposition of the control tasks within each of these strata, indicating the decomposition on the basis of layers of control.[8] The individual tasks are defined in those figures.

The two uses of decomposition in the power systems control problem do not include the obvious one, decomposition on the basis of structure, which can further simplify the problem.[8,20] In fact, each stratum might be decomposed on the basis of a different structural arrangement if this proves useful. It must be emphasized that this decomposition

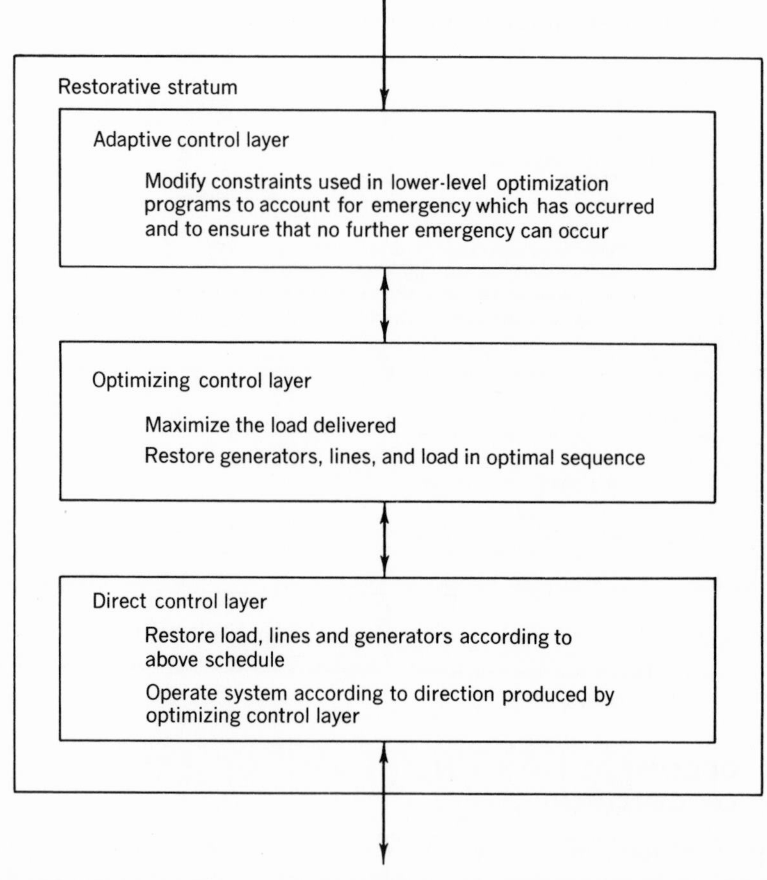

FIG. 8-11. Decomposition of restorative stratum into three layers of on-line control.

of the control problem does not presume an identical decomposition of the hardware which implements this system. The division of effort between one or more computers and analog or special-purpose hardware must be separately evaluated as outlined in Fig. 8-1. The complexity of the power system security and control problem clearly shows the advantage of decomposition into an on-line multilevel system.

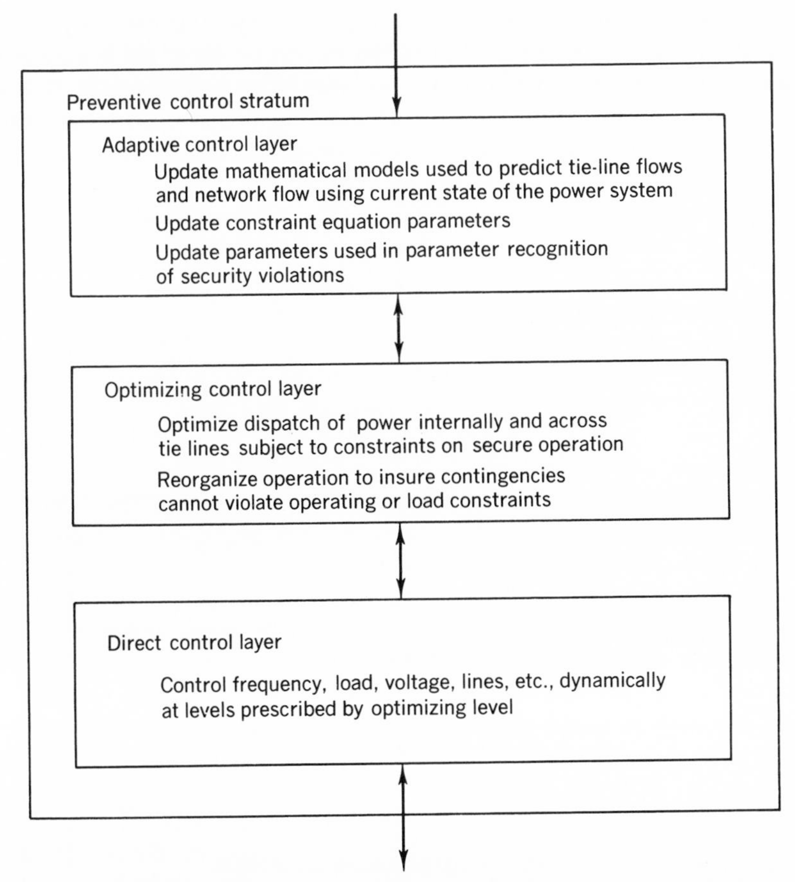

FIG. 8-12. Decomposition of preventive stratum into three layers of on-line control.

8-5 DECOMPOSITION ON THE BASIS OF LEVELS OF CONTROL

One of the most effective approaches to the design of computer control systems is decomposition on the basis of levels of control, called *layers of control* to differentiate the levels from those derived from decomposition on the basis of influence and structure.[18,22] The partitioning of the on-line control problem into layers leads immediately to simplification of the design problem, simplification of the programming problem, and an increase in generality since designs for a given layer can be used in may different situations. Moreover, coordination between layers becomes straightforward and can be carried out in either a single computer or a multicomputer on-line configuration.

The basis for partitioning the problem in this case is neither type of task nor physical structure but rather the *frequency of the disturbances* entering the system. Thus faster disturbances are assigned to one layer of the control and slower disturbances to another. Because of the nature of the process control problem, the resulting layers are significantly different but work together effectively.

The emphasis in this discussion is on the method for decomposition and *not* on the methods for the design of the control systems for each layer. In fact, however, one of the most significant results of this decomposition is that new techniques have been developed for the design of control systems on the assumption of a multilayer structure which significantly simplified the implementation of modern computer control.[1,6,7,13,30,33,28,3,4,29] The magnitude of the effect of a disturbance on the process and the extent to which this effect can be minimized through control is dependent in great measure upon the relative frequency of occurrence of the disturbance, and this provides a convenient method for classifying disturbance in general. Figure 8-13 indicates an arbitrary classification of disturbances into four frequency ranges.[31] The highest frequency range includes disturbances which directly affect system variables, whereas the lowest frequency range includes disturbances such as changing market conditions or system configuration. Intermediate to these ranges are the disturbances which affect mode of operation such as load or demand and those which affect the parameters of the system models but not the structure of the system itself (aging, drifts in raw material characteristics, seasonal changes, etc.). Although these classifications are arbitrary, they are very useful, because the techniques used to control the system in the presence of each of these disturbance classes are essentially different, and thus provide a decomposition of the control problem on the basis of *layers* of control.

Note that this classification is not intended to be all-inclusive but rather to be indicative of the types of disturbance and control systems used.

The highest frequency range of disturbances for which control is effective includes those disturbances which directly affect process variables. Undesired changes in steam pressure, ambient temperature, flows, compositions, and the like can cause a direct and immediate change in process variables which may be undesirable. Consequently, control systems are used to counteract the effects of these changes: feedback control in general and perhaps feed-forward control if the disturbance is measurable.

Disturbances which affect the way in which the processes are operated occur at a lower rate and fall in the second-highest frequency range. For example, load changes in a power system dictate a change in the set

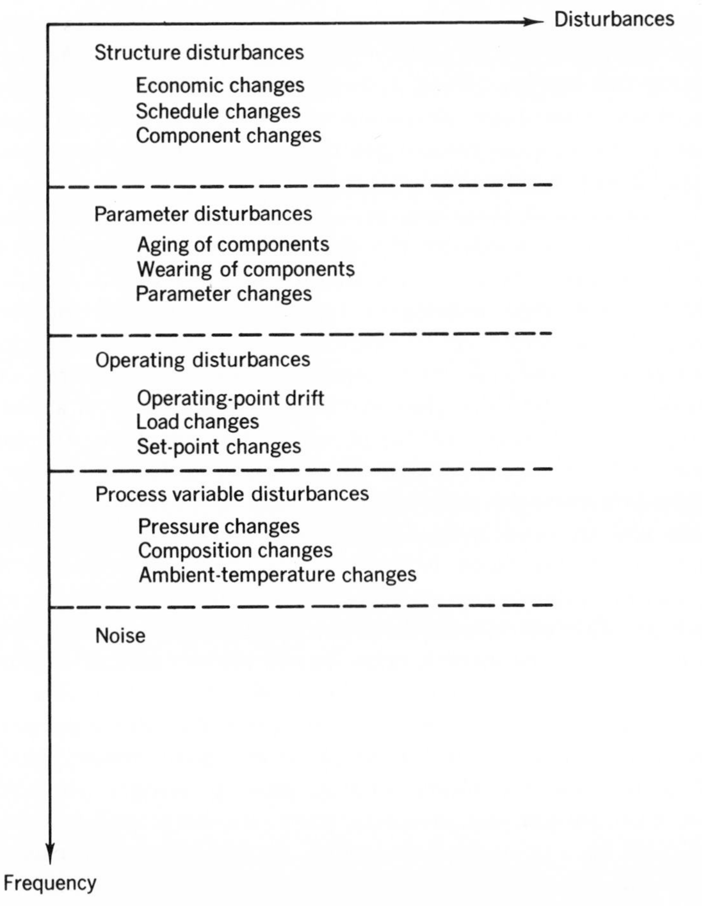

FIG. 8-13. Decomposition of disturbances according to frequency.

points of the turbine generator systems. In this case the load change is a disturbance and the purpose of the control system is to maintain optimal operation of the power generating system by changing the operating point.

The next-lower frequency range which can be singled out includes disturbances which affect the parameters of the process itself. In general, these disturbances or changes occur less often than do changes in the operating conditions. Examples are long-term drift in equilibrium points, changes in process dynamics with time and/or operating conditions, aging of components with attendant changes in process parameters, etc. If the successful operation of the process is dependent upon any of these process parameters, it is necessary to add a control system to counteract their effects.

The lowest frequency range of disturbances for which control is applied includes changes in the environment of the process which in general affects the actual *structure* of the process (in contrast to disturbances which affect the parameters of the process). Examples include changes in the economic conditions of the market for the process output, changes in the schedules for the process, addition or removal of a component in the process, and changes in the process objectives (as for example from maximum throughput to fixed throughput with maximal quality standards). In every case, the disturbances mentioned above are real in the sense that ignoring them will cause deterioration in performance of the process.

Although the assignment of a particular disturbance to one category or another may often be arbitrary, it is useful to consider these classes of disturbances because we can observe from Fig. 8-13 that the techniques available for the design of an appropriate control system are essentially different for each frequency range.

Control systems corresponding to the four frequency ranges are diagrammed in Fig. 8-14 in the form of a multilevel system, with one level of control devoted to each frequency range.

Consider first the fourth level of control, which is concerned with the lowest frequency disturbances, that is, those which affect the actual structure of the process. It is at this level that the overall process control problem is formulated and adjusted when disturbances occur, and consequently this level sets the goals for the levels below it. Functions such as production planning and scheduling based on received and expected orders are carried out at this level. Since this task has often been done by management with the aid of data obtained from the process operation and processed into the form of reports by computers, the fourth-layer control system is often called a *management information and control system.* Notice that it is inconsequential whether the control loop is automatic or contains a human being—the objective is the same.

The second highest layer of control (level 3) is concerned with disturbances which affect parameters of the process but not its actual structure. This layer is concerned with the actual numerical values of parameters used in the models of the process, limits of variables, goals, and controllers and hence the disturbances which affect these parameters. It is the objective of this layer of control to adapt or update parameters of mathematical models, values of operating limits, parameters in goals to be optimized by lower levels, and the like. In general, this is done by using past operating data or experience. This control is often called *adaptive control* but not simply in the sense of regulator controller tuning but rather in the broader sense of making use of any information gained from operating experience.

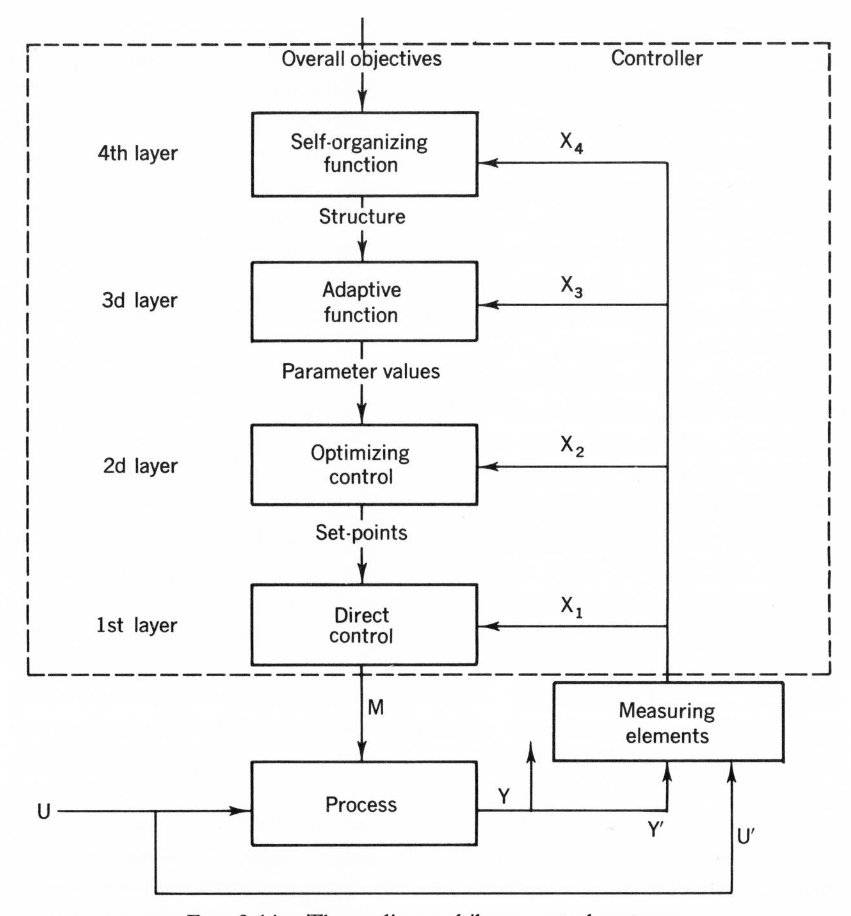

FIG. 8-14. The on-line multilayer control system.

Next in the descending hierarchy is the second layer of control, which responds to disturbances which affect the performance of the process as measured by some goal supplied by the upper layers. The controller compensates for disturbances by changing the mode of operation of the process. This may be done by optimizing some performance function subject to operating constraints, and for this reason, the level is often termed the *optimizing layer of control.* Actually in the general control problem, this name is not appropriate because no explicit performance function is used. For example, this layer might choose the operating condition of the process in order to achieve some satisfactory level of performance without actually bothering to optimize this performance. The start-up problem to be discussed is an example in point.

The first layer of control is concerned with compensating for the disturbances in the highest frequency range, those which upset process variables and hence drive the process away from the desired operating condition as determined by the higher levels. The first-layer control system must therefore enforce the decisions of the upper levels. It turns on and off appropriate variables upon demand, monitors process variables to ensure that they take on the expected values, regulates variables which the upper levels specify, checks limits of process variables and alarms the upper levels of control when they exceed safe values, and gathers data as required for successful operation of the upper control levels. Thus this level is the interface between the actual process and the upper levels of the control system.

Notice that the function of each layer of the control system is to use data supplied by all the lower layers together with specific information from the layer immediately above it in order to make decisions which affect the layer immediately below it. Thus this decomposition of the process control problem results naturally in a hierarchical structure. This is convenient not only from a system planning point of view, but also from a computer programming point of view since the programming tasks for each level are different and may be done by different individuals or groups at different times. Notice that the higher the level of control, the lower the frequency of the disturbances with which it is concerned and consequently the lower the rate at which the control program must be executed. However, it is also true that the higher the level of control, the more complex the control calculations become. Lefkowitz has observed that the product of computing load per calculation times the rate at which the calculation must be performed is essentially a constant for a properly designed multilevel control system.[18] In effect, this shares the control load equally among the four layers of control.

Notice that the decomposition into four layers of control is useful in the sense that the design of the overall control system has been reduced to the design of four separate but (slightly) interacting control systems, and that, moreover, the techniques used in the design of each level of control are essentially different.

The decomposition of the overall process control problem into this multilevel form is best illustrated by two examples: the control of an operating unit in a continuous process and the start-up of a process.

First consider the control of a single operating unit in a process.[10,11,17] Such a unit receives as its inputs raw material and the outputs of other process units. In general the mean characteristics of these inputs are known but their actual values vary considerably around the mean and constitute a source of disturbance. The output of the operating unit may be either an end product itself or an intermediate product which in turn is the input to another unit. In either case, the level of output is often constrained between close limits together with constraints on the quality of the output. Manipulated variables are to be chosen by the control system so as to meet the specifications of the output and so that the operating unit is operated at minimum cost or maximum performance in some prescribed sense.

The key level of control in this situation is the second level, for its objective is to optimize performance of the operating unit subject to constraints on its outputs. This is often set up in the form of a steady-state optimization problem with the output of the second level being the set points for the various controlled variables of the process. These set points are in turn transmitted to the first-level control, which may be direct digital control, whose objective is to maintain these process variables at their set points despite the variations in the process inputs (the high frequency disturbances). Of course, as the second level changes the set points for the process, the process dynamics may also change enough that the controllers on the first level need retuning. In general, this can partially be done at the first level using self-tuning techniques. These are most often successful if only a single parameter such as the loop gain is tuned. If other parameters must be tuned or updated, the second level is responsible. Thus the information transmitted from the second- to first-level controllers includes not only the set points for the controllers, but also new controller parameters if required. In addition, updated process variable limits may be transmitted.

The objective of the fourth level of control in this example is to set up the structure for the process optimization problem. That is, this level defines the optimization problem and the structure for the mathematical models of the process and constraints which the lower levels use to actually perform the optimization and enforce the result. The third level of control, the adaptive level, determines the appropriate parameters of the mathematical model and updates them as time progresses. In the general case, the control system would not function if any level were missing, although in specific instances, a level might not be needed (for example, the adaptive level might be called upon so seldom that its function can be handled by off-line calculation and the changes manually entered into the second-level control system).

In this example we notice an increase in computing complexity and a decrease in frequency with increasing levels. The overall computational load per level is approximately constant and this equal division of effort simplifies the design and implementation of the control system. We observe that if there is one such multilevel control system for each operating unit in a process, then not only is the control system partitioned in level, but also each level is itself decomposed into a number of parallel controllers. The result is a multilevel, multigoal control system.

A second example is the process start-up problem.[39] The complexity of the start-up problem varies considerably from process to process and with the conditions under which it must take place. One method of starting up a process is to derive a succession of states and to drive the process from one state to the next, holding it at a given state by regulatory control action until all process variables are within prescribed limits and the process is in or near equilibrium for that state. This control problem also lends itself to the multilevel description. The fourth layer of control defines the goal for the start-up. That is, it specifies on the basis of current values of the process measurements what the initial state of each component of the process may be and what criterion should be used to drive the process to its desired operating conditions. This function is similar to production planning.

The second layer or optimizing control chooses the best or at least a satisfactory sequence of states in which to operate the process. For example, in some processes it might be possible to drive the process controlled variables to a state corresponding to a small percentage of full load after turning on various pumps, valves, etc., and then when the process is in equilibrium, change all set points to their final value and allow the direct digital control system to drive the process to its full-load condition. In other processes this might cause variables to exceed safe operating conditions, and consequently several transition states must be defined. At any rate, when this layer has decided what changes are to be made, it transmits this information to the first-level control, which turns on various valves, pumps, etc., and causes appropriate direct digital control loops to begin functioning. In addition, the first layer checks that the valves and pumps actually respond and so notifies the second level. The first layer monitors process variables and alarms the second and higher layers when the variables exceed safe limits prescribed by the higher layers.

The function of the third layer in this situation is to take advantage of past experience to modify models, safe operating limits, and other parameters used by the second layer in determining the sequence of states. Although the function of this layer is clearly updating or adapting of parameter values, it is not model adaptation in the same sense as occurs in the control of the operating unit.

Decomposition of the start-up problem in this way simplifies the design of each of the four layers and makes possible the development of a general strategy for almost any class of process. The problem of implementing the control system in *software* is nontrivial, however, and a further decomposition on the basis of structure can lead to simplification in this area.[32,34]

REFERENCES:

1. Burghart, J. H.: An Adaptive Technique for On-line Steady-state Optimizing Control, *Case Western Reserve Univ. Syst. Res. Cen. Rept.* SRC 80-C-65-32, 1965.
2. Calahan, D. A.: "Computer-aided Network Design," McGraw-Hill, 1968.
3. Coviello, G. J.: Optimal Control of Multivariable Systems, *AIEE Conf. Paper* 62–1280, 1962.
4. Coviello, G. J.: An Organization Approach to the Optimization of Multivariate Systems, *Proc. Joint Autom. Control Conf.*, 1964.
5. Chestnut, H.: "Systems Engineering Methods," Wiley, 1967.
6. Durbeck, R. C.: Principles for Simplification of Optimizing Control Models, *Case Western Reserve Univ. Syst. Res. Cen. Rept.* SRC 66-C-64-27, 1964.
7. Durbeck, R. C., and L. S. Lasdon: Control Model Simplification Using a Two-level Decomposition Technique, *Case Western Reserve Univ. Syst. Res. Cen. Rept.* SRC 62-C-64-24, 1965.
8. Dy-Liacco, T. E.: Control of Power Systems via the Multi-level Concept, doctoral dissertation, Systems Research Center, Case Western Reserve University, Cleveland, 1968.
9. Fiedler, H. J., and L. K. Kirchmayer: Automation Developments in the Control of Interconnected Electric Utility Systems, *Proc. IFAC Conf. Computer Control*, Toronto, 1968.
10. Findeisen, W., and I. Lefkowitz: Design and Applications of Multilayer Control, *Proc. IFAC Congr., 4th*, Warsaw, Poland, 1969.
11. Findeisen, W., and J. Szymanowski: A Case Study in Multilayer Control, *Case Western Reserve Univ. Syst. Res. Cen. Rept.* SRC 112-C-67-45, 1967.
12. Hadley, G.: "Nonlinear and Dynamic Programming," Addison-Wesley, 1964.
13. Happ, H. H.: Multi-computer Configurations and Diakoptics: Dispatch of Real Power in Power Pools, *Proc. Power Ind. Computer Appl. Conf.*, 1967.
14. Hare, V. C., Jr.: "Systems Analysis: A Diagnostic Approach," Harcourt, Brace & World, 1967.
15. Lasdon, L. S., and J. D. Schoeffler: A Multi-level Technique for Optimization, *Proc. Joint Autom. Control Conf.*, 1965, pp. 85–91.
16. Lasdon, L. S., and J. D. Schoeffler: Decentralized Plant Control, *ISA Trans.*, vol. 5, no. 2, pp. 175–183, April, 1966.
17. Lefkowitz, I.: Multi-level Approach to the Design of Complex Control Systems, *Proc. Syst. Eng. Conf.*, Chicago, 1965.
18. Lefkowitz, I.: Multi-level Approach Applied to Control System Design, *Trans. Am. Soc. Mech. Engrs.*, 1966.
19. Lefkowitz, I., and J. Burghart: An Adaptive Technique for On-line Optimizing Control, *Proc. Joint Autom. Control Conf.*, 1967.
20. Mitter, S. K., and T. E. Dy-Liacco: Multi-level Approach to the Control of Interconnected Power Systems, *Proc. IFAC Conf. Computer Control*, Menton, France, 1967.
21. Mesarovic, M. D.: Multilevel Systems and Concepts in Process Control, *Proc. IEEE*, vol. 58, no. 1, pp. 111–125, January, 1970.
22. Mesarovic, M. D., I. Lefkowitz, and J. D. Pearson: Advances in Multi-level Control, *Proc. IFAC Symp. Eng. Control Syst. Design*, Tokyo, 1965.

23. Mesarovic, M. D.: A Conceptual Framework for the Studies of Multi-level Multi-goal Systems, *Case Western Reserve Univ. Syst. Res. Cen. Rept.* SRC 110-A-66-47, 1965.
24. Miller, A.: Automation in the Steel Industry, *Automation*, pp. 7–14, November, 1966.
25. Pender, H., and K. McIlwain: "Electrical Engineer's Handbook," pp. 5–12, Wiley, 1950.
26. Peschon, J.: Case Study on Optimum Control of Electric Power System, Case Studies in System Control, *Proc. Joint Autom. Control Conf.*, 1968, pp. 1–56.
27. Roth, J. F.: The Application of the Hierarchy System to On-line Process Control, *J. Brit. IRE*, pp. 117–126, August, 1962.
28. Shimizu, K.: Multi-layer Approach to Adaptive Optimizing Control under Disturbances, *Case Western Reserve Univ. Syst. Res. Cen. Rept.* SRC 118-C-67-51, 1967.
29. Sanders, J. L.: Multi-level Control, *Proc. Joint Autom. Control Conf.*, 1964.
30. Schoeffler, J. D., and I. Lefkowitz: Predictive Control, in "Encyclopedia of Linguistics Information and Control," Pergamon, 1968.
31. Schoeffler, J. D., T. Willmott, and J. Dedourek: Programming Languages for Industrial Process Control, *Proc. IFAC Conf. Computer Control*, Menton, France, 1967.
32. Schoeffler, J. D.: Process Control Software, *Datamation*, vol. 12, no. 2, pp. 33–42, February, 1966.
33. Schoeffler, J. D., and H. Fertik: The Design of Non-linear Multi-variable Control Systems from State Dependent Linear Models, *Am. Soc. Mech. Engrs. J. Basic Eng.*, ser. D, vol. 88, no. 2, pp. 355–361, June, 1966.
34. Schoeffler, J. D., and R. Temple: "RTL—A Real Time Programming Language," Systems Research Center, Case Western Reserve University, Cleveland, 1968.
35. Lasdon, L. S.: Duality and Decomposition in Mathematical Programming, *IEEE Trans. Sci. Cybernetics*, vol. SSC-4, no. 2, July, 1968.
36. Schoeffler, J. D.: Real-time Multi-level Systems, *Proc. Syst. Eng. Conf.*, Chicago, 1965.
37. Takahara, Y.: Multi-level Systems and Uncertainties, *Case Western Reserve Univ. Syst. Res. Cen. Rept.* SRC 99-A-66-42, 1966.
38. Yourdon, E.: "Real-time Systems Design," Information and Systems Institute, Cambridge, Mass., 1967.
39. Bloom, G. B.: "Programming Language for Process Start-up," master's thesis, SRC-68-15, Systems Research, Case Western Reserve University, Cleveland, 1968.
40. Schoeffler, J. D., Dynamic Simulation and Control of the Fourdrinier Papermaking Process, *Proc. IFAC Congr.*, 3d, London, 1966.

INDEX

INDEX